AF812858

Studies in Big Data

Volume 121

Series Editor

Janusz Kacprzyk, Polish Academy of Sciences, Warsaw, Poland

The series "Studies in Big Data" (SBD) publishes new developments and advances in the various areas of Big Data- quickly and with a high quality. The intent is to cover the theory, research, development, and applications of Big Data, as embedded in the fields of engineering, computer science, physics, economics and life sciences. The books of the series refer to the analysis and understanding of large, complex, and/or distributed data sets generated from recent digital sources coming from sensors or other physical instruments as well as simulations, crowd sourcing, social networks or other internet transactions, such as emails or video click streams and other. The series contains monographs, lecture notes and edited volumes in Big Data spanning the areas of computational intelligence including neural networks, evolutionary computation, soft computing, fuzzy systems, as well as artificial intelligence, data mining, modern statistics and Operations research, as well as self-organizing systems. Of particular value to both the contributors and the readership are the short publication timeframe and the world-wide distribution, which enable both wide and rapid dissemination of research output.

The books of this series are reviewed in a single blind peer review process.

Indexed by SCOPUS, EI Compendex, SCIMAGO and zbMATH.

All books published in the series are submitted for consideration in Web of Science.

Sanjay Chaudhary · Chandrashekhar M. Biradar ·
Srikrishnan Divakaran · Mehul S. Raval
Editors

Digital Ecosystem for Innovation in Agriculture

Springer

Editors
Sanjay Chaudhary
School of Engineering and Applied Science
Ahmedabad University
Ahmedabad, Gujarat, India

Chandrashekhar M. Biradar
CIFOR-ICRAF India
Asia Continental Program
New Delhi, Delhi, India

Srikrishnan Divakaran
School of Engineering and Applied Science
Ahmedabad University
Ahmedabad, Gujarat, India

Mehul S. Raval
School of Engineering and Applied Science
Ahmedabad University
Ahmedabad, Gujarat, India

ISSN 2197-6503 ISSN 2197-6511 (electronic)
Studies in Big Data
ISBN 978-981-99-0576-8 ISBN 978-981-99-0577-5 (eBook)
https://doi.org/10.1007/978-981-99-0577-5

© The Editor(s) (if applicable) and The Author(s), under exclusive license to Springer Nature Singapore Pte Ltd. 2023
This work is subject to copyright. All rights are solely and exclusively licensed by the Publisher, whether the whole or part of the material is concerned, specifically the rights of translation, reprinting, reuse of illustrations, recitation, broadcasting, reproduction on microfilms or in any other physical way, and transmission or information storage and retrieval, electronic adaptation, computer software, or by similar or dissimilar methodology now known or hereafter developed.
The use of general descriptive names, registered names, trademarks, service marks, etc. in this publication does not imply, even in the absence of a specific statement, that such names are exempt from the relevant protective laws and regulations and therefore free for general use.
The publisher, the authors, and the editors are safe to assume that the advice and information in this book are believed to be true and accurate at the date of publication. Neither the publisher nor the authors or the editors give a warranty, expressed or implied, with respect to the material contained herein or for any errors or omissions that may have been made. The publisher remains neutral with regard to jurisdictional claims in published maps and institutional affiliations.

This Springer imprint is published by the registered company Springer Nature Singapore Pte Ltd.
The registered company address is: 152 Beach Road, #21-01/04 Gateway East, Singapore 189721, Singapore

To my wife (Sunita), son (Mandar), and daughter (Anuradha)

by Sanjay Chaudhary

Dr. Ravi Prabhu, Director General,
The World Agroforestry, Dr. Javed Rizvi,
Dr. Maarten van Ginkel, and Dr. S. K. Dhyani

by Chandrashekhar Biradar

To my wife Vidya, son Adhithya, mother Vidya, father Divakaran, and my colleagues for their support and encouragement

by Srikrishnan Divakaran

In dedication to my wife—Hemal,
son—Hetav, in memory of my
mother—Charu, and
father—Shirish, for their constant encouragement and support through thick and thin

by Mehul Raval

Foreword

Meeting the food demands in the 21st century seems to be a continued story of catch-up. It has become increasingly clear that the breathing time Norman Borloug and colleagues afforded us with the Green Revolution has not been effectively used to harness food demand. Although over that past half century significant progress has been achieved in reducing hunger and poverty as well as improving food security and nutrition, in the process we have seriously degraded the natural resource base (soils, water, and biodiversity), on which our food production is based. Climate change is adding an extra threat to food security. Calls are getting louder to shift to sustainable and regenerative food production systems that will feed an ever-increasing population. The agricultural science and development communities are looking to expand the technologies and toolbox that will help create efficient agricultural systems that are economically, socially, and environmentally sustainable.

One of the modern technologies that might help address the complex problems facing agriculture may be found in the innovative application of data science, big data analytics, remote sensing, Internet of Things, computer vision, machine learning, cloud computing, artificial intelligence, etc. Technologies are in place to capture big data in real-time manner, and modern farmers in advanced economies are increasingly providing and using such data for farm and resource management. In the upcoming decade, an increasing number of farms will partake in this farm modernization process. Policymakers as well as farmers will benefit from it to make better decisions. The challenge will lie in involving the resource-poor farmers in this transformative process.

This book intends to provide the reader with an overview of the frameworks and technologies involved in the digitalization of agriculture, as well as the data processing methods, decision-making processes, and innovative services/applications for enabling digital transformations in agriculture. This book has two broad sections: (1) *Frameworks, Tools, and Technologies for Transforming Agriculture* and (2) *Problems and Applications of Digital Agricultural Transformations.*

The first part offers an overview of the challenges and opportunities in transforming agriculture through efficient and cost-effective digital services and applications. The chapters in this part discuss the key aspects of building a framework for allocating digital resources necessary for developing digital services and applications in agriculture. It also discusses some of the key principles and concepts as well as technologies and tools in AI and machine learning useful for the creation of resource-efficient services and applications on these platforms.

The second part presents key principles and concepts in computer vision, machine learning, remote sensing, and artificial intelligence (AI). They demonstrate their use in developing intelligent services and applications to solve agricultural problems that arise in the context of plant phonemics, sustainable agriculture, yield prediction, and farm data integration. It also presents some analytical tools for analyzing policies for allocating farm resources and measuring farm productivity.

The value and relevance of any book can be measured by the range and depth of topics and the quality of its chapters. This book meets those criteria. The editors of this book were successful in assembling a valuable collection of chapters on technological advances to help address a key and essential challenges in agriculture. The chapters are well written by competent authors associated with globally reputed organizations. This book provides qualitative and relevant reference content on digital infrastructure for innovation in sustainable agriculture. This book offers an excellent source of knowledge and information that will help a range of professionals, from policymakers to scientists to technology developers and end-users. I congratulate the editors and authors of this book for their commendable contribution.

Austin, TX, USA
December 2022

Paul Vlek
Professor Emeritus
Former Director
Center for Development Research (ZEF)
Division of Ecology and Natural Resources
University of Bonn
Bonn, Germany

Former Executive Director of WASCAL
West African Science Service Center on Climate
Change and Adapted Land Use
Accra, Ghana

Preface

This book on *Digital Ecosystem for Innovation in Agriculture* is an attempt to provide the reader with an accessible big picture of technologies and innovations for the digital transition of agriculture, as well as the key challenges and research trends in the development of data analytical frameworks, tools, and their applications in the context of Digital Augmentation in Agriculture. "Digital Agriculture," here we refer to leveraging the digital technological advances in agriculture and agroecosystems to improve the functional productivity of processes involved in agri-food systems. Furthermore, the transformation should be economically viable and ecologically sustainable.

The increased availability of sophisticated remote sensing satellite services, widespread use of unmanned aerial vehicles (UAVs), better access to quality of data, power of cyber infrastructure, and the easy deployment of inexpensive Internet of Things (IoT) sensors, standardized interfaces, operations, and programmable frameworks have propelled applications in agricultural ecosystems. It is coupled with significant advances in data science (tools and techniques for filtering, compressing, processing, and analyzing) and big data analytics (capturing, storing, retrieving, and visualizing big data). In addition, machine learning (tools and algorithms for building models, making predictions, and performing statistical analysis on data) has accelerated the development of various niche services in Digital Agriculture.

The main focus of this book is on (i) Frameworks and Systems: Handling Big Data, employing remote sensing technology, making provisions for providing and accessing computing, storage, and services over the cloud, and the Internet of Things for collecting, filtering, storing, retrieving, integrating, and visualizing farm data; (ii) Tools and Techniques in Data Science and Machine Learning for developing models, algorithms, and services for providing niche agricultural services that involve data analytics, making predictions and providing statistical guarantees on their predictions.

The book chapters are divided into two parts. The book's first part, *Frameworks, Tools, and Technologies for Transforming Agriculture*, focuses on the critical issues in developing platforms/frameworks for effectively allocating digital infrastructure for building digital services/applications in agriculture. One of the vital issues is climate change which is now touching our daily life. It is also threatening agriculture and needs mitigation. The first chapter "A Brief Review of Tools to Promote Transdisciplinary Collaboration for Addressing Climate Change Challenges in Agriculture by Model Coupling" suggests increasing agriculture efficiencies and making room for renewable bioenergy crops. The chapter summarizes the tools that promote collaboration for developing sustainable and climate-resilient agriculture and discusses model coupling about plant and agri-sciences. The second chapter "Machine Learning and Deep Learning in Crop Management—A Review" provides a survey of the tools and techniques of machine and deep learning employed in agriculture. It discusses algorithms for crop management activities like crop yield prediction, diseases, and pest and weed detection. Satellites, drones, and on-ground sensors are essential in providing data for the digital ecosystem for agriculture innovation. However, all three modes of data collection are executed in isolation. Therefore, the need for an orchestration platform to exploit the potential of remote sensing data is presented in the third chapter "Need for an Orchestration Platform to Unlock the Potential of Remote Sensing Data for Agriculture".

However, it is crucial to develop strategies to connect multimodal data. In that context, the fourth chapter "An Algorithmic Framework for Fusing Images from Satellites, Unmanned Aerial Vehicles (UAV), and Farm Internet of Things (IoT) Sensors" shows an algorithmic framework for constructing higher-dimensional maps through data fusion of satellite images and unmanned aerial vehicles with multi-sensor farm data. Remote sensing plays a critical role in mapping and monitoring crops on a large scale. The availability of open-source data and cloud resources play a significant role in developing remote sensing-based solution. Therefore, the fifth chapter "Globally Scalable and Locally Adaptable Solutions for Agriculture" focuses on using open-source high-resolution spectral, spatial, and temporal resolution satellite data, open-source cloud-based platforms, and big data algorithms for agriculture. Continuous knowledge management (KM) can trigger innovation in agriculture. Therefore, developing a theoretical framework to guide the KM process is crucial. The sixth chapter "A Theoretical Framework of Agricultural Knowledge Management Process in the Indian Agriculture Context", the last chapter of the part, uses the famous Indian milk cooperative sector as an example, derives various systemic factors, and guides agri-organization through KM processes.

The book's second part, *Problems and Applications of Digital Agricultural Transformations*, presents specific challenges for Digital Agriculture that employs computer vision, machine learning, and remote sensing tools and techniques. This part spans from chapter "Simple and Innovative Methods to Estimate Gross Primary Production and Transpiration of Crops: A Review" to "Computer Vision Approaches for Plant Phenotypic Parameter Determination". The most significant carbon and water fluxes in agroecosystems are gross primary production (GPP) and transpiration (TR) of crops. Crop yield estimate using GPP and transpiration measurement can improve irrigation in cropland. The seventh chapter "Simple and Innovative Methods to Estimate Gross Primary Production and Transpiration of Crops: A Review" reviews simple and innovative methods to estimate gross primary production and transpiration. It reviews the state of the science, including in situ and remote sensing methods, while focusing on the biophysical foundation. The growth of computation power has facilitated the creation of 3D models of plants or trees. Virtual plants can simulate crop growth (in silico) compared to the natural environment. The eighth chapter "Role of Virtual Plants in Digital Agriculture" overviews the role of virtual plants in Digital Agriculture, showcasing that in silico implementation is an alternative to time-consuming, labor-intensive actual field trials. Finally, this chapter covers the concept of VP modeling, its applications, and some challenges in its application. Anthropogenic activities can impact soil carbon pools and phytomass on a vast scale. Orchards and plantations can affect the carbon pool, which needs to be carefully studied. The ninth chapter "Remote Sensing for Mango and Rubber Mapping and Characterization for Carbon Stock Estimation—Case Study of Malihabad Tehsil (UP) and West Tripura District, India" provides a case study involving remote sensing for mango and rubber mapping and characterization for carbon stock estimation. It uses Sentinel-2 data and machine learning to classify tree species. It demonstrates that simultaneous high-resolution phytomass and soil mapping with geospatial techniques significantly enhances India's capability to monitor and model terrestrial carbon pools.

Deep learning (DL) and computer vision (CV) advances are penetrating agriculture and natural resource management. The next set of chapters showcases such use cases of DL. The tenth chapter "Impact of Vegetation Indices on Wheat Yield Prediction Using Spatio-Temporal Modeling" presents the use of spatial-temporal modeling to study the impact of vegetation indices on wheat yield prediction. It showcases the use of convolutional neural networks (CNNs) and long short-term memory (LSTM) for yield prediction. Irrigation is a crucial phase of crop cultivation, and its scheduling and water management play essential roles in arid regions. Therefore, estimating the crop-specific water requirement at the farm and a more significant level is necessary. The eleventh chapter "Farm-Wise Estimation of Crop Water Requirement of Major Crops Using Deep Learning Architecture" illustrates the use of deep learning in evaluating farm-wise crop water requirements of major crops. It showcases platform development for adequately managing water resources across states in India.

Usually, remote sensing satellite data is available as multispectral images. More spectral information can increase the accuracy of the machine learning algorithms used in Digital Agriculture. Hyperspectral sensing (HyS) provides very high spectral resolution and is useful in land use and land cover classification (LULC). The twelfth chapter "Hyperspectral Remote Sensing for Agriculture Land Use and Land Cover Classification" presents LULC using hyperspectral sensing. A review of current algorithms for processing HyS datasets is carried out in this article. It includes validating various atmospheric correction (AC) models, dimensionality reduction techniques, and classification methods. In plant breeding, phenotypic trait measurement, i.e., morphological and physiological characteristics, is necessary to develop improved crop varieties. Computer vision-based techniques have emerged as an efficient method for non-invasive and non-destructive plant phenotyping. The thirteenth chapter "Computer Vision Approaches for Plant Phenotypic Parameter Determination" presents Computer Vision Approaches for Plant Phenotypic Parameter Determination. A deep learning-based encoder-decoder network is developed to recognize and count the number of spikes from visual images of wheat plants.

Ahmedabad, India	Sanjay Chaudhary
New Delhi, India	Chandrashekhar M. Biradar
Ahmedabad, India	Srikrishnan Divakaran
Ahmedabad, India	Mehul S. Raval
December 2022	

Acknowledgements We are thankful to

- Contributing authors
- Springer
- Janusz Kacprzyk, Aninda Bose, and Nareshkumar Mani
- Ahmedabad University
- Center for International Forestry Research (CIFOR) and World Agroforestry (ICRAF)
- Family members

Contents

Editors and Contributors

About the Editors

Sanjay Chaudhary, Ph.D., is Dean of Students at Ahmedabad University as well as Professor and Associate Dean of School of Engineering and Applied Science of Ahmedabad University. During 2001–2013, he was Professor as well as Dean (Academic Programs) at Dhirubhai Ambani Institute of Information and Communication Technology (DA-IICT), Gandhinagar, India. His research areas are cloud computing, blockchain technology, big data analytics, and ICT applications in agriculture and rural development. He has authored nine books and nine book chapters. He has published more than 150 research papers in international conferences, workshops, and journals. He has received research grants from leading organizations including IBM, Microsoft, and Department of Science and Technology, Government of India. Seven Ph.D. candidates have completed their Ph.D. successfully under his supervision. He holds a doctorate degree in computer science from Gujarat Vidyapith. His literary articles are published regularly in leading Gujarati magazines.

Chandrashekhar M. Biradar, Ph.D., is Country Director, CIFOR-ICRAF-India with Asia Continental Program; and Chief of Party (CoP, TOFI Program-Trees outside Forests in India). Dr. Biradar is Landscape Ecologist with broad experience in executing agroecosystem research and outreach across the diverse landscapes in Asia, Africa, and the Americas. Dr. Biradar has a multidisciplinary educational background, with B.Sc. in Forestry; M.Sc. Forestry (Tree Improvement and Genetic Resources, specialization in Agroforestry), and a Ph.D. in Environmental Sciences and Earth Observation Systems (with a focus on forest ecology, biodiversity, and geoinformatics). His core expertise focuses on digital augmentation, geoinformatics, and regenerative agroecosystems. Before joining CIFOR-ICRFA, he worked with ICARDA, the University of New Hampshire, the University of Oklahoma, IWMI, and the IIRS-ISRO. Dr. Biradar has over 25 years of experience serving as Researcher, Principal Scientist, Manager, Head of Units, and Research Team Leader and has published over 400+ research articles, tools, and products and received several national and international awards. His current research for development focused on harnessing advances in technologies, agroforestry and forestry, agroecology, indigenous knowledge, and citizen science to restore functional agroecosystems for ecologically sustainable and economically viable landscapes and livelihoods.

Srikrishnan Divakaran completed his Ph.D. in Computer Science in 2002 from Rutgers University, New Brunswick, USA. Then, from 2002 to 2008, he worked as Assistant Professor in the computer science department at Hofstra University, Long Island, NY, as Associate Professor at DAIICT from 2009 to 2016 before joining School of Engineering and Applied Sciences at Ahmedabad University in 2017. Dr. Divakaran has nearly 20 years of research and over 15 years of teaching experience and over five years of industry experience at leading multinational companies in computing and finance. Dr. Divakaran has taught a wide range of courses in computer science as well as related disciplines like bioinformatics and operations research and has a strong research background in designing algorithms for problems with applications in bioinformatics/computational biology, distributed systems, and operations research. In terms of research, over the past 7 years, his interests have broadly been in the area of design and analysis of online and approximation algorithms for problems in bioinformatics/computational biology, distributed systems, and operations research. In bioinformatics, his current research focus is on the design and analysis of approximation algorithms and heuristics for the following problems: (1) constrained generalized tree alignment, (2) template-based methods for sequence alignment, and (3) fast heuristics for exact string matching. In distributed systems, his research focus is in the design and analysis of online and offline approximation algorithms for problems in resource allocation, load balancing, and list update. In operations research, his research interests have been in the design and analysis of online and approximation algorithms for bin packing and scheduling with setups.

Mehul S. Raval is an Associate Dean—Experiential Learning and Professor at the Ahmedabad University. His research interest is in computer vision and engineering education. He obtained a Bachelor's degree (ECE) in 1996, a Master's degree (EC) in 2002, and a Ph.D. (ECE) in 2008 from the College of Engineering Pune/University of Pune, India. He has 25+ years of experience as an academic with visits to Okayama University, Japan, under a Sakura Science fellowship, Argosy visiting associate professor at Olin College of Engineering, the US, during Fall 2016, and visiting Professor at Sacred Heart University, CT, in 2019. He publishes in journals, magazines, conferences, and workshops and reviews papers for leading publishers—IEEE, ACM, Springer, Elsevier, IET, and SPIE. He has received research funds from the Board of Research in Nuclear Science (BRNS) and the Department of Science and Technology, Government of India. He supervises doctoral, M.Tech., and B.Tech. students. In addition, he chairs programs and volunteers on technical program committees for conferences, workshops, and symposiums. He also serves on the Board of Studies (BoS) to develop Engineering curricula for various Universities in India. He is a senior member of IEEE, a Fellow of IETE, and a Fellow of The Institution of Engineers (India). Dr. Raval served the IEEE Gujarat section during 2008–2015 and from 2018–2020 as a Joint Secretary. In addition, he did the IEEE signal processing society (SPS) chapter—IEEE Gujarat Section as vice chair and executive committee member in 2014. Currently, he is helping the IEEE Computational Intelligence Society Chapter—IEEE Gujarat Section as Chair.

Contributors

Alka Arora ICAR-Indian Agricultural Statistics Research Institute (IASRI), New Delhi, India

Jeffrey Basara Department of Meteorology, University of Oklahoma, Norman, OK, USA

Jorge Celis Department of Microbiology and Plant Biology, University of Oklahoma, Norman, OK, USA

Magesh Chandramouli Computer Graphics Technology, Purdue University NW, Hammond, IN, USA

Sanjay Chaudhary School of Engineering and Applied Science, Ahmedabad University, Ahmedabad, India

Viswanathan Chinnusamy ICAR-Indian Agricultural Research Institute (IARI), New Delhi, India

Vipul K. Dabhi Department of Information Technology, Dharmsinh Desai University, Nadiad, Gujarat, India

V. K. Dadhwal School of Natural Science and Engineering, National Institute of Advanced Studies (NIAS), Bengaluru, India

Mihir Dakwala Amnex Infotechnologies Pvt. Ltd. Ahmedabad, Gujarat, India

Srikrishnan Divakaran School of Engineering and Applied Sciences, Ahmedabad University, Ahmedabad, India

Michael Henke School of Agronomy, Hunan Agricultural University, Changsha, People's Republic of China

MuraliKrishna Iyyanki DRDO, Delhi, India

Adinarayana Jagarlapudi Centre of Studies in Resources Engineering (CSRE), Indian Institute of Technology Bombay, Mumbai, India

Deepak Jaiswal Environmental Sciences and Sustainable Engineering Centre, IIT Palakkad, Palakkad, Kerala, India

Sanjiv Kumar Jha Principal Smart Infra—SA, Amazon Web Services (AWS), Bengaluru, India

Sneha S. Kulkarni Amnex Infotechnologies Pvt. Ltd. Ahmedabad, Gujarat, India

Jay Prakash Kumar Amnex Infotechnologies Pvt. Ltd. Ahmedabad, Gujarat, India

Mohit Kumar Computer Application, ICAR-Indian Agricultural Statistics Research Institute (IASRI), New Delhi, India

Pratyush Kumar Amnex Infotechnologies Pvt. Ltd. Ahmedabad, Gujarat, India

Sudhir Kumar ICAR-Indian Agricultural Research Institute (IARI), New Delhi, India

Sudeep Marwaha ICAR-Indian Agricultural Statistics Research Institute (IASRI), New Delhi, India

Heather McCarthy Department of Microbiology and Plant Biology, University of Oklahoma, Norman, OK, USA

Gaurav Mishra Development Management Institute (DMI), Patna, Bihar, India

Tanuj Misra Teaching Cum Research Associate, Rani Lakshmi Bai Central Agricultural University, Jhansi, UP, India

Hasit Parmar L. D. College of Engineering, Ahmedabad, India

S. V. Pasha School of Natural Science and Engineering, National Institute of Advanced Studies (NIAS), Bengaluru, India

Pragnesh Patel School of Engineering and Applied Science, Ahmedabad University, Ahmedabad, India

Suchitra M. Patil Centre of Studies in Resources Engineering (CSRE), Indian Institute of Technology Bombay, Mumbai, India

Gogumalla Pranuthi International Crops Research Institute for the Semi-Arid Tropics, Patancheru, Hyderabad, India

Mehul S. Raval School of Engineering and Applied Science, Ahmedabad University, Ahmedabad, India

K. Saketh School of Natural Science and Engineering, National Institute of Advanced Studies (NIAS), Bengaluru, India

Maitrik Shah School of Engineering and Applied Science, Ahmedabad University, Ahmedabad, India

Rupavatharam Srikanth International Crops Research Institute for the Semi-Arid Tropics, Patancheru, Hyderabad, India

Sruthi Surendran Environmental Sciences and Sustainable Engineering Centre, IIT Palakkad, Palakkad, Kerala, India

Ram Naresh Kumar Vangala Food and Agribusiness School (FABS), Hyderabad, Telangana, India

Satya Sahithi Veeramallu Jawaharlal Nehru Technological University, Hyderabad, India

Sunil K. Vithlani Department of Information Technology, Dharmsinh Desai University, Nadiad, Gujarat, India

Pradeep Wagle USDA-ARS, Oklahoma and Central Plains Agricultural Research Center, El Reno, OK, USA

Xiangming Xiao Department of Microbiology and Plant Biology, University of Oklahoma, Norman, OK, USA

Frameworks, Tools, and Technologies for Transforming Agriculture

A Brief Review of Tools to Promote Transdisciplinary Collaboration for Addressing Climate Change Challenges in Agriculture by Model Coupling

Sruthi Surendran and Deepak Jaiswal

Abstract Climate Change threatens agriculture, and agriculture can also play an essential role in climate change mitigation and adaptation through increasing agricultural efficiencies and greening the energy sector by making room for sustainable renewable bioenergy crops. Efforts for climate change mitigation and adaptation with a focus on agriculture must come from transdisciplinary collaboration, which is often not easy and require one to come out of one's comfort zone. However, at the same time, plenty of tools can facilitate and make such collaboration easy, especially for those who are not modellers or modelling with a focus on a specific aspect of climate change and agriculture. This chapter summarizes such tools and their application in accelerating transdisciplinary collaboration for more sustainable and climate-resilient agriculture.

Keywords yggdrasil · Model Coupling · Crop modelling · BioCro · Climate change

1 Introduction

The Green revolution led by Normal Borlaug resulted in mitigating the problem of hunger and malnutrition by introducing high-yielding varieties of various staple crops (Evenson & Gollin, 2003). One of the reasons for the high yield of these newly developed varieties was dwarfism, which reduced lodging losses and increased grain yield (harvest index) since less biomass was partitioned to stem and other unusable plant parts. However, excess demand for inputs (fertilizers, pesticides, and water) by these high-yielding dwarf varieties results in many inadvertent consequences, such as polluted surface water bodies, soil degradation, and unsustainable exploitation of

S. Surendran · D. Jaiswal (✉)
Environmental Sciences and Sustainable Engineering Centre, IIT Palakkad, Palakkad, Kerala 678557, India
e-mail: dj@iitpkd.ac.in

S. Surendran
e-mail: 162114001@smail.iitpkd.ac.in

© The Author(s), under exclusive license to Springer Nature Singapore Pte Ltd. 2023
S. Chaudhary et al. (eds.), *Digital Ecosystem for Innovation in Agriculture*, Studies in Big Data 121, https://doi.org/10.1007/978-981-99-0577-5_1

groundwater resources for irrigation (Kolbert, 2021). Additionally, recent trends of major staple crops are showing signs of slowing rates of yield increase, which will not be able to meet the future demand of the growing population (Ray et al., 2013). In fact, excess pressure on agriculture to meet the demand of a growing population may increase the global emission of greenhouse gases (GHGs) responsible for climate change (Bajželj & Richards, 2014). Climate change has been mentioned as one of the factors responsible for slowing down the rate of yield increase of major staple crops (Ray et al., 2019). The positive feedback loop between the adverse impact of climate change on agriculture and the expansion of agriculture to meet the growing demand (Bajželj & Richards, 2014) requires disruptive innovation to create a more sustainable and equitable world for the future generation. Agriculture driven by digital technologies provides an opportunity to build upon the existing science and technology and create innovative solutions consisting of interdisciplinary and transdisciplinary approaches. The gradual increase in investment in the AgTech industry in recent years seems an encouraging step toward data-driven sustainable agriculture (De Clercq et al., 2018). To make use of the recent thrust on data-driven agriculture from the perspective of a biophysical crop modeller, this chapter is organized to provide a brief scientific summary of the intimate cross-link between climate and agriculture, various approaches to address the climate change challenges in agriculture, the necessity of transdisciplinary approaches and available digital tools for the same towards climate resilient agriculture with emphasis on water availability and direct impact of climate change on plant growth.

2 How Does Agriculture Contribute to Climate Change?

The global food system contributes ~ 24% of global GHGs emissions (CO_2 equivalent), of which 10% comes from agricultural activities and the remaining 14% from land-use change (Lynch, 2021). Energy usages in various farming operations (ploughing, harvesting, irrigation, transportation), production of inputs (fertilizers, pesticides), atmospheric loss of excess fertilizers in the form of N_2O, emission of CH_4 under anaerobic conditions, loss of soil organic carbon (SOC) are some of the key pathways responsible for the emissions of GHGs from agriculture (Lenka et al., 2015). Animal agriculture (especially beef) is also a major contributor to CH_4 emissions (Subak, 1999). Indirect land-use changes can also cause an increase in GHGs emissions. Indirect land-use changes are often associated with the change in the production or typical usage of a product in one place (e.g. corn production in the USA mid-west going for biofuel production) that may trigger the clearing of forest areas elsewhere (e.g. tropical forest in Africa) and release of huge amount of CO_2 stored in below-ground and above-ground biomass.

3 How Does Climate Change Threaten Agriculture?

Climate change is often associated with the change in climate variables such as rising temperature, increasing vapour pressure deficit, changing rainfall patterns, etc. Such changes are mostly predicted to cause enhanced yield losses due to pathogens (Tonnang et al., 2022), changed water demand by crops (Ort & Long, 2014), depleted nutrient contents of crops (Ainsworth & Long, 2005), and greater sensitivity to drought (Lobell et al., 2014). Both qualitative and quantitative losses are expected to occur especially in densely populated areas (Porter et al., 2014).

4 Ongoing Approaches for a Future-Proof Agriculture

One can imagine two extreme scenarios for a future-proof agriculture. One of the scenarios is accepting the worst and adapting to climate change by either developing improved and new varieties of crops that would be able to withstand projected climate change or having an assessment of how climate change would alter crops' ability to grow in different longitudes/latitudes and outside their typical environment and potential replacement of the existing crops. Another extreme of the scenarios would now be to actually restrict global warming to 1.5 °C to minimize the irreversible impact of climate change; however recent trends of GHGs emissions suggest that this is unlikely to happen (Jaiswal et al., 2019).

Recently successful efforts to develop improved crop varieties resulted from collaboration among people from diverse backgrounds such as genetic and bioengineering, plant biology, crop modelling, and agronomy. The development of an improved variety of soybean that yields > 20% of traditional varieties (De Souza et al., 2022) was assisted by the identification of specific genes responsible for improving photosynthesis using computer simulations. Using IoT's (Internet of Things) and related digital technologies for rapid phenotyping and screening of new plant varieties, in conjunction with genetic engineering, have potential to significantly speed up the slow process of the traditional breeding approach and is only beginning to show its promise in developing crop varieties for future-proof agriculture. Another pathway to climate change adaptation comes via the integration of geospatial technologies, crop modelling, process-engineering and climate modelling aided by remote sensing products. This approach (Fig. 1) was used in assessing the potential of expanding the production of Brazilian ethanol and the role of the same to reduce global greenhouse gas emissions without disturbing forests and other protected areas (Jaiswal et al., 2017). A biophysical crop model BioCro (Lochocki et al., 2022) was parameterized for a popular Brazilian sugarcane variety, and high performance computing resources were used to simulate the yield of sugarcane using current and future projections of climate all over Brazil. The differences in the predicted future yield and past yield of sugarcane were used to identify regions where climate change will favour or be detrimental to sugarcane cultivations in the future. These regional yield maps were further integrated with the satellite-based land-cover products and outputs from sugarcane biorefinery industries to assess CO_2 offset potential. While climate change surely impacts agriculture, a large-scale change in land-use pattern (e.g., planting fast-growing perennials on suitable marginal land) can alter weather patterns also. A collaboration between crop modeller and climate modeller (He et al., 2022) resulted in a coupled tool demonstrating that large-scale land use changes (planting perennial grasses on the marginal land) can alter key climate variables including rainfall and temperature. Such transdisciplinary approaches are essential to tackle the complex challenges in the field of climate-change, agriculture, and their mutual interaction. Attacking such a complex problem from a transdisciplinary perspective often require coming out of one's comfort zone but is the necessity of current time. The next section of this chapter uses an example to discuss why a unidimensional approach may not be sufficient to deal with the issue of climate change and agriculture.

5 Why Will a Unidimensional Approach Not Work? An Example of Disconnect Between Hydrologists and Crop Modelling Communities

5.1 Crop Models and Agriculture

Crop models provide numerous benefits for agriculture. They support crop agronomy, pest management, breeding, and the management of natural resources, the interactions between the atmosphere, the crop, and the soil and also help to study the effects of climate change on plants (Asseng et al., 2014). Plant growth models have been acknowledged as a very useful tool for guiding agronomic management and have been widely used in decision support systems (Guo, 2006). It integrates multidisciplinary knowledge including botany, agronomy, plant physiology, meteorology, soil science, mathematics, and computer science. Crop model outputs help with preseason planning, in-season management choices, and future forecasts of field, regional, and national yield. Models can also be used to maximize profitability, enhance grain quality, decrease environmental impacts and minimize crop failure risks in addition to optimizing yields (Chenu et al., 2017).

5.2 Hydrology and Agriculture

Agriculture is also a major user of water (Fig. 2). Agricultural water use for crops depends on a number of factors, including climate, topography, lithology, soil, management techniques, crop type etc. Knowing these parameters enables estimating agricultural water demand and developing crop management practices. Given the significance of water in agriculture, it is crucial to have a better understanding

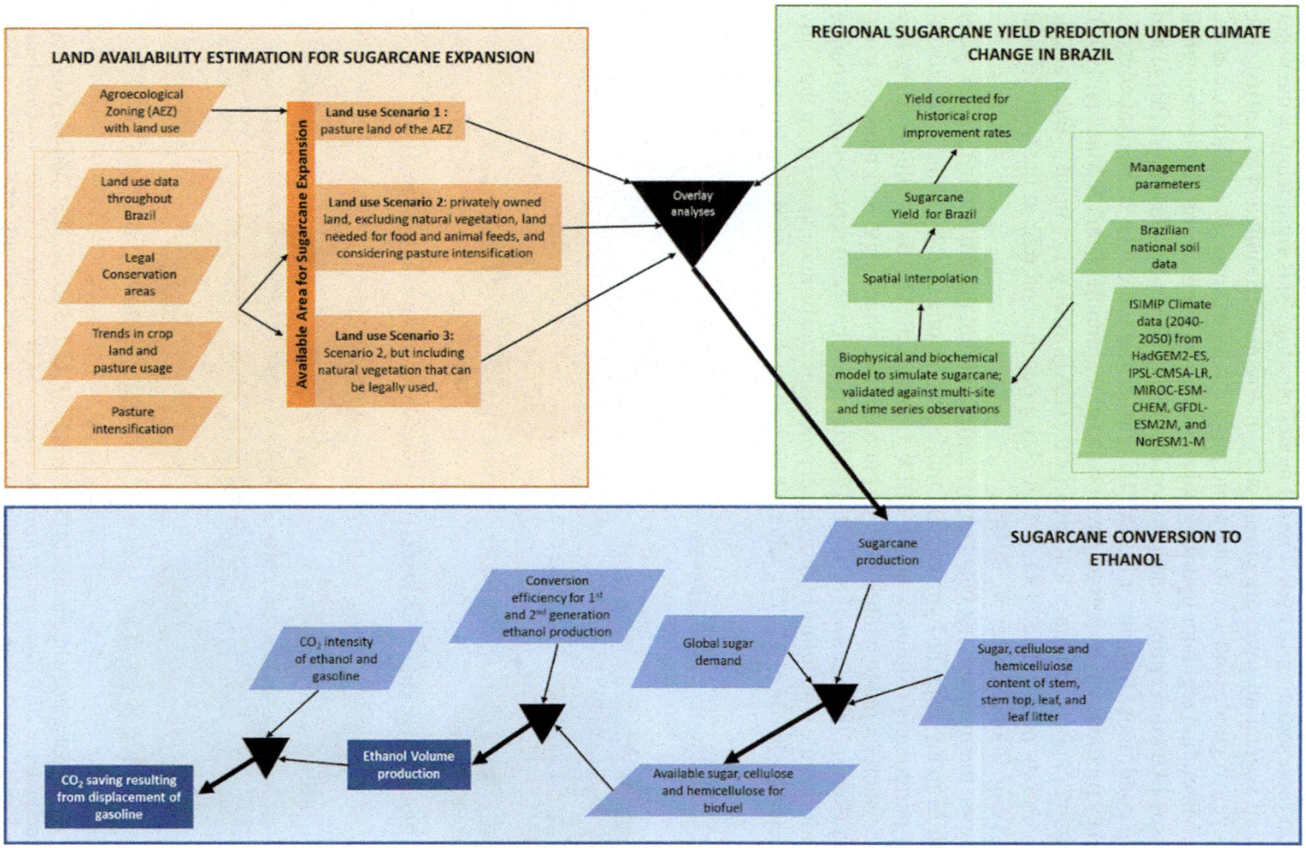

Fig. 1 Overview flow diagram of the method used to determine the future potential of sugarcane ethanol production and resulting CO_2 savings in Brazil. Jaiswal et al., (2017)

of the hydrological condition in crop models in order to enhance their simulation capabilities.

Plant representation in hydrological modelling (Fig. 3) is significant as the water cycle and vegetation cover are intrinsically interrelated (Beaulieu et al., 2016). The water cycle regulates the distribution and productivity of vegetation (Churkina et al., 1999). Vegetation regulates hydrological processes through several eco-hydrological processes like interception, transpiration etc. (Schlesinger and Jasechko, 2014) and surface conditions (Murray et al., 2011). Furthermore, the type of vegetal cover is also a key predictor of evapotranspiration and global runoff (Dunn & Mackay, 1995). Vegetation canopy structure, leaf area index (LAI), root water uptake and transpiration are few plant related factors that play a detrimental role in the hydrological cycle (Peng, 2000; Peng et al., 2014).

Despite the intimate link between dynamics of plant growth and hydrology, plant growth in many popular hydrological models is represented by simple root distribution parameters (Simunek et al., 2013; Liang et al., 2016), while ignoring key features such as photosynthesis, light interception, phenology, and biomass partitioning (Peng, 2000; Jaiswal et al., 2017). The significance of representing dynamic vegetation while simulating hydrological processes can be understood from the study conducted by Jiao et al. (2017), where they derived dynamic vegetation from remote

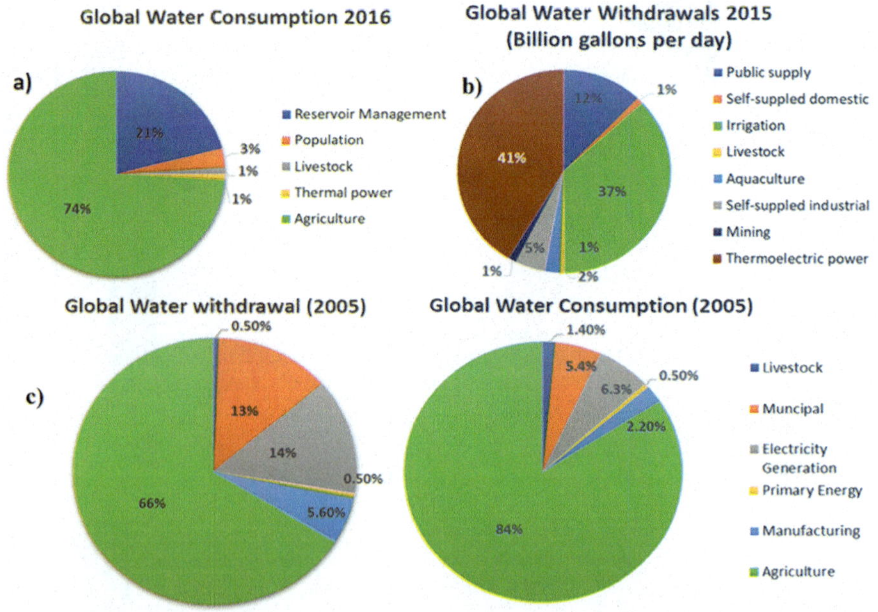

Fig. 2 (**a**) Global water withdrawals 2015 (*Courtesy of the U.S. Geological Survey*), (**b**) Global water consumption 2016 [*Data for the figure has been used from* Qin et al. (2019)] (**c**) Distribution of global water consumption and withdrawals by sector in the year 2005. (*Adapted from* Hejazi et al. (2014))

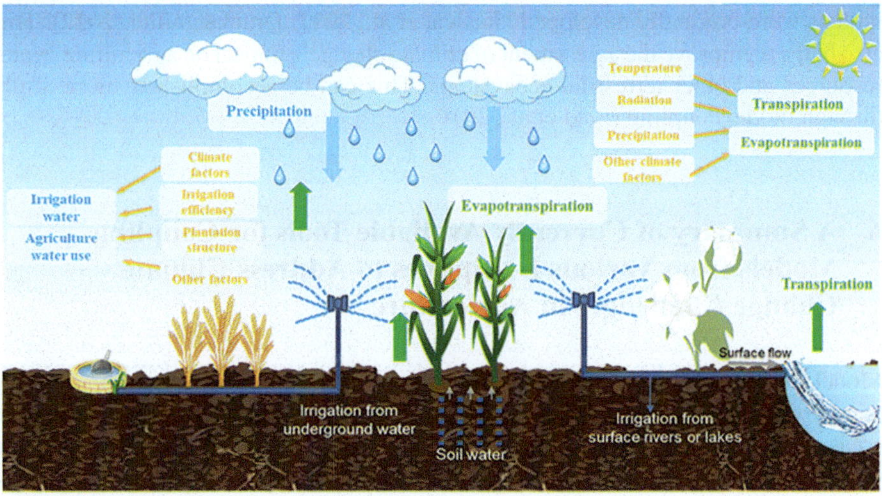

Fig. 3 Role of agricultural water use in hydrological processes by (Wang et al. 2022) is licensed under CC BY 4.0

sensing observations and incorporated it in predictive model simulations which resulted in significantly variable trends in the results of evapotranspiration, runoff, soil surface moisture, and river discharge compared to the results produced using static vegetation conditions.

On the other hand, mechanistically rich crop growth models emphasize biophysical processes responsible for vegetation dynamics but often employ a simple representation of hydrological processes particularly in the soil water domain (Jones et al., 2003; Jaiswal et al., 2017). Ma et al. (2009) simulated the responses of various crops for different methods of estimating soil hydraulic properties and root distribution under various water management conditions and concluded that accurate simulation of plant growth depends not only on plant parameters, but also on soil parameters and accurate estimation of them. Even while soil-crop models include a well-established description of soil and water, they frequently lack an accurate representation of weather variables (Lopez-Jimenez et al., 2022).

As a result, a gap exists between hydrologists and plant scientists in terms of adapting hydrological model simulations to vegetation dynamics and vice versa (Blöschl et al., 2019). This gap often results in inappropriate assumptions in making climate change impact assessment and also finding ways for climate change adaptation.

Hence, it is essential to bring the mechanistically rich features of hydrological and crop growth models together especially in the context of climate change and its impact on hydrological systems (Beaulieu et al., 2016), vegetation growth (Jaiswal et al., 2017) and sustainable agriculture. Not only is it important to bring them together, it is equally important to integrate process-rich physical based models that can make reliable predictions beyond the spatial and temporal scales under which models were originally developed (Jaiswal et al., 2017; Druckenmiller, 2022). This also helps to improve the assessment of climate change's impact on agriculture, water resources and other terrestrial ecosystems. This can be accomplished successfully through model coupling (Siad et al., 2019).

6 A Summary of Currently Available Tools for Coupling Models from Various Disciplines to Address Climate Change Challenges in Agriculture

Model coupling can make it easier to analyse the relationships inside and among complex systems, revealing new realities about the world and shedding light on areas of science that are yet to be explored. The necessity for integrated simulation models to address concerns about sustainable agriculture production in relation to resource scarcity and climate stresses has been further highlighted by challenges like food security, environmental degradation, and climate change.

For example, an accurate simulation of water flow distribution can help crop models by providing a more accurate estimate of evapotranspiration rates. Since hydrological models include representations of all these processes, connecting hydrologic and crop growth models is expected to be advantageous for both simulations (Manfreda et al., 2010). There are numerous studies that have integrated hydrological and crop growth models for a variety of agricultural applications, such as the simulation of furrow irrigation and crop yield (Wang et al., 2014), to study the effect of CO_2 on grassland (Kellner et al., 2017), Simulations of Soil Water Dynamics in the Soil–Plant-Atmosphere system (Shelia et al., 2018), Irrigation water salinity impacts assessment (Wang et al., 2017), Impacts of groundwater balance on cotton growth (Han et al., 2015), Water and nitrogen management (Liang et al., 2016), Agricultural water management (Zhang et al., 2012), Irrigation modelling of wheat cultivation (Zhou et al., 2012), and many more.

There are various ways to couple models, from simple manual data exchanges to automated frameworks for integration (Table 1).

According to Siad et al. (2019), the degree of interdependence between model variables is referred to as the level of coupling. In high-level coupling (e.g. embedded or integrated), each component and its linked one must be presented in such a way that the code or framework can be executed. At the same time, low level coupling (e.g. shared and loose coupling) enables communication and independent management among components. In decoupled coupling (sequential coupling), the components function separately and independently.

Table 1 Types of coupling

<table>
<tr><th>Sr. No</th><th>Type of coupling</th><th colspan="2">Description</th></tr>
<tr><td>1</td><td>Sequential coupling</td><td colspan="2">Models are completely decoupled</td></tr>
<tr><td>2</td><td>Loose coupling</td><td colspan="2">Models exchange I/O data</td></tr>
<tr><td rowspan="2">3</td><td rowspan="2">Shared coupling</td><td>Unified GUI</td><td>Models share graphical user interfaces</td></tr>
<tr><td>Shared Data</td><td>Models share I/O database</td></tr>
<tr><td rowspan="2">4</td><td rowspan="2">Tight coupling</td><td>Embedded</td><td>One model is completely contained in the other (usually as a subroutine)</td></tr>
<tr><td>Integrated</td><td>Models are merged at the source code level in one coherent model</td></tr>
<tr><td>5</td><td>Framework</td><td colspan="2">Using an overall modelling framework, where the models are coupled using a third-party tool commonly called "Coupler" based on a combination of the previous methods. The original code can be wrapped using a specialized software such as OMS3 (David et al., 2016), OpenMI (Gregersen et al., 2007), or PYTHON (www.python.org)</td></tr>
</table>

Source Siad et al. (2019), Peña-Haro et al. (2012)

6.1 Coupled Models for Agricultural and Eco-Hydrological Simulations

6.1.1 Code Wrapping and Modelling Framework

The Object Modelling System (OMS) is a framework and development tool for designing, building, validating, and deploying agro-environmental models (David et al., 2016). It is a framework for component-based environmental modelling. Although based on the Java platform, it can communicate with C/C++ and FORTRAN through the Java Native Access (JNA) technology on all significant operating systems and architectures. It has been employed in many agricultural applications, including enhancing the use of water resources and predicting crop production in local agricultural management (Zhang et al., 2012). Similar to this, Peña-Haro et al. (2012) combined the crop growth and production model WOFOST (Van Diepen et al., 1989) with the unsaturated flow model HYDRUS-1D and the saturated flow model MODFLOW. External coupling (via input/output data manipulation) and code wrapping were used in combination to achieve the coupling. For wrapping WOFOST, OMS3 was used and for writing scripts, PYTHON was used. The importance of a coupled model was demonstrated during the model's testing in a synthetic case, and crop development based on ground water level was studied.

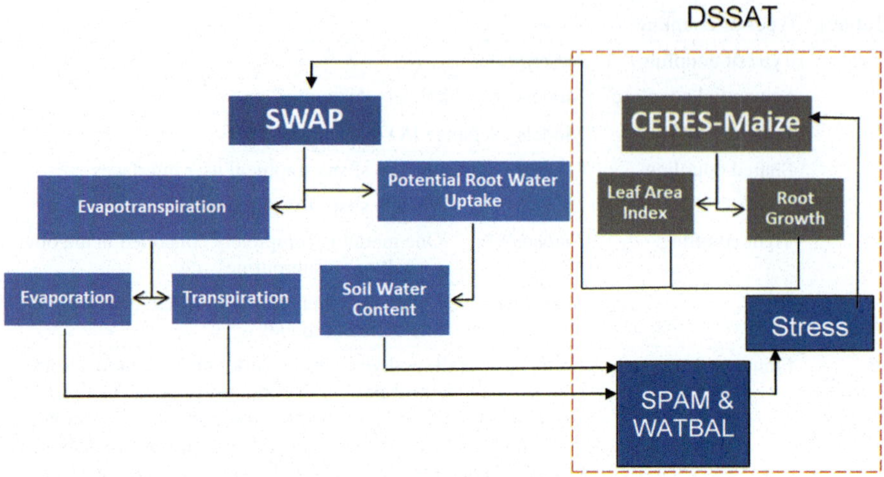

Fig. 4 CERES-Maize-based algorithm. Dokoohaki et al., (2016)

6.1.2 Source Code Modification and Embedded Coupling

Dokoohaki et al., (2016) coupled the Cropping System Model (CSM)-Crop Environment Resource Synthesis (CERES)-Maize (CSM-CERES-Maize) with the Soil Water Atmosphere and Plant (SWAP) model in order to benefit from the advantages of both models for the purpose of biomass simulation.

Prior to combining the CSM-CERES-Maize and SWAP models, the structure and algorithm of each model's component was examined. SWAP model was made simpler by removing certain modules and converting required input data format into DSSAT format. The WATBAL and SPAM modules in the DSSAT V4.0 package were then used to replace the obtained SWAP model (Fig. 4).

6.1.3 Using Python

In this study by Kellner et al. (2017), the Python programming language was used to couple a hydrological model, Catchment Modelling Framework-CMF (Kraft et al., 2011) and the plant growth modelling framework-PMF (Multsch et al., 2011) as recommended by Perkel (2015) for the investigation of CO_2 effects on a permanent, temperate grassland system. This coupled model has also been used to simulate wheat development under different management strategies (Houska et al., 2014).

6.2 Coupled Models for Agricultural and Plant Biological Applications

The role of plant genomes and phenotypes is critical to ensure food security and make agriculture more efficient. Agrigenomics or the use of genomics in agriculture, has and will continue to boost sustainable productivity and provide answers to the growing problems associated with feeding the world's population. The genetic markers connected to desirable traits can help develop new varieties which care climate resilience, highly productive, require lesser resources, and are easy to manage. To better understand the genetic variety influencing phenotypes, researchers are studying agricultural species using genotyping and next-generation sequencing technology. Crop modelling along with GIS analysis is expected to play an important role in identifying specific varieties and traits suitable for a specific site and climate (Long et al., 2015) while avoiding expensive multi-year and multi-site field trials.

Plant models range from empirical to mechanistic. Then there are data-driven machine learning models that are completely predictive. Although empirical models are an excellent tool to reflect plant conditions accurately, they are not very good at forecasting emergent features. Machine learning models have its predictive capability constrained to the data used to train them (Baker et al., 2018). Mechanistic models, are based on physical processes and have the ability to predict using mathematical representations, but their predictive capabilities are also frequently limited to the range of observed data (Baker et al., 2018).

To comprehend how plants react to a changing climate or to investigate how plants can be engineered to achieve particular phenotypic goals that could have its wide range of application in agriculture, it is necessary to integrate empirical and mechanistic modelling approaches and create models that span across biological levels through integrative multiscale modelling (Baker et al. 2018; Gazestani & Lewis, 2019; Zhang et al., 2020). Plant models can also be microscale and macroscale. Coupling models that are specific to cells, genes and tissues to organ, plant growth, physiological and ecological models are equally important.

It is in such cases that modelling framework coupling tools like yggdrasil (Lang, 2019) and OpenAlea (Pradal et al., 2008) can be used.

OpenAlea follows a top to down approach focusing on integrating ecophysiological and agronomic processes with plant architectural models. Pradal et al. (2015) has also demonstrated a well explained workflow in phenotyping, phenomics and environmental control designed to study the interaction between plant architecture and climate change. In a different study conducted by Yang et al. (2016), OpenAlea was used as a component of work to examine how changes in the transitions between vegetative and reproductive growth in a variety of apple cultivar are related to the impact of long-term water stress on tree architecture and production. It was concluded that this study could be used to open new avenues for in silico studies of agronomical scenarios.

Virtual Laboratory (VLab)/L-studio environment is another prominent platform for running, connecting, and integrating models within the plant community

(Prusinkiewicz et al., 2007). It has commonly been used to describe the growth of plants at scales ranging from individual cells and tissues to entire plant ecosystems. Recent versions of LStudio/VLab, which is based on L-systems, was enhanced with the compiled language L+C (Prusinkiewicz et al., 2007). L+C, in conjunction with the programming language XL (extended L-Systems), enables more flexible integration by facilitating the specification of complex plant models. Based on real-world parameters, this was used to generate 3D structural tomato modelling (He et al., 2010). L-Studio can also be used to visualize data and modify plant architecture to create architectures resulting from applying different management practices (Auzmendi & Hanan, 2018).

The OpenMI interface is another interface that helps in connecting hydrological and environmental models (Moore & Tindall, 2005). Marshall-Colon et al. (2017) has also mentioned about various modelling environments and frameworks like Cactus (Goodale et al., 2003), SemGen (Gennari et al,. 2011), Swift (Wilde et al. 2011) and many others whose problem solving frameworks could be used to develop an integrated environment for plant community that can help with agriculture.

6.3 Coupling Challenges

Although model coupling can help overcome barriers in connecting complex systems, it is not without its own set of challenges.

- As individual process-based model components have evolved in complexity and specialization over time, their accuracy has increased but it has become increasingly difficult to integrate them into useful multiscale models.
- The other challenge faced during combining models is the strategy of hard-coding interlinked parameter values, equations, solution algorithms and user interfaces while model development, rather than treating these each as separate components.
- Additionally, it is challenging to integrate independently developed computational models because of compatibility issues brought on by different programming languages or data formats.
- Both spatial and temporal scales must be compatible with models and the corresponding data. Even though the models may share information, the coupled system's results are meaningless if the models' scales are different.
- In the context of agriculture, ecology and hydrology, there are eco-hydrological models that can be used to combine complex interactions between agro-ecological and hydrological cycles, such as APEX (Izaurralde, 2010), APSIM (Keating et al., 2003), AquaCrop (Steduto et al., 2009), CropSyst (Stöckle et al., 2003), DAISY (Abrahamsen & Hansen, 2000), DNDC (Gilhespy et al., 2014), DSSAT (Jones et al., 2003), STIC (Brisson et al., 1998), and WOFOST (Van Diepen et al., 1989). But they are constrained to specific data, specific purpose, and specific scales.

6.4 An Introduction to yggdrasil

A majority of the coupling challenges can be overcome by using a framework with the help of third-party coupling tools. yggdrasil is one such tool (Lang, 2019). Although OMS3 provides a framework for component-based environmental modelling, it can only interface with C/C++ and FORTRAN, giving yggdrasil the upper hand because yggdrasil is capable of handling and coupling models developed in different programming languages.

yggdrasil is an open-source Python coupling tool that can connect existing computational models written in different languages and execute integration network. yggdrasil was primarily created with the goal of integrating models of plant biology, but it may also be used for other models with programming languages that yggdrasil supports. yggdrasil as of now supports models written in R, Python, FORTRAN, MATLAB, C/C++ , and Java. yggdrasil was created as a part of the Crops in Silico program, which aims to construct an entire crop in silico, from the level of genes to field level (Marshall-Colon et al., 2017). The entire documentation can be found in GitHub (https://github.com/cropsinsilico/yggdrasil).

6.4.1 Application of yggdrasil

Below are a couple of studies that have used yggdrasil for research in plant biology.

Shameer et al. (2022) used yggdrasil to couple a flux balance analysis (FBA) model with a dynamic photosynthesis model named e-photosynthesis capable of capturing the kinetics of photosynthetic reactions and associated chloroplast carbon metabolism. FBA models are capable of simulating large metabolic networks in a computationally efficient way as they are composed of linear equations. Although this makes them computationally efficient, most of the biological processes are nonlinear and FBA models cannot represent them directly. It is for this that they coupled the FBA model with e-photosynthesis, which is a kinetic model composed of ODE's that can capture nonlinear processes also. FBA and ODE modules were formulated, and they were coupled in two ways, a general loosely coupled approach and a tailored tightly coupled approach. The values between both the modules were passed using yggdrasil while coordinating their parallel execution.

In another study by Kannan et al. (2019), the yggdrasil framework has been used to one-way couple a protein translation model (Python) and a leaf photosynthesis model (MATLAB). Scaling of modelling processes from gene expression to photosynthetic metabolism to predict leaf physiology has been done in order to assess how well photosynthesis adapts to increasing atmospheric CO_2 concentrations.

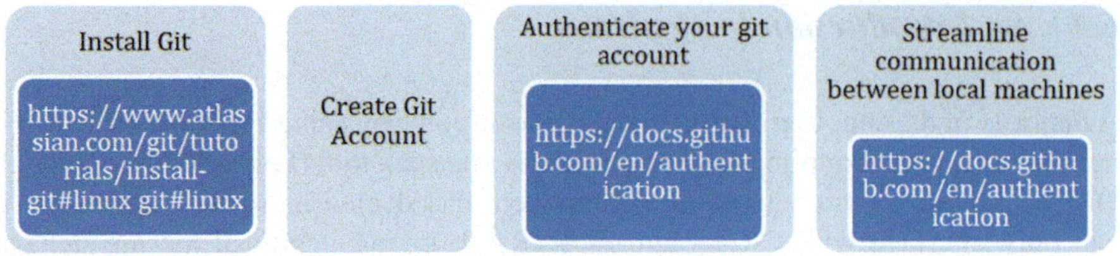

Fig. 5 Git installation and connecting to local machine

6.4.2 Installation and Working with yggdrasil

As a first step to working with yggdrasil, we reviewed the material from the 5th Annual Crops in Silico Symposium and Hackathon, which took place in June 2021, to learn more about yggdrasil. (https://cropsinsilico.org/2021-annual-meeting/). These resources go through some of the fundamentals of using yggdrasil to link models in Jupyter notebooks (https://cropsinsilico.github.io/yggdrasil/hackathon2021/index.html). We will work in the environment provided by Cis2021 Hackathon, which requires a different installation process to access the materials and use them, since the hackathon offers a variety of materials, such as simple models to learn using yggdrasil. Alternately, once we are familiar with how yggdrasil functions, we can install it on a local machine using the instructions provided here. (https://cropsinsilico.github.io/yggdrasil/).

Installation Using Cis2021 Hackathon Instructions (Ubuntu)

1. Install GitHub in order to access the yggdrasil files by carrying out the following steps (Fig. 5):
2. Materials from Jupyter notebooks stored in the repository https://github.com/cropsinsilico/CiS2021-hackathon forms the basis for this hackathon. We can use MyBinder, a local install, or a Docker Container to run them. In order to access and use yggdrasil, we used the Docker Container (https://cropsinsilico.github.io/yggdrasil/hackathon2021/setup.html#docker-container) to run the files.

Using Docker Container

(a) Open the Docker Desktop after installing according to the directions provided in the above link (Docker Container). Choose Images and the image named langmm/hackathon2021:0.0.8 and run it (Fig. 6).
(b) In the terminal, navigate to the directory where the Docker Container folder was created. Then type the command docker run -it --rm -p 8888:8888 -e NB_UID=$(id -u) --user root -v **<some directory name>**:/tmp langmm/hackathon2021:0.0.8 (Fig. 7).
(c) Thereafter, navigate to the link http://localhost:8888/tree to access the Jupyter notebook to work with the Cis2021 Hackathon materials.

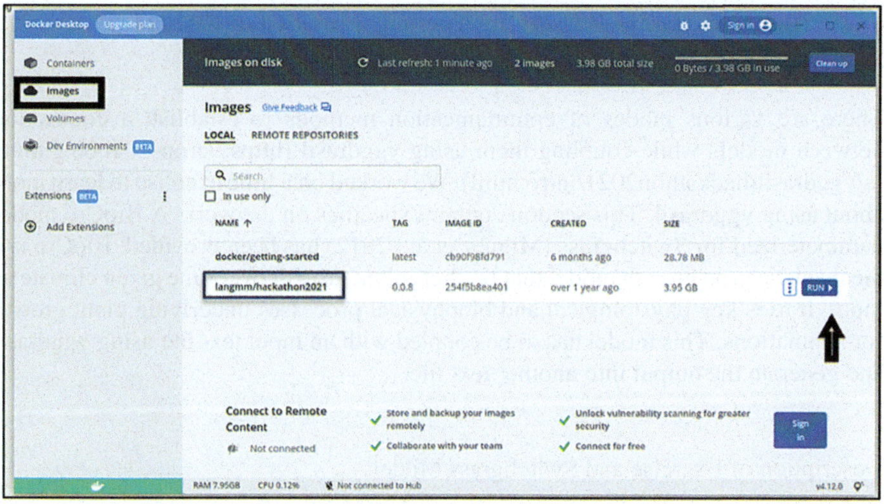

Fig. 6 Screenshot Docker image

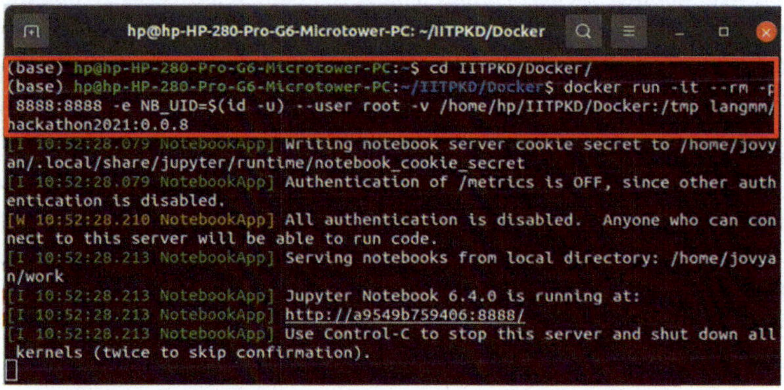

Fig. 7 To open Jupyter notebook using terminal

(d) Alternatively we can also run the command $ sudo docker run -it --rm -p 8888:8888 -e NB_UID=$(id -u) –user root --v<**some directory name**>:/tmplangmm/hackathon2021:0.0.8 in the terminal and open the link http://localhost:8888/tree to access the Jupyter notebook to work with the Cis2021 Hackathon materials.

6.4.3 Running yggdrasil Using Instructions Provided in Cis2021 Hackathon

There are various modes of communication methods to establish a connection between models while coupling them using yggdrasil (https://cropsinsilico.github.io/yggdrasil/hackathon2021/intro.html). We worked on a little exercise to learn more about using yggdrasil. This section contains specifics on the work. A BioCro model parameterized for Switchgrass (Miguez et al., 2012) has been provided. BioCro is a mechanistic crop growth model that predicts plant growth over time given climate as input. It uses key physiological and biophysical processes underlying plant growth for simulations. This model has to be coupled with an input text file using yggdrasil and generate the output into another text file.

Description of Exercise and Switchgrass Model

In this particular example, we will be expressing our model as a function. While integrating models as functions, yggdrasil wraps a model that can be written as a function with another model without having to create a new code to integrate both the models (https://cropsinsilico.github.io/yggdrasil/autowrap.html).

The BioCro model provided here is a model parameterized to simulate the growth of Switchgrass (Miguez et al., 2012) and will be referred to as the Switchgrass model henceforth. The online documentation (https://ebimodeling.github.io/biocro-documentation/) and *A Practical Guide to BioCro* provided whose details have been provided in Lochocki et al. (2022) both provide more information on the functions and library entries present in the BioCro II R package. *A Practical Guide to BioCro* provides basic examples of using BioCro to conduct a simulation and calculate a response curve, along with R essentials and other information that is beneficial to beginners using BioCro.

Case 1. Run the Switchgrass model with one input text file.

The Switchgrass model was connected to an input text file that serve as the output from some model to be integrated with BioCro (Fig. 8). This text file will contain one parameter value from the switchgrass model being run, with the exception that the parameter values are different from what is used in the switchgrass model. In order to run a model using yggdrasil, a YAML configuration file is required. This YAML file is a file that is written to help yggdrasil understand the model. It provides the name of the model, location, language, function (model) that needs to be wrapped. Not only is a YAML configuration file required to describe the model, but a YAML connection file with details of an input text file and an output file is also required. The input and output files are put into channels so that yggdrasil can use them to integrate with models. The YAML file connects the inputs/output and the model files.

Firstly a new folder "Switchgrass" is created in the file browser opened via the link http://localhost:8888/tree (Fig. 9).

Fig. 8 Switchgrass model integrated using channel connections

Fig. 9 Creating folder “Switchgrass” in Jupyter

The necessary files that need to be worked with like the R Script (plot_optimization_1input.R), weather input file (growing_season.rds), BioCro package to be installed (BioCro_1.00.tar.gz), input text file (alphaStem.txt), and YAML file (BioCro_1input) are uploaded (Fig. 10).

The R Script plot_optimization_1input.R is the Switchgrass model code in R. The input parameters are provided as shown in Fig. 11.

There are several other modules used to simulate Switchgrass as shown below in Fig. 12.

In this case we have called the model as a function of the input parameter alphaStem. The major input driver that BioCro uses is the weather input file (Fig. 13). Figure 14 shows the Switchgrass model modified to perform as a function. As shown in Fig. 14, the first line installs the BioCro package, the fifth line calls Switchgrass model as a function (Fig. 14), and the eighth line assigns the weather parameters from the RDS file (Fig. 13). The value of alphaStem (the input parameter

Fig. 10 Files uploaded in Jupyter notebook to run the model

```
parameters_list_switchgrass = with(list(), {
  datalines =
    "symbol value
    lat                26.639      # latitude, needed in many modules
    soil_clod_size     0.04        # needed in two_layer_soil_profile module
    soil_reflectance   0.2         # needed in two_layer_soil_profile module
    soil_transmission  0.01        # needed in two_layer_soil_profile module
    specific_heat      1010        # needed in two_layer_soil_profile module
    stefan_boltzman    5.67e-8     # needed in two_layer_soil_profile module
    iSp                1.9         # needed in parameter_calculator (lai /leaf 
    Sp_thermal_time_decay  0.00038 # neeeded in parameter_calculator, linear de
    mrc1             0.0008333     # needed in no_leaf_resp_partitioning_growth_
    mrc2             0.000025      # needed in no_leaf_resp_partitioning_growth_
    nlayers            10          # needed in c4_canopy module , number of can
    chil               1           # needed in c4_canopy module , leaf angle di
    kd                 0.3         # needed in c4_canopy module , leaf angle di
    heightf            3           # needed in c4_canopy module
    leafwidth          0.04        # needed in c4_canopy module
```

Fig. 11 Parameter list for Switchgrass model

```
initial_state_switchgrass = with(list(), {
  datalines =
    "symbol value
    Rhizome 0.02                          # Mg/ha
    Leaf    0.01                          # Mg/ha, 80% of total seed mass per land area
    Stem    0.01                          # Mg/ha, 10% of total seed mass per land area
    Root    0.02                          # Mg/ha, 10% of total seed mass per land area
    Grain   0.0                           # Mg/ha # treating this as the Pod/reproductive (minus seed)
    soil_water_content 0.37               # Needed in two_layer_soil_profile module
    LeafN 2
    TTc 0                                 # Accumulated thermal time at the start
    LeafLitter 0                          # Leaf litter at the start
    RootLitter 0                          # Root Litter at the start
    RhizomeLitter 0                       # Rhizome litter at the start
    StemLitter 0                          # Stem Litter at the start
     rhizome_senescence_index 0           # Not used but needed  in the module 'partitioning_growth'
    ## needed for two layer soil profile
    cws1    0.185                         # equal to initial soil_water_content, from Justins sorghum parameters (see 10/3/19 email -m
    cws2    0.185                         # equal to initial soil_water_content, from Justins sorghum parameters (see 10/3/19 email -m
    DVI -1                                # Used in development_index.h,miscanthus_partitioning_coefficient_logistic,thermaltime_devel
    "
  data_frame = read.table(textConnection(datalines), header=TRUE)
  values = as.list(data_frame$value)
  names(values) = data_frame$symbol
  values
})


ss_modules_list_swithcgrass = c('soil_type_selector','stomata_water_stress_linear',
                  'leaf_water_stress_exponential','parameter_calculator',
                                  'solar_zenith_angle', 'shortwave_atmospheric_scattering', 'incident_shortwave_par_from_cwrf',
                                  'ten_layer_canopy_properties','ten_layer_c4_canopy', 'ten_layer_canopy_integrator',
                                  'thermaltime_development_rate_calculator',
                                  'logistic_partitioning_with_rhizome_renobilization',
                                  'partitioning_growth_with_rhizome_as_reserved_c_storage','delta_TT'
)


deriv_modules_list_switchgrass = c('miscanthus_senescence',
                              'partitioning_growth',
                              'two_layer_soil_profile',
                              'development_index',
                              'thermal_time_accumulator')
```

Fig. 12 Various modules to run Switchgrass model

	year	doy	hour	solar	temp	rh	windspeed	precip	solardiff
2257	2002	95	0	0.00000	1.72395477	1.0000000	1.955702	2.236739e-04	0.00000
2258	2002	95	1	0.00000	0.98976793	1.0000000	1.955702	2.236739e-04	0.00000
2259	2002	95	2	0.00000	0.42640656	0.3855464	1.955702	2.236739e-04	0.00000
2260	2002	95	3	0.00000	0.07226279	0.6666055	1.955702	2.236739e-04	0.00000
2261	2002	95	4	0.00000	-0.04852905	1.0000000	1.955702	2.236739e-04	0.00000
2262	2002	95	5	0.00000	0.07226279	1.0000000	1.955702	2.236739e-04	0.00000
2263	2002	95	6	130.89521	0.42640656	1.0000000	1.955702	2.236739e-04	25.22248
2264	2002	95	7	819.24407	0.98976793	1.0000000	1.955702	2.236739e-04	52.03545
2265	2002	95	8	1050.76715	1.72395477	0.9834020	1.955702	2.236739e-04	58.45403
2266	2002	95	9	1152.60573	2.57893346	1.0000000	1.955702	2.236739e-04	61.11284
2267	2002	95	10	1203.97880	3.49643860	1.0000000	1.955702	2.236739e-04	62.42264
2268	2002	95	11	1229.09722	4.41394374	1.0000000	1.955702	2.236739e-04	63.05596

Fig. 13 Weather input file growing_season.rds

```
install.packages("BioCro_1.00.tar.gz", repos = NULL, type = 'source')
# BioCro build from the branch "direct_radiation_input" of biocro-dev
library(BioCro)

switchgrass = function(alphaStem){

# parameters and weather inputs
growing_season = readRDS(file = "../Switchgrass/growing_season.rds")

```

Fig. 14 Calling Switchgrass model as function

```
# alphaStem
# %g
10.0012

```

Fig. 15 Input text file alphaStem.txt

```
168  parameters_list_switchgrass$alphaStem = alphaStem
169
170  growing_season = get_growing_season_climate(growing_season, threshold_temperature = 0)
171
172  result = Gro_solver(initial_state =  initial_state_switchgrass,
173                      parameters = parameters_list_switchgrass,
174                      varying_parameters = growing_season,
175                      steady_state_module_names = ss_modules_list_swithcgrass,
176                      derivative_module_names = deriv_modules_list_switchgrass, verbose = TRUE)
177    stem = max(result$Stem)
178  }
```

Fig. 16 Model solver command

for which the model is called as a function for) is assigned using the input text file (Fig. 15). All the remaining modules used to run the Switchgrass model comes inside the called function "switchgrass".

The parameter alphaStem is assigned as shown in line 168 of Fig. 16 and the model is solved using the commands shown in Fig. 16.

Fig. 17 YAML file to run model and connect the model with the input/output text file

```
model:
  name: Switchgrass
  language: R
  args: ../Switchgrass/plot_optimization_linput.R
  function: switchgrass

connections:
  - input:
      name: ../Switchgrass/alphaStem.txt
      filetype: table
    output: Switchgrass:input
  - input: Switchgrass:output
    output:
      name: ../Switchgrass/result.txt
      filetype: table
      field_names: [stem]

```

The output that needs to be written into an output text file, the maximum biomass of the stem, is provided in the code as shown in line 177 of Fig. 16.

The YAML file required to configure and connect the model with the input/output files and channels is as shown in Fig. 17.

Once all the files are ready to run, the YAML file is run in Jupyter notebook as shown in Fig. 18.

Output generated is written into a text file (Fig. 19).

Alternatively, yggdrasil can also be run in the terminal as shown in Figs. 20 and 21 using the yggrun command.

Case 2. Run the Switchgrass model with two input text files and obtain two values of output.

In this case the Switchgrass model was connected to two input text files. And the output consists of two values (Fig. 22).

The code is modified as shown in Figs. 23, 24, and 25. The function includes two arguments, alphaStem and betaStem line 6 of Fig. 23. The parameter is assigned as shown in line 182 and 183 of Fig. 24 and the results to be written into an output text file are called for as shown in line 193 of Fig. 24. The input text files are shown in Fig. 26.

The YAML file is modified as shown in Fig. 26.

The output obtained is as shown in Fig. 27.

6.4.4 Running yggdrasil in the Local Machine (Ubuntu)

To show how yggdrasil operates in a local machine, we used an example problem light model from the Cis2021 Hackathon materials. This problem uses a simple model to describe how to use yggdrasil to run a model as a function with YAML configuration and a connection file.

After the installation of yggdrasil in the local machine (Refer sec 6.4.2), a directory [Model_asfunction (Fig. 28)] was created to download the model script (Fig. 29),

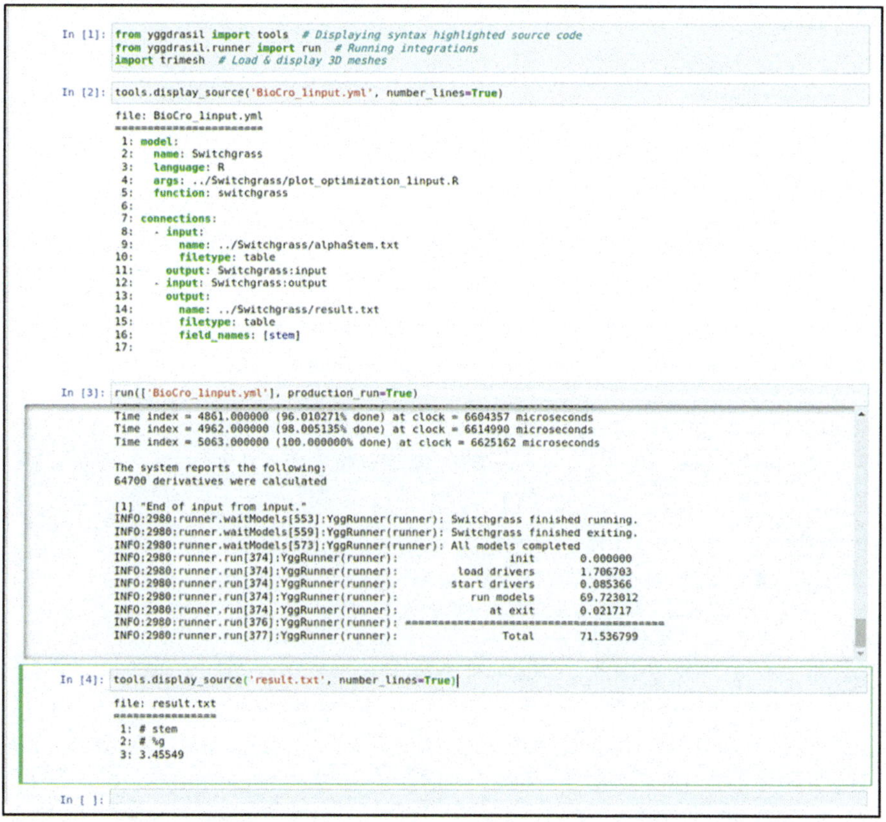

Fig. 18 Notebook with commands to run model using yggdrasil

Fig. 19 Output text file

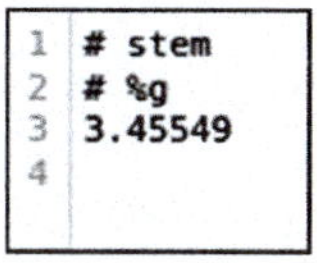

input text file (Fig. 30) and the YAML configuration and connection file (Fig. 31) from the Cis2021 Hackathon materials into the local machine, inorder to run the model using yggdrasil.

The model is run using the command yggrun <YAMLFILE> as shown in Fig. 32. The output obtained is shown in Fig. 33.

Several other simple examples are also available here (https://cropsinsilico.github.io/yggdrasil/hackathon2021/notebooks.html) to work with yggdrasil's operation. These are examples demonstrating integrating models as functions in Python, C++ and FORTRAN, integrating models using interfaces, one way model to model connection, Remote Procedure Call (RPC) and many more.

Fig. 20 To run in terminal similar to running in Ubuntu terminal

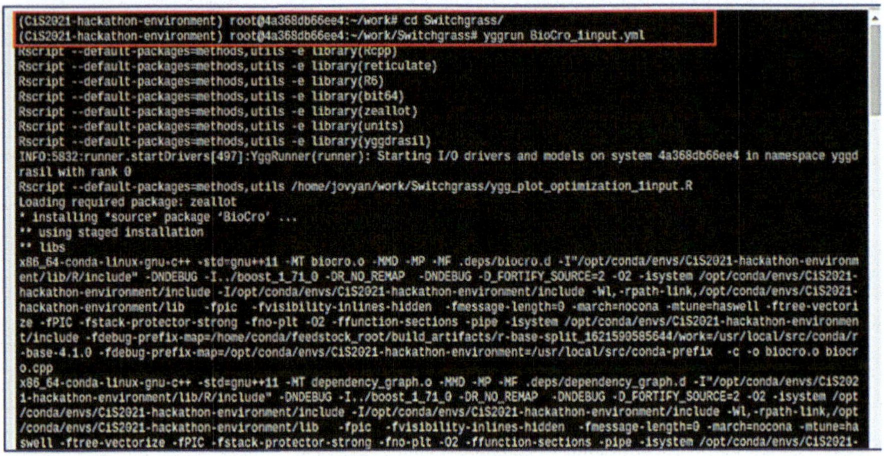

Fig. 21 Command to run YAML file for the model in terminal

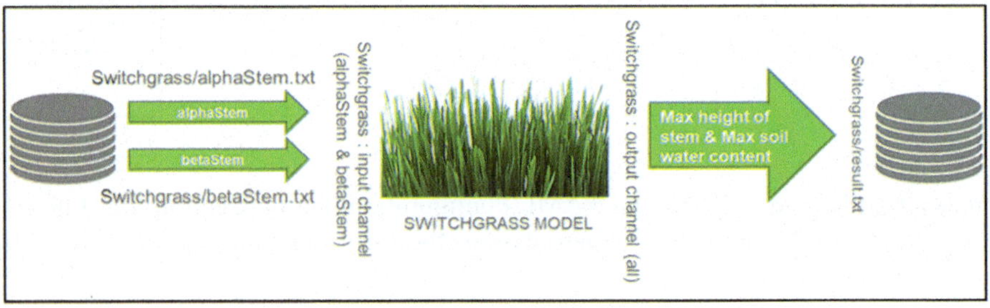

Fig. 22 Switchgrass model integrated using 2 input channel connections

```
install.packages("BioCro_1.00.tar.gz", repos = NULL, type = 'source')
# Biocro build from the branch "direct_radiation_input" of biocro-dev
library(BioCro)
#library(lattice)

switchgrass = function(alphaStem, betaStem){
```

Fig. 23 Calling function with two arguments

```
parameters_list_switchgrass$alphaStem = alphaStem
parameters_list_switchgrass$betaStem = betaStem

growing_season =  get_growing_season_climate(growing_season, threshold_temperature = 0)
result = Gro_solver(initial_state =  initial_state_switchgrass,
                     parameters = parameters_list_switchgrass,
                     varying_parameters = growing_season,
                     steady_state_module_names = ss_modules_list_swithcgrass,
                     derivative_module_names = deriv_modules_list_switchgrass, verbose = TRUE)
  stem = max(result$Stem)
  soilwater = max(result$soil_water_content)
  return(list(stem,soilwater))
}
```

Fig. 24 Assigning parameters and two values of output

```
# alphaStem
# %g
10.0012

```

```
# betaStem
# %g
-5.555

```

Fig. 25 Input files

```
model:
  name: Switchgrass
  language: R
  args: ../Switchgrass/plot_optimization_2input2result.R
  function: switchgrass
  inputs:
      - name: alphaStem
      - name: betaStem
  output:
      - name: all

connections:
  - input: ../Switchgrass/alphaStem.txt
    output: alphaStem
    filetype: table
  - inputs: ../Switchgrass/betaStem.txt
    output: betaStem
    filetype: table
  - input: all
    output: ../Switchgrass/result.txt
    filetype: table
    field_names: stem,soilwater

```

Fig. 26 YAML file

```
# stem	soilwater
# %g	%g
4.28824	0.400079
```

Fig. 27 Output text file result.txt

Fig. 28 Directory to download light model

```
#' Compute the intensity of light.
#'
#' @param doy Day of year.
#' @param height Distance from ground in cm.
#'
#' @return Intensity of light in ergs cm^-2 s^-1.
light <- function(doy, height) {
  # Define parameters that are static across a run
  amplitude <- units::set_units(80.0, 'ergs cm-3 s-1')
  doy_offset <- units::set_units(0.0, 'days')

  # Calculate intensity
  intensity <- amplitude * height * (1.0 + sin(2.0 * pi *
    units::drop_units((doy - doy_offset) / units::set_units(365, 'days'))))
  return(intensity)
}
```

Fig. 29 Light model R Script

7 Conclusion

Here, we provided a brief review of the tools that are available for model coupling with emphasis on agriculture and plant sciences. We have also discussed yggdrasil in details and used very simple exercises to demonstrate two different ways to couple models using yggdrasil. We believe that this chapter will be useful for those who

```
# doy    height
# days   cm
# %f     %f
0        0
1        0.5
2        1.0
3        2.0
4        4.0
5        5.0
6        7.0
7        10.0
```

Fig. 30 Input light model

```yaml
model:
  name: light
  language: R
  args: ../Model_asfunction/light_v0.R
  function: light

connections:
  - input:
      name: ../Model_asfunction/light_v0.txt
      filetype: table
    output: light:input
  - input: light:output
    output:
      name: ../Model_asfunction/output.txt
      filetype: table
      field_names: [intensity]
```

Fig. 31 YAML configuration and connection file for light model

```
(base) hp@hp-HP-280-Pro-G6-Microtower-PC:~/IITPKD/Book_Chap_yggdrasil/Trial_yggdrasil/Model_asfunct
ion$ yggrun light_v0_R.yml
Rscript --default-packages=methods,utils -e library(Rcpp)
Rscript --default-packages=methods,utils -e library(reticulate)
Rscript --default-packages=methods,utils -e library(R6)
Rscript --default-packages=methods,utils -e library(units)
Rscript --default-packages=methods,utils -e library(bit64)
Rscript --default-packages=methods,utils -e library(zeallot)
Rscript --default-packages=methods,utils -e library(yggdrasil)
INFO:5682:runner.startDrivers[800]:YggRunner(runner): Starting I/O drivers and models on system hp-
HP-280-Pro-G6-Microtower-PC in namespace yggdrasil with rank 0
Rscript --default-packages=methods,utils /home/hp/IITPKD/Book_Chap_yggdrasil/Trial_yggdrasil/Model_
asfunction/ygg_light_v0_light.R
Loading required package: zeallot
/usr/bin/bash: /home/hp/miniconda3/lib/libtinfo.so.6: no version information available (required by
 /usr/bin/bash)
/usr/bin/bash: /home/hp/miniconda3/lib/libtinfo.so.6: no version information available (required by
 /usr/bin/bash)
[1] "End of input from temp_doy."
INFO:5682:runner.waitModels[867]:YggRunner(runner): light finished running.
```

Fig. 32 Running light model in terminal using yggdrasil command yggrun

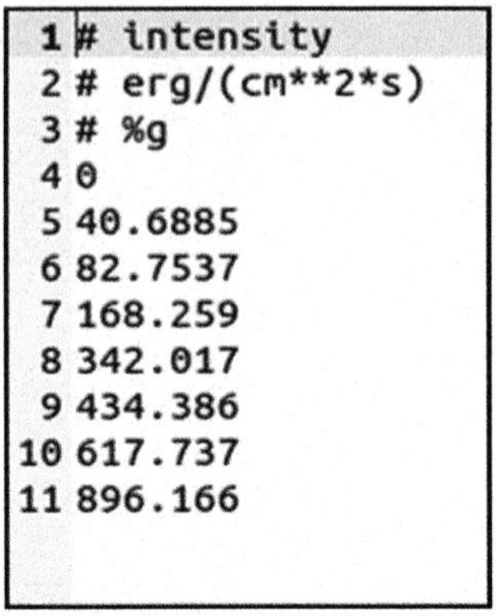

```
# intensity
# erg/(cm**2*s)
# %g
0
40.6885
82.7537
168.259
342.017
434.386
617.737
896.166
```

Fig. 33 Output of light model

are starting to work on model coupling and particularly those who are interested in coupling models that were originally developed using different programming languages.

8 Future Direction

We expect that tools like yggdrasil can be used to couple two or more complicated models to address both scientific and applied challenges. Two ideal models for coupling would be HYDRUS 1D (Simunek et al., 2013) and BioCro II (Lochocki et al., 2022) to improve the assessment of climate change impact on crop growth and productivity.

9 Open Challenges

Coupling tools such as yggdrasil provides an efficient way to combine process-based crop growth models and ML/DL based models. For example, crop growth models mostly lack yield loss due to pathogens. Integration of ML/DL based models to predict plant diseases with the crop growth models can be beneficial for research as well as farming communities. Integration of crop growth models with live data from sensors can help development of smart farming techniques for climate change adaptation and ensuring future food security. In addition to climate change, agriculture directly connects to many other disciplines such as economics, policy making, social sciences, poverty alleviation, etc. Ability to easily integrate crop growth models with policy models can provide insight into future action plan to meet the UN's sustainable development goals (Viana et al., 2022) and remains one of the many open challenges.

References

2015 water-use withdrawals by category | U.S. Geological Survey. https://www.usgs.gov/media/images/2015-water-use-withdrawals-category. Accessed 24 October 2022.

Abrahamsen, P., & Hansen, S. (2000). Daisy: An open soil-crop-atmosphere system model. *Environmental Modelling & Software, 15*, 313–330. https://doi.org/10.1016/S1364-8152(00)00003-7

Ainsworth, E. A., & Long, S. (2005). What have we learned from 15 years of free-air CO_2 enrichment (FACE)? A meta-analytic review of the responses of photosynthesis, canopy properties and plant production to rising CO_2. *New Phytologist, 165*, 351–372.

Asseng, S., Zhu, Y., Basso, B., et al. (2014). Simulation modeling: Applications in cropping systems. In N. K. Van Alfen (Ed.), *Encyclopedia of agriculture and food systems* (pp. 102–112). Academic Press.

Auzmendi, I., & Hanan, J. (2018). Using L-studio to visualize data and modify plant architecture for agronomic purposes: Visualization and modification of plant architecture with L-studio. In *2018 6th International Symposium on Plant Growth Modeling, Simulation, Visualization and Applications (PMA)* (pp. 43–49)

Bajželj, B., & Richards, K. (2014). The Positive feedback loop between the impacts of climate change and agricultural expansion and relocation. *Land, 3*, 898–916. https://doi.org/10.3390/land3030898

Baker, R. E., Pena, J.-M., Jayamohan, J., & Jérusalem, A. (2018). Mechanistic models versus machine learning, a fight worth fighting for the biological community? *Biology Letters, 14*, 20170660.

Beaulieu, E., Lucas, Y., Viville, D., et al. (2016). Hydrological and vegetation response to climate change in a forested mountainous catchment. *Model Earth Syst Environ, 2*, 1–15. https://doi.org/10.1007/s40808-016-0244-1

Blöschl, G., Bierkens, M. F. P., Chambel, A., et al. (2019). Twenty-three unsolved problems in hydrology (UPH)—a community perspective. *Hydrological Sciences Journal, 64*, 1141–1158. https://doi.org/10.1080/02626667.2019.1620507

Brisson, N., Mary, B., Ripoche, D., et al. (1998). STICS: A generic model for the simulation of crops and their water and nitrogen balances. I. Theory and parameterization applied to wheat and corn. *Agronomie, 18*, 311–346. https://doi.org/10.1051/agro:19980501

Chenu, K., Porter, J. R., Martre, P., et al. (2017). Contribution of crop models to adaptation in wheat. *Trends in Plant Science, 22*, 472–490. https://doi.org/10.1016/j.tplants.2017.02.003

Churkina, G., Running, S. W., Schloss, A. L., & ThE. Participants OF. ThE. Potsdam NpP. Model Intercomparison. (1999). Comparing global models of terrestrial net primary productivity (NPP): The importance of water availability. *Global Change Biology, 5*, 46–55. https://doi.org/10.1046/j.1365-2486.1999.00006.x

David O, Markstrom, S. L., Rojas, K. W., et al. (2016). The object modeling system. In: *Agricultural system models in field research and technology transfer* (pp. 317–330). CRC Press.

De Clercq, M., Vats, A., & Biel, A. (2018). Agriculture 4.0: The future of farming technology. In *Proceedings of the World Government Summit*, Dubai, UAE (pp. 11–13).

De Souza, A. P., Burgess, S. J., Doran, L., et al. (2022). Soybean photosynthesis and crop yield are improved by accelerating recovery from photoprotection. *Science, 377*, 851–854. https://doi.org/10.1126/science.adc9831

Dokoohaki, H., Gheysari, M., Mousavi, S.-F., Zand-Parsa, S., Miguez, F. E., Archontoulis, S. V., & Hoogenboom, G. (2016). Coupling and testing a new soil water module in DSSAT CERES-Maize model for maize production under semi-arid condition. *Agricultural Water Management, 163*, 90–99. https://doi.org/10.1016/j.agwat.2015.09.002

Druckenmiller, H. (2022). Accounting for ecosystem service values in climate policy. *Nature Clinical Practice Endocrinology & Metabolism, 12*, 596–598. https://doi.org/10.1038/s41558-022-01362-0

Dunn, S. M., & Mackay, R. (1995). Spatial variation in evapotranspiration and the influence of land use on catchment hydrology. *Journal of Hydrology, 171*, 49–73. https://doi.org/10.1016/0022-1694(95)02733-6
Evenson, R. E., & Gollin, D. (2003). Assessing the impact of the green revolution, 1960 to 2000. *Science, 300*, 758–762. https://doi.org/10.1126/science.1078710
Gazestani, V. H., & Lewis, N. E. (2019). From genotype to phenotype: Augmenting deep learning with networks and systems biology. *Current Opinion in Systems Biology, 15*, 68–73. https://doi.org/10.1016/j.coisb.2019.04.001
Gennari, J. H., Neal, M. L., Galdzicki, M., & Cook, D. L. (2011). Multiple ontologies in action: Composite annotations for biosimulation models. *Journal of Biomedical Informatics, 44*, 146–154. https://doi.org/10.1016/j.jbi.2010.06.007
Gilhespy, S. L., Anthony, S., Cardenas, L., et al. (2014). First 20 years of DNDC (DeNitrification DeComposition): Model evolution. *Ecological Modelling, 292*, 51–62. https://doi.org/10.1016/j.ecolmodel.2014.09.004
Goodale, T., Allen, G., Lanfermann, G., et al. (2003). The Cactus framework and toolkit: Design and applications. In J. M. L. M. Palma, A. A. Sousa, J. Dongarra, & V. Hernández (Eds.), *High performance computing for computational science—VECPAR 2002* (pp. 197–227). Springer.
Gregersen, J. B., Gijsbers, P. J. A., & Westen, S. J. P. (2007). OpenMI: Open modelling interface. *Journal of Hydroinformatics, 9*, 175–191. https://doi.org/10.2166/hydro.2007.023
Guo, Y. (2006). Plant modeling and its applications to agriculture. In: *2006 second international symposium on plant growth modeling and applications* (pp. 135–141).
Han, M., Zhao, C., Šimůnek, J., & Feng, G. (2015). Evaluating the impact of groundwater on cotton growth and root zone water balance using Hydrus-1D coupled with a crop growth model. *Agricultural Water Management, 160*, 64–75. https://doi.org/10.1016/j.agwat.2015.06.028
He, R., Hu, J., He, Y., & Fang, H. (2010). Structural plant modelling based on real 3d structural parameters, resulting simulation system and rule-based language xl. Presented at *XVIIth World Congress of the International Commission of Agricultural and Biosystems Engineering (CIGR)*, Canadian Society for Bioengineering (CSBE/SCGAB) Québec City, Canada, June 13–17 2010.
He, Y., Jaiswal, D., Liang, X., et al. (2022). Perennial biomass crops on marginal land improve both regional climate and agricultural productivity. *GCB Bioenergy*. https://doi.org/10.1111/gcbb.12937
Hejazi, M., Edmonds, J., Clarke, L., et al. (2014). Long-term global water projections using six socioeconomic scenarios in an integrated assessment modeling framework. *Technological Forecasting and Social Change, 81*, 205–226. https://doi.org/10.1016/j.techfore.2013.05.006
Houska, T., Multsch, S., Kraft, P., et al. (2014). Monte Carlo-based calibration and uncertainty analysis of a coupled plant growth and hydrological model. *Biogeosciences, 11*, 2069–2082. https://doi.org/10.5194/bg-11-2069-2014
Izaurralde JRW and RC (2010) The APEX model. In: *Watershed models* (pp. 461–506). CRC Press.
Jaiswal, D., De Souza, A. P., Larsen, S., et al. (2019). Reply to: Brazilian ethanol expansion subject to limitations. *Nature Clinical Practice Endocrinology & Metabolism, 9*, 211–212. https://doi.org/10.1038/s41558-019-0423-y
Jaiswal, D., De Souza, A. P., Larsen, S., et al. (2017). Brazilian sugarcane ethanol as an expandable green alternative to crude oil use. *Nature Clim Change, 7*, 788–792. https://doi.org/10.1038/nclimate3410
Jiao, Y., Lei, H., Yang, D., et al. (2017). Impact of vegetation dynamics on hydrological processes in a semi-arid basin by using a land surface-hydrology coupled model. *Journal of Hydrology, 551*, 116–131. https://doi.org/10.1016/j.jhydrol.2017.05.060
Jones, J. W., Hoogenboom, G., Porter, C. H., et al. (2003). The DSSAT cropping system model. *European Journal of Agronomy, 18*, 235–265. https://doi.org/10.1016/S1161-0301(02)00107-7
Kannan, K., Wang, Y., Lang, M., et al. (2019). Combining gene network, metabolic and leaf-level models shows means to future-proof soybean photosynthesis under rising CO_2 *in silico Plants, 1*, diz008. https://doi.org/10.1093/insilicoplants/diz008

Keating, B. A., Carberry, P. S., Hammer, G. L., et al. (2003). An overview of APSIM, a model designed for farming systems simulation. *European Journal of Agronomy, 18*, 267–288. https://doi.org/10.1016/S1161-0301(02)00108-9

Kellner, J., Multsch, S., Houska, T., et al. (2017). A coupled hydrological-plant growth model for simulating the effect of elevated CO_2 on a temperate grassland. *Agricultural and Forest Meteorology, 246*, 42–50. https://doi.org/10.1016/j.agrformet.2017.05.017

Kolbert, E. (2021). Creating a better leaf could tinkering with photosynthesis prevent a global food crisis? In *The New Yorker*. https://www.newyorker.com/magazine/2021/12/13/creating-a-better-leaf. Accessed 14 September 2022.

Kraft, P., Vaché, K. B., Frede, H.-G., & Breuer, L. (2011). CMF: A hydrological programming language extension for integrated catchment models. *Environmental Modelling & Software, 26*, 828–830. https://doi.org/10.1016/j.envsoft.2010.12.009

Lang, M. (2019). yggdrasil: A Python package for integrating computational models across languages and scales. *in silico Plants, 1*. https://doi.org/10.1093/insilicoplants/diz001

Lenka, N., Lenka, N., Sejian, V., & Mohanty, M. (2015). Contribution of agriculture sector to climate change. In *Climate change impact on livestock: Adaptation and mitigation* (pp. 37–48).

Liang, H., Hu, K., Batchelor, W. D., et al. (2016). An integrated soil-crop system model for water and nitrogen management in North China. *Science and Reports, 6*, 25755. https://doi.org/10.1038/srep25755

Lobell, D. B., Roberts, M.J., Schlenker, W., et al. (2014). Greater sensitivity to drought accompanies maize yield increase in the U.S. *Midwest, 344*, 5.

Lochocki, E. B., Rohde, S., Jaiswal, D., et al. (2022). BioCro II: A software package for modular crop growth simulations. *In silico Plants, 4*, diac003. https://doi.org/10.1093/insilicoplants/diac003

Long, S. P., Karp, A., Buckeridge, M. S., et al. (2015). Feedstocks for biofuels and bioenergy. *Bioenergy & Sustainability: Bridging the Gaps* 302–347

Lopez-Jimenez, J., Vande Wouwer, A., & Quijano, N. (2022). Dynamic modeling of crop-soil systems to design monitoring and automatic irrigation processes: A review with worked examples. *Water, 14*, 889. https://doi.org/10.3390/w14060889

Lynch, J. (2021). Agriculture's contribution to climate change and role in mitigation is distinct from predominantly fossil CO_2-Emitting Sectors. *Frontiers in Sustainable Food Systems, 4*, 9.

Ma, L., Hoogenboom, G., Saseendran, S. A., et al. (2009). Effects of estimating soil hydraulic properties and root growth factor on soil water balance and crop production. *Agronomy Journal, 101*, 572–583. https://doi.org/10.2134/agronj2008.0206x

Manfreda, S., Smettem, K., Iacobellis, V., et al. (2010). Coupled ecological–hydrological processes. *Ecohydrology, 3*, 131–132. https://doi.org/10.1002/eco.131

Marshall-Colon, A., Long, S. P., Allen, D. K., et al. (2017). Crops In Silico: generating virtual crops using an integrative and multi-scale modeling platform. *Frontiers in Plant Science, 8*, 786. https://doi.org/10.3389/fpls.2017.00786

Miguez, F. E., Maughan, M., Bollero, G. A., & Long, S. P. (2012). Modeling spatial and dynamic variation in growth, yield, and yield stability of the bioenergy crops Miscanthus × giganteus and P anicum virgatum across the conterminous United States. *Gcb Bioenergy, 4*, 509–520.

Moore, R. V., & Tindall, C. I. (2005). An overview of the open modelling interface and environment (the OpenMI). *Environmental Science & Policy, 8*, 279–286. https://doi.org/10.1016/j.envsci.2005.03.009

Multsch, S., Kraft, P., Frede, H. G., & Breuer, L. (2011). Development and application of the generic Plant growth Modeling Framework (PMF). In: Chan, F., Marinova, D. and Anderssen, R.S. (eds) *MODSIM2011, 19th International Congress on Modelling and Simulation. Modelling and Simulation Society of Australia and New Zealand,* December 2011.

Murray, S. J., Foster, P. N., & Prentice, I. C. (2011). Evaluation of global continental hydrology as simulated by the Land-surface Processes and eXchanges Dynamic Global Vegetation Model. *Hydrology and Earth System Sciences, 15*, 91–105. https://doi.org/10.5194/hess-15-91-2011

Ort, D., & Long., (2014). Limits on yields in the corn belt. *Science, 344*, 484–485.

Peña-Haro, S., Zhou, J., Zhang, G., et al. (2012). A multi-approach framework to couple independent models for simulating the interaction between crop growth and unsaturated-saturated flow processes. In: R. Seppelt, A.A. Voinov, S. Lange, D. Bankamp (Eds.), International Environmental Modelling and Software Society (iEMSs) 2012, *International Congress on Environmental Modelling and Software, Managing Resources of a Limited Planet: Pathways and Visions under Uncertainty, Sixth Biennial Meeting*, Leipzig, Germany, July 2012, pp. 1224–1231.

Peng, C. (2000). From static biogeographical model to dynamic global vegetation model: A global perspective on modelling vegetation dynamics. *Ecological Modelling, 135*, 33–54. https://doi.org/10.1016/S0304-3800(00)00348-3

Peng, H., Zhao, C., Feng, Z., et al. (2014). Canopy interception by a spruce forest in the upper reach of Heihe River basin, Northwestern China. *Hydrological Processes, 28*, 1734–1741. https://doi.org/10.1002/hyp.9713

Perkel, J. M. (2015). Programming: Pick up Python. *Nature, 518*, 125–126. https://doi.org/10.1038/518125a

Porter, J. R., Xie, L., Challinor, A. J., et al. (2014). Food security and food production systems In: *Climate Change 2014: Impacts, Adaptation, and Vulnerability. Part A: Global and Sectoral Aspects. Contribution of Working Group II to the Fifth Assessment Report of the Intergovernmental Panel on Climate Change* [Field, C.B., V.R. Barros, D.J. Dokken, K.J. Mach, M.D. Mastrandrea, T.E. Bilir, M. Chatterjee, K.L. Ebi, Y.O. Estrada, R.C. Genova, B. Girma, E.S. Kissel, A.N. Levy, S. MacCracken, P.R. Mastrandrea, and L.L.White (eds.)]. Cambridge University Press, Cambridge, United Kingdom and New York, NY, USA, pp. 485–533.

Pradal, C., Dufour-Kowalski, S., Boudon, F., et al. (2008). OpenAlea: A visual programming and component-based software platform for plant modelling. *Functional Plant Biol, 35*, 751–760. https://doi.org/10.1071/FP08084

Pradal, C., Fournier, C., Valduriez, P., & Cohen-Boulakia, S. (2015) OpenAlea: scientific workflows combining data analysis and simulation. In *Proceedings of the 27th International Conference on Scientific and Statistical Database Management* (pp. 1–6). Association for Computing Machinery, New York, NY, USA.

Prusinkiewicz, P., Karwowski, R., & Lane, B., et al. (2007). The L+C plant-modelling language. In J. Vos, L. F. M. Marcelis, & P. H. B. De Visser (Eds.), *Functional-structural plant modelling in crop production* (pp. 27–42). Springer.

Qin, Y., Mueller, N. D., Siebert, S., et al. (2019). Flexibility and intensity of global water use. *Nature Sustainability, 2*, 515–523. https://doi.org/10.1038/s41893-019-0294-2

Ray, D. K., Mueller, N. D., West, P. C., & Foley, J. A. (2013). Yield trends are insufficient to double global crop production by 2050. *PLoS ONE, 8*, e66428.

Ray, D. K., West, P. C., Clark, M., et al. (2019). Climate change has likely already affected global food production. *PLoS ONE, 14*, e0217148. https://doi.org/10.1371/journal.pone.0217148

Schlesinger, W. H., & Jasechko, S. (2014). Transpiration in the global water cycle. *Agricultural and Forest Meteorology, 189–190*, 115–117. https://doi.org/10.1016/j.agrformet.2014.01.011

Shameer, S., Wang, Y., Bota, P., et al. (2022). A hybrid kinetic and constraint-based model of leaf metabolism allows predictions of metabolic fluxes in different environments. *The Plant Journal, 109*, 295–313. https://doi.org/10.1111/tpj.15551

Shelia, V., Šimůnek, J., Boote, K., & Hoogenbooom, G. (2018). Coupling DSSAT and HYDRUS-1D for simulations of soil water dynamics in the soil-plant-atmosphere system. *Journal of Hydrology and Hydromechanics, 66*, 232–245. https://doi.org/10.1515/johh-2017-0055

Siad, S. M., Iacobellis, V., Zdruli, P., et al. (2019). A review of coupled hydrologic and crop growth models. *Agricultural Water Management, 224*, 105746. https://doi.org/10.1016/j.agwat.2019.105746

Simunek, J. J., Šejna, M., Saito, H., et al. (2013). The Hydrus-1D software package for simulating the movement of water, heat, and multiple solutes in variably saturated media, Version 4.17, HYDRUS Software Series 3, Department of Environmental Sciences. USA 342

Steduto, P., Hsiao, T. C., Raes, D., & Fereres, E. (2009). AquaCrop—The FAO crop model to simulate yield response to water: I. *Concepts and Underlying Principles. Agronomy Journal, 101*, 426–437. https://doi.org/10.2134/agronj2008.0139s

Stöckle, C. O., Donatelli, M., & Nelson, R. (2003). CropSyst, a cropping systems simulation model. *European Journal of Agronomy, 18*, 289–307. https://doi.org/10.1016/S1161-0301(02)00109-0

Subak, S. (1999). Global environmental costs of beef production. *Ecological Economics, 13*.

Tonnang, H. E., Sokame, B. M., Abdel-Rahman, E. M., Dubois, T. (2022). Measuring and modelling crop yield losses due to invasive insect pests under climate change. *Current Opinion in Insect Science*, 100873.

Van Diepen, C. A., Wolf, J., van Keulen H, Rappoldt C (1989) WOFOST: a simulation model of crop production. *Soil Use and Management, 5*, 16–24. https://doi.org/10.1111/j.1475-2743.1989.tb00755.x

Viana, C. M., Freire, D., Abrantes, P., et al. (2022). Agricultural land systems importance for supporting food security and sustainable development goals: A systematic review. *Science of the Total Environment, 806*, 150718.

Wang, J., Huang, G., Zhan, H., et al. (2014). Evaluation of soil water dynamics and crop yield under furrow irrigation with a two-dimensional flow and crop growth coupled model. *Agricultural Water Management, 141*, 10–22. https://doi.org/10.1016/j.agwat.2014.04.007

Wang, K., Sun, S., Li, Y., et al. (2022). Response of regional agricultural water use to the change of climate and plantation structure in the typical agricultural region of China. *Journal of Water and Climate Change, 13*, 1370–1388. https://doi.org/10.2166/wcc.2022.416

Wang, X., Liu, G., Yang, J., et al. (2017). Evaluating the effects of irrigation water salinity on water movement, crop yield and water use efficiency by means of a coupled hydrologic/crop growth model. *Agricultural Water Management, 185*, 13–26. https://doi.org/10.1016/j.agwat.2017.01.012

Wilde, M., Hategan, M., Wozniak, J. M., et al. (2011). Swift: A language for distributed parallel scripting. *Parallel Computing, 37*, 633–652.

Yang, W., Pallas, B., Durand, J.-B., et al. (2016). The impact of long-term water stress on tree architecture and production is related to changes in transitions between vegetative and reproductive growth in the 'Granny Smith' apple cultivar. *Tree Physiology, 36*, 1369–1381. https://doi.org/10.1093/treephys/tpw068

Zhang, G., Zhou, J., & Zhou, Q., et al. (2012). Integrated eco-hydrological modelling by a combination of coupled-model and algorithm using OMS3. International Congress on Environmental Modelling and Software.

Zhang, J., Petersen, S. D., Radivojevic, T., et al. (2020). Combining mechanistic and machine learning models for predictive engineering and optimization of tryptophan metabolism. *Nature Communications, 11*, 4880. https://doi.org/10.1038/s41467-020-17910-1

Zhou, J., Cheng, G., Li, X., et al. (2012). Numerical modeling of wheat irrigation using coupled HYDRUS and WOFOST models. *Soil Science Society of America Journal, 76*, 648–662. https://doi.org/10.2136/sssaj2010.0467

Machine Learning and Deep Learning in Crop Management—A Review

Sunil K. Vithlani and Vipul K. Dabhi

Abstract The use of computer vision techniques based on machine learning (ML) and deep learning (DL) algorithms has increased in order to improve agricultural output in a cost-effective manner. Researchers have used ML and DL techniques for different agriculture applications such as crop classification, automatic crop harvesting, pest and disease detection from the plant, weed detection, land cover classification, soil profiling, and animal welfare. This chapter summarizes and analyzes the applications of these algorithms for crop management activities like crop yield prediction, diseases and pest detection, and weed detection. The study presents advantages and disadvantages of various ML and DL models. We have also discussed the issues and challenges faced while applying the ML and DL algorithms to different crop management activities. Moreover, the available agriculture data sources, data preprocessing techniques, ML algorithms, and DL models employed by researchers and the metrics used for measuring the performance of models are also discussed.

Keywords Crop yield prediction · Deep learning · Disease detection · Machine learning · Weed management

1 Introduction

Agricultural practices consist of various activities like soil preparation, sowing, manuring, irrigation, weeding, harvesting, and storage. Traditional farmers perform all of these activities manually, which are both time-consuming and costly. It is also necessary to have indigenous knowledge and experience with these activities to gain the maximum yield (Hamadani et al., 2021). The global economy depends heavily on the agricultural sector (Adriano Cruz 2014). The agricultural system has

S. K. Vithlani (✉) · V. K. Dabhi
Department of Information Technology, Dharmsinh Desai University, Nadiad, Gujarat, India
e-mail: sunilvithlani.it@ddu.ac.in

V. K. Dabhi
e-mail: vipuldabhi.it@ddu.ac.in

© The Author(s), under exclusive license to Springer Nature Singapore Pte Ltd. 2023
S. Chaudhary et al. (eds.), *Digital Ecosystem for Innovation in Agriculture*,
Studies in Big Data 121, https://doi.org/10.1007/978-981-99-0577-5_2

to produce vast quantities of high-quality food that will increase as the human population continues to grow. Conventional agriculture has a number of issues, such as a surge in food demand as a result of the world's rapid urbanization, scarcity of natural resources, climate variability, and safety and wellness concerns.

Adoption of modern tools and technologies like sensors, drones, satellites, artificial intelligence (AI), and data analytics helps farmers, policymakers, and researchers to tackle the issues of conventional farming more efficiently. According to a survey presented by L. Benos et al. (Benos et al., 2021), the application of ML/DL in agriculture domain can be classified into four major categories: (1) Crop Management, (2) Water Management, (3) Soil Management, and (4) Livestock Management. Crop management can be further sub-divided into crop yield prediction, disease and pest's detection, and weed management. Water is crucial for plant survival, so proper irrigation system is required for proper growth of a crop. Soil management deals with the problems like maintaining the fertility of the soil after each crop season and how to deal with soil erosion problem? Animal welfare and its tracking are the key activities in livestock management. About 15–25 percent of India's potential crop output is to be lost due to pests, weeds, and diseases (Wolanin et al., 2020). To increase the crop yield, it is very important to protect the crop from the pests, weeds, and diseases and provide the appropriate nutrient required for the crop growth. In this survey, we have focused only the recent development in the crop management.

Together with big data technologies and high-performance computing, ML and DL have emerged to open up new possibilities for unraveling, quantifying, and comprehending data-intensive processes in agricultural operational environments. Machine learning is subset of AI. In ML, features are manually extracted from the dataset and provided as input to train a model. A model learns form the past experience and identify the relation between the features. The trained model is then tested using previously unseen data and is evaluated based on the performance of the model. DL is subset of ML. It does not require hand-crafted features as input. DL model takes raw data as input, and features are extracted gradually by each layer of the network. The brain's neural networks in humans served as an inspiration for DL models. DL models consist of input layer, various number of hidden layers, and output layer, where each hidden layer utilizes the input data to extract particular features at various scales or resolutions, combine them into a higher-level feature, and then use this higher-level feature to predict the future. DL models can be categorized into three categories: reinforcement learning, unsupervised learning, and supervised learning (Nguyen et al., 2019). The process of learning a function from labeled training data is known as supervised learning. Each pair of data in a supervised learning dataset consists of an input object and the desired output value. In a supervised learning model, the training data is analyzed to create a function that can be applied to prediction (Nguyen et al., 2019).

An in-depth analysis of the ML and DL models' use in agriculture is provided in this paper. Several pertinent papers are presented that highlight the important and distinctive characteristics of well-known ML and DL models. The state of the art for ML and DL used in various agricultural activities is presented in Sect. 3. Section 4 presents the summary and conclusion.

2 Literature Review

In today's world due to the scarcity of the precious resources like agriculture land and irrigation water, crop management plays a crucial role to cope up with the increasing food requirement (Rashid et al., 2021). Crop management helps for efficient utilization of natural resources and improve the income of the farmer by providing better crop yield. Various farming costs can be also reduced using proper crop management.

Crop management can help in various agricultural practices, such as determining the best crop for a specific location and identifying elements that would destroy the crops, such as weeds, insects, and crop diseases. It also helps to gain insights about the various growth stages of the crop, water management, and pesticide and herbicide management. IoT, data science, ML, and DL are a few of the technologies that deal mostly with data and are really helpful for comprehending and offering fantastic insights into data. In Fig. 1, we have represented the dataset used, ML and DL models applied, and problems/issues as well as future directions reported by the authors in their literature. In this section, we have presented the brief literature review of ML and DL models used for crop yield prediction, crop diseases detection, and weed management.

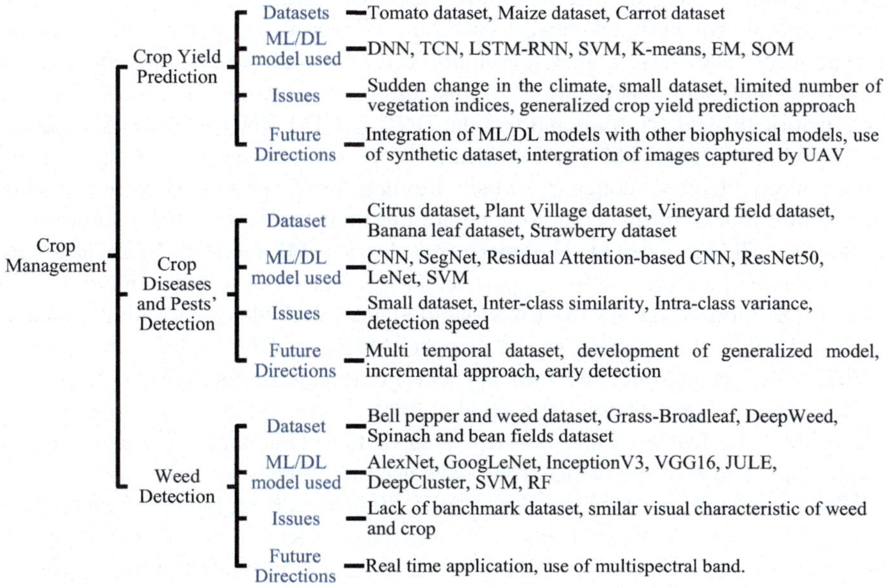

Fig. 1 Dataset, ML/DL models, issues, and future direction reported in the literature survey for crop management

2.1 Crop Yield Prediction

Crop yield predication plays an important role in estimating food availability. Crop growth is affected by various parameters like weather condition, type of soil, nutrient information, humidity, usage of pesticides, fertilizers, water, and other complex conditions (Rashid et al., 2021). Crop yield prediction can be done at both the macro-level and the micro-level (Senthilnath et al., 2016). In macro-level crop yield prediction, remote sensing data is used to measure various vegetation indices and predict the growth at the farm level or field level. Plant-level yield prediction is done at the micro-level of crop yield prediction. Information related to all these parameters can be collected using various methods like remote sensing, unmanned aerial vehicle (UAV), field survey, and sensors mounted on the field. Different vegetation indices are calculated using such data and used to predict the yield. Some of the vegetation indices are the normalized difference vegetation index (NDVI), green vegetation index (GVI), chlorophyll absorption ratio index (CARI), enhanced vegetation index (EVI), and many others. LSTM and CNN are widely used DL models for crop yield prediction using remote sensing data. Vegetation indices and meteorological data are mostly used by researchers for yield prediction (Muruganantham et al., 2022).

Gong et al. (Gong et al., 2021) have employed deep neural networks (DNN), which are composed of long short-term memory-recurrent neural networks (LSTM-RNN) and temporal convolutional networks (TCN), to forecast crop productivity in tomato greenhouses. They have gathered temporal information from three greenhouses on environmental factors including CO2 concentration, temperature, relative humidity, and historical yield. To train and test the DNN (LSTM-RNN + TCN), the dataset is normalized. The LSTM-RNN is used to preprocess the original input to extract representative feature sequences, which are then further processed by a sequential of residual blocks in the TCN to generate the final features used for future yield prediction. The proposed DNN outperforms classical ML models like linear regression (LR), random forest (RF), support vector machine (SVM), decision tree (DT) as well as DL models like LSTM-RNN (single layer), LSTM-RNN (multiple layers), LSTM-RNN, with root mean square errors (RMSEs) of 10.45 ± 0.94 for dataset 1, 6.76 ± 0.45 for dataset 2, and 7.40 ± 1.88 for dataset 3, respectively.

The authors (Rico-Fernández et al., 2019) presented a novel approach for segmenting the leaf area and non-leaf area of the plant image and using that to predict the growth of the plants. They have used 60 images from the carrot dataset, CWFID dataset (Haug & Ostermann, 2015), 50 images from the maize dataset (Lu et al., 2015), and 43 images from the tomato dataset (Rico-Fernández et al., 2019). One of the sample images from the CWFID dataset is presented in Fig. 2. Images were preprocessed, and color feature was extracted by transforming an image into different color spaces like RGB, CIE Lab, CIE Luv, HSV, HSL, YCrCb, and 2G-R-B. The color feature was also calculated using different color models like ExG, 2G-R-B, VEG, CIVE, MExG, COM1, and COM2. This color feature vector was provided to SVM for the classification of the leaf area and non-leaf area of the plant image. Three techniques were discussed for crop segmentation: CIE Luv + SVM, CIVE + SVM, and COM2 + SVM. They have compared these three approaches for the quality of segmentation and the accuracy of the models. CIE Luv + SVM performs better with an average quality of segmentation $Q_{se} = 0.89$ in the tomato dataset, $Q_{seg} = 0.88$ in the maize dataset, and $Q_{seg} = 0.91$ in the carrot dataset.

Senthilnath et al. (2016) have compared three clustering methods to detect tomatoes from remotely sensed RGB images captured by a UAV. Video has been recorded by the camera mounted on a UAV. The video was reviewed offline, and images were extracted from it. Three unsupervised spectral clustering methods, namely K-means, expectation maximization (EM), and self-organizing map (SOM), are used for grouping pixels into tomatoes and non-tomatoes. After clustering, spatial segmentation has been applied to overcome the misclassified tomato region. ROC parameters like precision, recall, and F1-score were compared and concluded that EM proved to be better (precision: 0.97) than K-means (precision: 0.66) and SOM (precision: 0.89).

In the present time, computer vision with machine learning and deep learning is getting attraction. To train the ML and DL models using computer vision concepts, it requires image datasets. Table 1 presents few of the image datasets available for the crop yield prediction.

Issues in Crop Yield Prediction:

- Gong et al. (2021) have considered historical environmental and yield information for crop yield prediction.
- It is easy to develop a ML/DL model for a single crop type, but it is very complex to design a generalized model which will work well for multi-crop and variety of environment (Rico-Fernández et al., 2019).
- Crop yield prediction can be also done by identifying the fruit size or fruit count, but it is very tedious task to identify the fruit from the aerial images in the early growth stage of the plant (Senthilnath et al., 2016).
- Earlier researchers have only performed correlation analyses between vegetation indices, soil data, and yield data, and these were mostly carried out for particular crop types, specific years, and a restricted number of vegetation indices (Muruganantham et al., 2022).

Fig. 2 Sample image from CWFID dataset (Haug & Ostermann, 2015). The original collected image in RGB (left), annotated image (middle), and masked image (right)

Table 1 Summary of image datasets used for crop yield prediction

Datasets	Type of image	Camera mounted	Images	Annotation	Environment	Accessibility	URL
CWFID (Haug & Ostermann, 2015)	Multispectral	Field robot	60	Pixel level	Organic carrot farm	Public	https://github.com/cwfid
Tomato dataset (Rico-Fernández et al., 2019)	RGB	Hand-holding	43	Pixel level	Greenhouse	Private	https://goo.gl/eMk3DE
Maize dataset (Lu et al., 2015)	RGB	Hand-holding	50	Pixel level	Maize farm	Private	https://sites.google.com/site/poppinace/
Tomato dataset (Senthilnath et al., 2016)	RGB	UAV	2	Frame level	Tomato farm	Private	–

- Any change in the climatic conditions during crop growth stages can drastically affect plant growth, which eventually leads to variability in crop yield estimation (Muruganantham et al., 2022) which was not taken into consideration by many researchers.

Future Direction in Crop Yield Prediction:

- To achieve more robust and accurate crop yield prediction, ML and DL models can be integrated with biophysical models (Gong et al., 2021).
- A more generalized approach can be developed using ML/DL algorithms and tested on the synthetic datasets to verify whether the synthetic images can reduce the need of huge amount of data to train the model or not (Rico-Fernández et al., 2019).
- Images captured using UAV can be stitched together to cover the larger cultivated area for yield prediction and fruit identification (Senthilnath et al., 2016).

2.2 Crop Diseases and Pests' Detection

One of the reasons for the decline in the quality and output of agricultural crops is plant disease. Plants are damaged by disease for a variety of reasons, including an imbalance of moisture in the soil and environment, a deficiency of necessary substances in the soil, the presence of harmful substances in the atmosphere, or an attack by some pests, fungi, or bacteria. Plant disease and pest detection techniques using ML and DL can be divided into three categories (Liu & Wang, 2021):

1. Classification techniques (What?)
2. Detection techniques (Where?)
3. Segmentation techniques (How?).

Classification techniques are used to determine which disease or pest is present in the plant. Models based on ML and DL can be used as feature extractors and classifiers. Images are fed into DL-based feature extractor models, which return an extracted feature vector. SVM or any other classifier can use these feature vectors to classify pests and diseases into single and multiclass categories. Researchers have used a variety of approaches for classification, including classification after locating the region of interest (RoI), a sliding window-based algorithm, and heat map-based classification, among others.

Detection techniques are used to answer questions like "Where is the infection or pest?" It is used to determine the exact location of the infected plant's part. There are two types of detection networks: two-stage detection and one-stage detection. The first stage of a two-stage detection model extracts regions of lesion areas from plant images and defines a bounding box around them. The second stage identifies and fine-tunes the area's localization. Two-stage methods are slower than one-stage networks, but they are more accurate. Two-stage detection networks include R-CNN, Faster R-CNN, and other models (Du et al., 2020). One-stage detection networks use only one stage to detect the infected area. They are faster as compared to two-stage methods, but the accuracy is also low. You Only Look Once (YOLO) and Single Shot Detector (SDD) have been widely used as one-stage detectors (Li et al., 2020).

Segmentation techniques are used to highlight the infected part, healthy part, and other regions from the plan images. Mask FCN and Mask R-CNN (Fuentes et al., 2017) are widely used DL models for segmentation. Table 2 presents few of the image datasets available for crop diseases and pest detection. Below we have presented a survey of a few recent research papers for plant disease and pest detection.

Khattak et al. (2021) recommended using a two-layer CNN model to classify five different diseases of citrus fruit and leaves. From the citrus dataset (Rauf et al., 2019) and the PlantVillage dataset (Hughes & Salathe, 2015; Arun pandian & Geetharamani, 2019), they have collected a total of 2293 images of healthy and diseased citrus fruits and leaves. Figure 3 displays sample images from the citrus dataset's disease classes and healthy class. The ImageDataGenerator class and API were used

Table 2 Summary of image datasets used for disease detection

Datasets	Type of image	Camera mounted	Images	Annotation	Environment	Accessibility	URL
Citrus dataset (Rauf et al., 2019)	RGB	Hand-holding	759	Image level	Citrus farm	Public	https://data.mendeley.com/datasets/3f83gxmv57/2
PlantVillage dataset (Hughes & Salathe, 2015)	RGB	Multiple methods	61,486	Image level	Multiple plant	Public	https://data.mendeley.com/datasets/tywbtsjrjv/1
Vineyard (Kerkech et al., 2020)	RGB	UAV	> 10,000	Bounding box	Vineyard	Private	–
Multi-crop and disease dataset (Picon et al., 2019)	RGB	Hand-holding	121,955	Image level	Multiple crop	Private	–
Strawberry dataset (Ebrahimi et al., 2017)	RGB	Field robot	100	Image level	Greenhouse	Private	–

Fig. 3 Sample images from the citrus dataset (Rauf et al., 2019). **a** Healthy fruit image, **b** black spot disease, **c** canker disease, **d** greening disease, and **e** scab disease

to process the dataset. The CNN model is trained using 80% of the images. They carried out ten experiments with different combinations of the number of convolution layers, the number of filters, the size of the filters, and the number of epochs. When compared to other models, the performance of the CNN model with two convolution layers, sixteen filters, and two filter dimensions with eight or more epochs is good. For all ten experiments, performance evaluation metrics like recall and precision are also contrasted. Additionally, the proposed CNN-based approach is compared to traditional ML and DL-based modes, with the results showing that the proposed model has higher accuracy (94.55%) as compared to others.

Kerkech et al. (Kerkech et al., 2020) proposed an approach for image segmentation to detect the disease-affected areas in the vineyard field using RGB and infrared images. The images are captured using a UAV having RGB and infrared sensors mounted on it. The captured images are labeled properly using a semi-automatic method. To align the images, the infrared images are registered over the RGB images. The SegNet model (Badrinarayanan et al., 2017) is used to segment the image into four classes, i.e., shaded areas, soil, healthy, and symptomatic vines. Two SegNet models were trained separately for the RGB images and for the infrared images. Both models' outputs are combined in two ways: "fusion AND" and "fusion OR". In the "fusion AND" approach, the disease is considered to be detected if it is present in both the RGB and infrared images. In the “fusion OR” approach, it is considered that the disease is detected if it is present in the RGB or in the infrared image. The performance of models was also compared using various performance metrics like precision, recall, F1-score, and accuracy. The "fusion OR” approach provides better accuracy (95.02%) than the “fusion AND” approach (88.14%), the visible image-based model (94.41%), and the infrared image-based model (89.16%).

Kartik et al. (2020) implemented the traditional CNN, CNN with residual layer, and attention-based residual CNN (Wang et al., 2017) architectures for disease detection in tomato leaves. A dataset was collected from the PlantVillage (Hughes & Salathe, 2015; Arun pandian & Geetharamani, 2019), an open-source platform which contains images of three disease classes (i.e., early blight, late blight, and leaf mold) and one healthy class. Data augmentation techniques like central zoom, random crop and zoom, and contrast images were used to generate multiple images and increase the dataset size. A total of 95,999 images were used for training, and 24,001 images were used for validation purposes. In CNN with a residual layer, three residual layers were used to concatenate the useful features from the earlier layers to the deeper layers. In addition to that, the attention technique was used to focus on only relevant features. A fivefold cross-validation technique was implemented for validating each model. The residual attention-based CNN model performed well as compared to the residual-based CNN model without an attention mechanism and the traditional CNN model, with an average classification accuracy of 98%, 95%, and 84%, respectively.

Picon et al. (2019) presented single-crop and multi-crop plant disease classification models over the images taken by cell phones in real field conditions. The dataset used in the paper consists of a total of 1,21,955 images of multiple crops like wheat, corn, rapeseed, barley, and rice. The dataset contains a total of seventeen disease classes and five healthy classes for each of these five crops. They used ResNet-50, a CNN-based model, as a baseline model to develop an independent (single) crop disease detection model and train it using the images of a particular crop dataset only. For multi-crop plant disease classification, they have trained the CNN model using the entire dataset. The multi-crop plant disease classification model provides a slightly better result (0.93 average balanced accuracy) than the single-crop classification model (0.92 average balanced accuracy). In another approach, they integrated CropID-crop metadata into the multi-crop classification model to provide the crop specification information and reduce the classification error by 71%. They got 0.98 average balanced accuracy for this CropID-based multi-crop plant disease classification model. Intra-class similarity and inter-class differences can be handled by training the model over multiple crops.

Amara et al. (2017) presented a deep learning-based strategy for categorizing banana leaf diseases. They made use of a dataset from the PlantVillage dataset (Arun pandian & Geetharamani, 2019) that contained 3700 images of banana leaves. The deep learning model is trained using images that have been resized to 60×60. To categorize the three banana leaf diseases—healthy, banana Sigatoka, and banana speckle—a LeNet-CNN-based deep learning model is used. When the model is trained using half of the images from the dataset and tested using the other half, its accuracy is 99.72%.

Ebrahimi et al. (2017) have suggested an SVM-based method for vision-based pest detection. Hundred images were obtained from a strawberry greenhouse using a camera mounted over a robot arm. The gamma operator is used to remove the background from captured images. Any remaining background was eliminated using contrast stretching and histogram equalization. SVM with different kernel functions of region index and color index is used for classification. The performance of the SVM is assessed using the MSE, RMSE, MAE, and MPE parameters. SVM is used to identify pests from images with an MPE of less than 2.25%.

Issues in Crop Diseases and Pests' Detection:

- Small datasets used for model training can lead to significant failure of the developed model (Kerkech et al., 2020).
- It is too challenging to precisely identify small-sized lesions during early detection stage (Picon et al., 2019).
- Disease and pest growth is variable in different crop development stages (Karthik et al., 2020).
- Presence of multiple diseases on a plant/crop is also a very challenging task (Karthik et al., 2020).
- Intra-class variance (Picon et al., 2019; Venkataramanan et al., 2021), i.e., diseases from same classes having different characteristics leads to miss-classification.
- Inter-class similarity (Picon et al., 2019; Venkataramanan et al., 2021), i.e., diseases from different classes having similar characteristics leads to miss-classification.
- Background disturbance present in the image will also effect the accuracy of the model (Picon et al., 2019).
- Uneven illumination (usually in the corners of the image) also creates a hurdle in development of more accurate model (Amara et al., 2017).
- In some scenarios, one crop piece may be occluded by another component and cannot be separated without human intervention (Amara et al., 2017).
- Disease detection speed should be also high to take the immediate remedy action (Amara et al., 2017).

Future Direction in Crop Diseases and Pests' Detection:

- Prepare a concrete dataset which should be multi-temporal and captured during various stages of the crop life cycle (Khattak et al., 2021).
- Performance of ML/DL model can be improved by training them on multiple plant disease datasets of varying size (Khattak et al., 2021).
- Use other spectrums to capture images. In literature, mostly use visible light to capture photographs, but we can also use near-infrared or multispectral images. (Kerkech et al., 2020)
- Use crop metadata, 3D information from the canopy, and other information as supplementary input to ML/DL models to reduce false detection (Kerkech et al., 2020).
- Generalized ML/Dl model can be developed by using the incremental learning approach for new crop only (Picon et al., 2019).

- Automatic severity estimation model for the detected disease can be also developed to stop the spread of the disease (Amara et al., 2017).

2.3 Weed Detection

In addition to diseases, weeds are a major threat to crop production. The biggest problem with weed control is that they are difficult to spot and distinguish from crops. Herbicides are used to control the weed in the farm, but excessive use of herbicides is harmful for humans. Computer vision, ML, and DL algorithms can improve weed detection and identification at low cost and without concern for the environment or side effects.

Subeesh et al. (2022) have compared four deep convolutional neural network models for weed detection in playhouse grown bell peppers. The image dataset has been captured using digital camera of mobile phone. A total of 1106 images were captured out of which 421 images of weed and 685 images of bell paper. The images are preprocessed and augmented to increase the dataset size. From the augmented dataset, 80% of the images was used to train 4 different models, i.e., AlexNet, GoogLeNet, Inception-V3, Xception, and their performance was compared using accuracy, precision, recall, and F1-score matrices. Inception-V3 performs well with accuracy of 97.7% as compared to other 3 models.

For the purpose of weed detection in the field, dos Santos et al. (2019) implemented the joint unsupervised learning of deep representations and image clusters (JULE) and deep clustering for unsupervised learning of visual features (DeepCluster). The JULE is made up of several stacked configurations of convolutional, batch normalization, ReLU, and pooling layers. The DeepCluster model was developed using AlexNet and VGG16 as its foundation to extract features, while K-means was utilized as the clustering technique. They have used two datasets, Grass-Broadleaf (dos Santos et al. 2017) and DeepWeeds. Figure 4 presents sample images of different classes of weeds from the DeepWeeds dataset. The DeepCluster model outperformed the JULE model. Using pre-trained VGG and AlexNet on ImageNet, it was possible to examine the impact of transfer learning on unsupervised learning. However, the accuracy of DeepWeed was enhanced by the pre-trained model, whereas Grass-accuracy Broadleaf's remained unchanged. Inception-V3, VGG16, and ResNet were trained using labeled images from the DeepCluster model in the semi-supervised approach. Inception-V3 proved better with 0.884 precision as compared to 0.860 precision of ResNet on the Grass-Broadleaf dataset, and VGG16 (precision:0.646) proved better than the ResNet (precision: 0.640) on the DeepWeed dataset.

For the purpose of weed detection in UAV photographs of spinach and bean fields, Bah et al. (2018) have suggested a CNN model with an unsupervised training dataset annotation collection. They believed that crops were planted in orderly rows and that weeds were plants that grew in between the rows. To determine the place of plant rows, the skeleton was subjected to the Hough transform. A blob coloring

Fig. 4 Sample images of different classes of weeds from the DeepWeeds (Olsen et al., 2019)

method was employed to identify the weeds after row detection. ResNet-18 was trained to detect weed in the images using the unsupervised training dataset. ResNet-18 was also compared with SVM and RF. AUC has been used for performance evaluation. ResNet-18 outperforms SVM and RF in supervised and unsupervised learning techniques. AUCs in the bean field are 91.37% for unsupervised data labeling and 93.25% for supervised data labeling. In the spinach field, they are 82.70% and 94.34%, respectively. Table 3 presents few of the image datasets available for weed detection.

Issues in Weed Detection:

- There is a lack of benchmark dataset which contains a variety of crops/weeds at various growth stages of crops and weeds (Bah et al., 2018).
- Requirement of large amount of labeled data for supervised learning algorithms (dos Santos et al., 2019).
- Weed and crop exhibit similar visual characteristic, so to train the model for higher accuracy, it requires other information also (Subeesh et al., 2022).

Future Direction in Weed Detection:

- Development of real-time applications like intelligent weeder and/or site-specific herbicide applicators based on the decision made by the DL models (Subeesh et al., 2022).
- Use of multispectral bands such as red edge or near-infrared images to distinguish plants, weed, and background and help to take decision to identify the required spraying area (Bah et al., 2018).

Table 3 Summary of image datasets used for weed detection

Datasets	Type of image	Camera mounted	Images	Annotation	Environment	Accessibility	URL
Bell pepper and weed dataset (Subeesh et al., 2022)	RGB	Smartphone	1106	Image level	Polyhouse	Private	–
Grass-Broadleaf (dos Santos et al. 2017)	RGB	UAV	15,336	Patch level	Soybean plantation	Public	https://data.mendeley.com/datasets/3fmjm7ncc6/2
DeepWeeds (Olsen et al., 2019)	RGB	Field robot	17,509	Image level	Multiclass weed	Public	https://github.com/AlexOlsen/DeepWeeds
WeedNet (Sa et al., 2018)	Multispectral	UAV	456	Frame level	Sugar beet farm	Public	https://github.com/inkyusa/weedNet
Bean and spinach field (Bah et al., 2018)	RGB	UAV	> 10,000	Patch level	Bean and spinach farm	Private	–

3 Discussion and Conclusions

Farmers have gained increasing insights for agricultural activities such as crop management, water management, and weed management through the use of new technologies in the recent years. While ML and DL are currently being used in agriculture, its applications remain far from ubiquitous. The first and most important stage in developing an AI-based agriculture ecosystem is to collect real-world field data and images under a variety of conditions. After the images are acquired, we need human professionals to precisely categorize them. Manually labeling images is also a time-consuming task that requires further attention from the scientific community. According to the study, the majority of researchers use smart (sensor-based) machines or fly unmanned aerial vehicles over farms to collect the necessary dataset on their own. Because the models trained using their own gathered dataset have more detailed information about the local environment, they perform well in similar types of environmental conditions but not in unfavorable environmental situations. PlantVillage and ImageNet are two popular open-source datasets for training models.

Choosing an effective model for developing an agriculture application involves balancing between accuracy and detection speed. The very accurate model (two-stage detection models) takes longer time but is more accurate. Single-stage detectors will produce spontaneous results, although their accuracy may be limited. It is dependent on the agricultural activity whether a real-time decision with moderate precision is required or an accurate outcome with some latency is required. For example, with an automatic fruit harvesting system, quick response is required, whereas a disease prediction system requires more exact class of the disease or healthy. SVM is a popular ML approach because it achieves high accuracy even when training with little amounts of data. Convolutional neural networks and recurrent neural networks are the most often utilized DL models because they perform well in variety of scenarios. The environmental conditions are different in different scenarios for agriculture activities. A model developed using one dataset to perform an agricultural activity cannot be universally applicable to all such activities or may not perform well. Crops like tomatoes, maize, wheat, rice, and soybean were widely studied by using ML.

The most commonly used performance metrics of the reviewed papers are briefly described in Table 4. Confusion matrix, accuracy, precision, and AUC are the major four performance measure criteria among others to evaluate the performance of the models. Mean square error (MSE) and root mean square error (RMSE) are also used to evaluate the performance of regression based problem. Adoption of machine learning and deep learning-based models in agriculture becomes more difficult and takes more care because a trained model may not be transferred to different environments because of the variety of situations in agriculture.

Based on the study, we can infer that ML in agriculture has a great potential to meet the emerging challenges. On the other hand, high infrastructure investment expenses restrict farmers from adopting these technologies. All the stack holders must adopt a changed perspective and start learning and spreading awareness about potential of AI in agriculture. Overall, given the growing awareness of the potential of artificial intelligence in agriculture, machine learning will become a behind-the-scenes facilitator for the construction of a more sustainable and productive farm.

Table 4 Commonly used metrics for performance measurement of ML/DL models

No	Performance metrics	Description
1	True positive (TP)	An outcome where the model correctly predicts the positive class
2	True negative (TN)	An outcome where the model correctly predicts the negative class
3	False positive (FP)	An outcome where the model incorrectly predicts the positive class
4	False negative (FN)	An outcome where the model incorrectly predicts the negative class
5	Confusion matrix	An NxN table that summarizes the number of correct and incorrect predictions
6	Accuracy	(TP + TN)/(TP + FP + FN + TN)
7	Recall	TP/(TP + FN)
8	Precision	TP/(TP + FP)
9	Specificity	TN/(TN + FP)
10	F1-score	(2 × Recall × Precision)/(Recall + Precision)
11	AUC	A number between 0.0 and 1.0 representing a binary classification model's ability to separate positive classes from negative classes. The closer the AUC is to 1.0, the better the model's ability to separate classes from each other
12	Correlation coefficient	It is a statistical measure of the strength of a linear relationship between two variables. Its values can range from −1 to 1
13	Coefficient of determination (R^2)	It is the proportion of the variance in the dependent variable that is predicted from the independent variable
14	Mean absolute error (MAE)	Average of all absolute errors between predicted value and actual value
15	Mean squared error (MSE)	It shows how far predictions fall from measured true values using Euclidean distance

It is expected that the current systematic effort will serve as a useful guide for academics, manufacturers, engineers, ICT system developers, policymakers, and farmers, contributing to additional systematic study on ML in agriculture. We hope that this state-of-the-art survey would be useful to beginners and researchers to find promising directions of research in the area of agriculture transformation through ML and DL.

Appendix: Literature Review Papers

#	Paper title	Dataset	Size of the dataset	Preprocessing steps	Main approach	Algorithm/model used	Performance metric used	Conclusion	Ref.
1	Detection of tomatoes using spectral-spatial methods in remotely sensed RGB images captured by UAV	Video has been recorded by a camera mounted on a UAV	Based on region of interest (ROI), images are extracted from the video	Image resizing	Three unsupervised spectral clustering methods are compared for grouping pixels into tomatoes and non-tomatoes	K-means, expectation maximization (EM), self-organizing map (SOM)	ROC parameters: precision, recall, and F1-score	EM proved to be better (precision: 0.97) than K-means (precision:0.66) and SOM (precision:0.89)	(Senthilnath et al., 2016)
2	A contextualized approach for segmentation of foliage in different crop species	1. Carrot dataset 2. Maize dataset 3. Tomato dataset	1. 60 images of carrot 2. 50 images of maize 3. 43 images of tomato	Color features were extracted by transforming the image into different color spaces like RGB, CIE Lab, CIE Luv, HSV, HSL, YCrCb, and 2G-R-B. The color feature was also calculated using different color indices like ExG, 2G-R-B, VEG, CIVE, MExG, COM1, and COM2	The color feature vector was provided to the SVM classifier to detect leaf area and non-leaf area in an image. Three different approaches were compared 1. CIE Luv + SVM, 2. CIVE + SVM, 3. COM2 + SVM	SVM	Quality of segmentation, accuracy of models	CIE Luv + SVM performs better as compared to others	(Rico-Fernández et al., 2019)

(continued)

(continued)

#	Paper title	Dataset	Size of the dataset	Preprocessing steps	Main approach	Algorithm/model used	Performance metric used	Conclusion	Ref.
3	Deep learning Based prediction on greenhouse crop yield Combined TCN and RNN	Environmental parameters (CO_2 concentration, relative humidity, etc.) and historical yield information from three different greenhouses of tomato		A temporal sequence of data containing both historical yield and environmental information is normalized and provided to the RNN	Representative features are extracted using the LSTM + RNN layer and fed into the temporal convolutional network	LSTM-RNN and TCN	MSE, RMSE	Mean and standard deviation of RMSEs 10.45 ± 0.94 for the dataset from greenhouse1, 6.76 ± 0.45 for dataset from greenhouse2, and 7.40 ± 1.88 for dataset from greenhouse3	(Gong et al., 2021)

(continued)

(continued)

#	Paper title	Dataset	Size of the dataset	Preprocessing steps	Main approach	Algorithm/model used	Performance metric used	Conclusion	Ref.
4	Vine disease detection in UAV multispectral images using optimized image registration and deep learning segmentation approach	RGB and infrared images collected using UAV	4 classes shadow, ground, healthy, and symptomatic class 17,640 samples for each class, among them 14,994 used for training and 2,646 for validation	The dataset was labeled using a semi-automatic method (a sliding window). Each block was classified by a LeNet5 network for pre-labeling. Labeled images are corrected manually, and labeled images are used for segmentation	Two SegNet models were trained separately for the RGB images and for the infrared images. Both models' outputs are combined in two ways "fusion AND" and "fusion OR"	SegNet	Precision, recall, F1-score, accuracy	"Fusion OR" approach provides a better accuracy (95.02%) over the "fusion AND" approach (88.14%), RGB image-based model (94.41%), and infrared image-based model (89.16%)	(Kerkech et al., 2020)
5	Attention embedded residual CNN for disease detection in tomato leaves	PlantVillage and augmented datasets	95,999 tomato leaf images for training and 24,001 images were used for validation	Data augmentation techniques like central zoom, random crop and zoom, and contrast images	CNN-based multiclass classification into three diseases classes (i.e., early blight, late blight, and leaf mold) and 1 healthy class for tomato leaves	CNN and modified CNN	Accuracy	Accuracy of baseline CNN model: 84%; residual CNN model: 95%; attention embedded residual CNN model: 98%	(Karthik et al., 2020)

(continued)

(continued)

#	Paper title	Dataset	Size of the dataset	Preprocessing steps	Main approach	Algorithm/model used	Performance metric used	Conclusion	Ref.
6	Crop conditional Convolutional Neural Networks for massive multi-crop plant disease classification over cell phone acquired images taken on real field conditions	Own dataset, collected images using a mobile phone	A total of 1,21,955 images of multiple crops like wheat, corn, rapeseed, barley, and rice are collected	Image resize	Three approaches were proposed to detect seventeen diseases of five healthy classes from five different crops 1. Independent model for each of the five crops 2. Single model, i.e., multi-crop, for the entire dataset 3. Use of crop metadata (CropID) information along with a multi-crop model	ResNet-50-CNN	AUC, sensitivity, specificity, balanced accuracy (BAC)	Independent single-crop models showed an average BAC of 0.92, whereas the baseline multi-crop model showed an average BAC of 0.93. Crop conditional CNN architecture performs well with 0.98 average BAC	(Picon et al., 2019)

(continued)

(continued)

#	Paper title	Dataset	Size of the dataset	Preprocessing steps	Main approach	Algorithm/model used	Performance metric used	Conclusion	Ref.
7	Automatic detection of Citrus fruit and leaves diseases using deep neural network model	Citrus dataset and PlantVillage dataset	2293 images	Images are preprocessed like normalizing and scaling and then used for training, validation, and testing to classify diseases into five classes	80% of preprocessed images are provided as input to CNN to train it. The remaining 20% of the images are used for validation and testing of the model. This proposed model is also compared with other ML/DL-based models	CNN (two layers)	Test accuracy, training loss, training time, precision, recall	The proposed CNN model has 95.65% accuracy	(Khattak et al., 2021)
8	A deep learning-based approach for banana leaf diseases classification	Images of banana leaves (healthy and diseased) were obtained from the PlantVillage dataset	3700 images	Images were resized to 60 × 60 pix. and converted to grayscale images and used for the classification process	Classification of the leaf images into three classes using LeNet architecture	LeNet architecture-CNN	Accuracy, precision, recall, F1-score	The model performs well for color images as compared to grayscale images. Accuracy is 99.72% for 50–50, train-test split	(Amara et al., 2017)

(continued)

(continued)

#	Paper title	Dataset	Size of the dataset	Preprocessing steps	Main approach	Algorithm/model used	Performance metric used	Conclusion	Ref.
9	Vision-based pest detection based on SVM classification method	Images were obtained from a strawberry greenhouse using a camera mounted on a robot arm	100 images	The non-flower regions are considered as background and removed by applying the gamma correction. Histogram equalization and contrast stretching were used to remove any remaining background	SVM with different kernel functions of region index and color index is used for classification	SVM	MSE, RMSE, MAE, MPE	Pests are detected from images using SVM with a mean percentage error less than 2.25%	(Ebrahimi et al., 2017)
10	Deep learning with Unsupervised Data Labeling for Weed Detection in Line Crops in UAV Images	Images were collected by UAV from two farm fields	Total: 17,044 (bean field), 15,858 (spinach field)	Background removal, skeleton, Hough transformation for (Crow Row) line detection	Images were labeled using unsupervised methods and supervised methods and used for crop/weed discrimination using CNN	ResNet-18, SVM, RF	AUC	AUCs in the bean field are 91.37% for unsupervised data labeling and 93.25% for supervised data labeling. In the spinach field, they are 82.70% and 94.34%, respectively	(Bah et al., 2018)

(continued)

(continued)

#	Paper title	Dataset	Size of the dataset	Preprocessing steps	Main approach	Algorithm/model used	Performance metric used	Conclusion	Ref.
11	Unsupervised deep learning and semi-automatic data labeling in weed discrimination	Grass-Broadleaf and DeepWeeds datasets	Grass-Broadleaf: Total 15,536 segments (3249 of soil, 7376 of soybean, 3520 of grass, and 1191 of broadleaf weeds). DeepWeeds: 17,509 images	Segmentation and image resize	Joint unsupervised learning of deep representations and image clusters (JULE) and deep clustering for unsupervised learning of visual features (DeepCluster)	Inception-V3, VGG16, ResNet	Precision	Inception-V3 has better precision 0.884 for Grass-Broadleaf dataset, and VGG16 has better precision 0.646 for DeepWeeds dataset as compared to ResNet	(dos Santos et al. 2019)
12	Deep convolutional neural network models for weed detection in polyhouse grown bell peppers	Images are captured using digital camera of mobile	Total of 1106 images collected and augmented to increase the size of dataset	Data augmentation, outlier detection, standardization, normalization	Four models based on CNN are compared for classification of image into bell paper or weed class	AlexNet, GoogLeNet, Inception-V3, Xception	precision, accuracy, recall, F1-score	Inception-V3 performance well as compared to other three models	(Subeesh et al., 2022)

References

Adriano Cruz, J. (2014). Enhancement of growth and yield of upland rice (*Oryza sativa* L.) by Actinomycetes. Agrotechnol s1. https://doi.org/10.4172/2168-9881.S1.008

Amara, J., Bouaziz, B., & Algergawy, A. (2017). A deep learning-based approach for banana leaf diseases classification. *(BTW 2017)-Workshopband.*

Arun Pandian, J., & Geetharamani, G. (2019). Data for: Identification of plant leaf diseases using a 9-layer deep convolutional neural network. *Mendeley Data, V1.* https://doi.org/10.17632/tywbtsjrjv.1

Badrinarayanan, V., Kendall, A., & Cipolla, R. (2017) Segnet: A deep convolutional encoder-decoder architecture for image segmentation. *IEEE transactions on pattern analysis and machine intelligence, 39*(12), 2481–2495.

Bah, M. D., Hafiane, A., & Canals, R. (2018). Deep learning with unsupervised data labeling for weed detection in line crops in UAV images. *Remote Sensing, 10*, 1690. https://doi.org/10.3390/rs10111690

Benos, L., Tagarakis, A. C., Dolias, G., et al. (2021). Machine learning in agriculture: A comprehensive updated review. *Sensors, 21*, 3758. https://doi.org/10.3390/s21113758

dos Santos, F. A., Freitas, D. M., da Silva, G. G., et al. (2019). Unsupervised deep learning and semi-automatic data labeling in weed discrimination. *Computers and Electronics in Agriculture, 165*, 104963. https://doi.org/10.1016/j.compag.2019.104963

dos Santos, F. A., Matte Freitas, D., Gonçalves da Silva, G., et al. (2017). Weed detection in soybean crops using ConvNets. *Computers and Electronics in Agriculture, 143*, 314–324. https://doi.org/10.1016/j.compag.2017.10.027

Du, L., Zhang, R., & Wang, X. (2020). Overview of two-stage object detection algorithms. *Journal of Physics: Conference Series, 1544*, 012033. https://doi.org/10.1088/1742-6596/1544/1/012033

Ebrahimi, M. A., Khoshtaghaza, M. H., Minaei, S., & Jamshidi, B. (2017). Vision-based pest detection based on SVM classification method. *Computers and Electronics in Agriculture, 137*, 52–58. https://doi.org/10.1016/j.compag.2017.03.016

Fuentes, A., Yoon, S., Kim, S. C., & Park, D. S. (2017). A Robust deep-learning-based detector for real-time tomato plant diseases and pests recognition. *Sensors, 17*, 2022. https://doi.org/10.3390/s17092022

Gong, L., Yu, M., Jiang, S., et al. (2021). Deep learning based prediction on greenhouse crop yield combined TCN and RNN. *Sensors, 21*, 4537. https://doi.org/10.3390/s21134537

Hamadani, H., Rashid,S. M., Parrah, J. D., et al. (2021). Traditional farming practices and its consequences. In Dar, G. H., Bhat, R. A., Mehmood, M. A., & Hakeem, .K. R. (Eds.), *Microbiota and biofertilizers, Vol 2: Ecofriendly tools for reclamation of degraded soil environs* (pp. 119–128). Springer International Publishing.

Haug, S., & Ostermann, J. (2015). A Crop/weed field image dataset for the evaluation of computer vision based precision agriculture tasks. In L. Agapito, M. M. Bronstein, & C. Rother (Eds.), *Computer vision—ECCV 2014 workshops* (pp. 105–116). Springer International Publishing.

Hughes, D. P., & Salathe, M. (2015). An open access repository of images on plant health to enable the development of mobile disease diagnostics. *arXiv preprint arXiv:1511.08060.*

Karthik, M. H., Anand, S., et al. (2020). Attention embedded residual CNN for disease detection in tomato leaves. *Applied Soft Computing, 86*, 105933. https://doi.org/10.1016/j.asoc.2019.105933

Kerkech, M., Hafiane, A., & Canals, R. (2020). Vine disease detection in UAV multispectral images using optimized image registration and deep learning segmentation approach. *Computers and Electronics in Agriculture, 174*, 105446. https://doi.org/10.1016/j.compag.2020.105446

Khattak, A., Asghar, M. U., Batool, U., et al. (2021). Automatic detection of citrus fruit and leaves diseases using deep neural network model. *IEEE Access*, 1–1. https://doi.org/10.1109/ACCESS.2021.3096895

Li, M., Zhang, Z., Lei, L., et al. (2020). agricultural greenhouses detection in high-resolution satellite images based on convolutional neural networks: Comparison of faster R-CNN, YOLO v3 and SSD. *Sensors, 20*, 4938. https://doi.org/10.3390/s20174938

Liu, J., & Wang, X. (2021). Plant diseases and pests detection based on deep learning: A review. *Plant Methods, 17*, 22. https://doi.org/10.1186/s13007-021-00722-9

Lu, H., Cao, Z., & Xiao, Y., et al. (2015). Joint crop and tassel segmentation in the wild. In *2015 Chinese Automation Congress (CAC)* (pp. 474–479).

Muruganantham, P., Wibowo, S., Grandhi, S., et al. (2022). A systematic literature review on crop yield prediction with deep learning and remote sensing. *Remote Sensing, 14*, 1990. https://doi.org/10.3390/rs14091990

Nguyen, G., Dlugolinsky, S., Bobák, M., et al. (2019). Machine learning and deep learning frameworks and libraries for large-scale data mining: A survey. *Artificial Intelligence Review, 52*, 77–124. https://doi.org/10.1007/s10462-018-09679-z

Olsen, A., Konovalov, D. A., Philippa, B., et al. (2019). DeepWeeds: A multiclass weed species image dataset for deep learning. *Science and Reports, 9*, 2058. https://doi.org/10.1038/s41598-018-38343-3

Picon, A., Seitz, M., Alvarez-Gila, A., et al. (2019). Crop conditional convolutional neural networks for massive multi-crop plant disease classification over cell phone acquired images taken on real field conditions. *Computers and Electronics in Agriculture, 167*, 105093. https://doi.org/10.1016/j.compag.2019.105093

Rashid, M., Bari, B. S., Yusup, Y., et al. (2021). A comprehensive review of crop yield prediction using machine learning approaches with special emphasis on palm oil yield prediction. *IEEE Access, 9*, 63406–63439. https://doi.org/10.1109/ACCESS.2021.3075159

Rauf, H. T., Saleem, B. A., Lali, M. I. U., et al. (2019). A citrus fruits and leaves dataset for detection and classification of citrus diseases through machine learning. *Data in Brief, 26*, 104340. https://doi.org/10.1016/j.dib.2019.104340

Rico-Fernández, M. P., Rios-Cabrera, R., Castelán, M., et al. (2019). A contextualized approach for segmentation of foliage in different crop species. *Computers and Electronics in Agriculture, 156*, 378–386. https://doi.org/10.1016/j.compag.2018.11.033

Sa, I., Chen, Z., Popović, M., et al. (2018). weedNet: Dense semantic weed classification using multispectral images and MAV for smart farming. *IEEE Robotics and Automation Letters, 3*, 588–595. https://doi.org/10.1109/LRA.2017.2774979

Senthilnath, J., Dokania, A., Kandukuri, M., et al. (2016). Detection of tomatoes using spectral-spatial methods in remotely sensed RGB images captured by UAV. *Biosystems Engineering, 146*, 16–32. https://doi.org/10.1016/j.biosystemseng.2015.12.003

Subeesh, A., Bhole, S., Singh, K., et al. (2022). Deep convolutional neural network models for weed detection in polyhouse grown bell peppers. *Artificial Intelligence in Agriculture, 6*, 47–54. https://doi.org/10.1016/j.aiia.2022.01.002

Venkataramanan, A, Laviale, M., Figus, C., et al. (2021). Tackling inter-class similarity and intra-class variance for microscopic image-based classification. In *Computer Vision Systems: 13th International Conference, ICVS 2021, Virtual Event, September 22–24, 2021, Proceedings 13* (pp. 93–103). Springer International Publishing.

Wang, F., Jiang, M., Qian, C., et al. (2017). Residual attention network for image classification. In *Proceedings of the IEEE conference on computer vision and pattern recognition* (pp. 3156–3164).

Wolanin, A., Mateo-García, G., Camps-Valls, G., et al. (2020). Estimating and understanding crop yields with explainable deep learning in the Indian Wheat Belt. *Environmental Research Letters, 15*, 024019. https://doi.org/10.1088/1748-9326/ab68ac

Need for an Orchestration Platform to Unlock the Potential of Remote Sensing Data for Agriculture

Sanjiv Kumar Jha

Abstract Satellites, drones, and on-ground sensors are essential in providing data for the digital ecosystem for agriculture innovation. Usually, all three modes of data collection are executed in isolation; however, they should work in coordination to leverage the strength of remote sensing. For example, making a classification model using remote sensing and ground data is mandatory. The current approach is to collect data through mobile phones by field officers at a specific time of crop harvesting. However, a more efficient method would be using an on-ground sensor to provide ground truth data continuously. Frequent training of the machine learning model will produce a time-sensitive digital crop signature on the ground. It will significantly impact how we do crop classification and yield estimation. This chapter evaluates and proposes a reference architecture for the orchestration platform that provides a coordination mechanism for the data collected from on-ground sensors, drones, and remote sensing satellites. The benefits of agriculture are examined through several use cases in this chapter.

Keywords Agriculture innovation · Crop classification · Drone · Remote sensing data · Satellite · Sensor · Orchestration platform

1 Background

In India, more than 50% of the population is employed in agriculture and related activities (DAC&FW, 2021), contributing only 17.4% to the country's Gross Value Added (GVA). However, more than 50% of the country's resources contribute to only one-fifth of the GVA. About 85% of farmers operate less than 5 acres, and the average size of farm holding is estimated to be less than 1 acre (DAC&FW, 2021). Due to that, access and affordability to technology for farming are limited. The most significant pain points for the farming sector today are:

S. K. Jha (✉)
Principal Smart Infra—SA, Amazon Web Services, Bengaluru, India
e-mail: sanjij@amazon.com

© The Author(s), under exclusive license to Springer Nature Singapore Pte Ltd. 2023
S. Chaudhary et al. (eds.), *Digital Ecosystem for Innovation in Agriculture*,
Studies in Big Data 121, https://doi.org/10.1007/978-981-99-0577-5_3

1. High volatility of input prices and sub-optimal selection of agriculture inputs.
2. Lack of access to technology for efficient cropping results in poor yields.
3. The uneven quality of produce and lack of large-scale testing.
4. Inefficient post-harvesting supply chains lead to wastage and principal losses.
5. Lack of access to credible financial solutions and working capital.

Three factors can transform Indian agriculture:

- Provision of fair prices to the farmer for farming input and yield. It involves solving accessibility and affordability of information and creating sustainable demand for agri-produce with minimum wastage.
- Improve yield and products quality—enabling farmers to sow and grow better as well as providing crop advisory to increase farming efficiency.
- Solving for financial inclusion—providing farmer-centric loans and insurance.

A key reason for the farmer not getting a fair input price is that the farmer lacks accessibility to quality inputs at the right time and during crop cycles. The use or overuse of fertilizers and pesticides is widespread. Many farmers still grow crops based on the rule of thumb and traditional practices. It points to a pressing need for agronomy and crop advisory services.

Vital for improving farmer income post-harvest is to ensure they get a better price for the product with the minimum waste. The farmer sows based on past information and knowledge, unaware of the market demand. There is massive wastage of food because of poor grading of the product leading to poor prices. There is high price volatility because of oversupply or insufficient necessary items.

Only 30% of farmers have access to institutional credit, and the remaining 70% remain dependent on informal credit channels (DAC&FW, 2021). Critical information like input use, crop and soil health, and quality of products aims to build the risk profile and creditworthiness of small farmers. Improved access to quality data and digitalization of records can increase the banker's ability to lend many-fold.

2 AgriTech as Emerging Sector

Many emerging AgriTech startups are trying to solve this problem using precision farming (Solomon, 2020a) technologies. There are about 1500 active Agri startups in India, of which more than 120 have raised an institutional investment of over $1.6B (AgriTech, 2022). Despite many innovations in this space, startups cannot unlock scale and raise growth funds, as it is evident that there is not a single unicorn in this segment. The main challenge is the economy of scale, as 85% of farmers operate less than 5 acres, and the average size of farm holding is over 1 acre (DAC&FW, 2021). The average farmer cannot afford expensive technology like drones or IoT. That was why the first wave of drone-based precision farming could not scale. Recently, remote sensing or satellite-based advisory services are gaining popularity as they can provide geo-expansion to scale.

2.1 *Remote Sensing in Agriculture*

Using satellite images, one can analyze agricultural and related economic activities. It helps to fix several essential parameters related to agriculture. Moreover, satellite images can check a large geographic area and accurately predict the field and crop situation without physically visiting the location. It allows real-time input of the data for decision-making. The remotely sensed data has several advantages over classical on-field approaches; it facilitates quick decision-making, performs spatial and temporal analysis, provides global coverage, and yields several tangible economic benefits. Many estimates are possible through remote sensing, like planting acreage, crop yield, and actual production. Moreover, it can support measuring the health of the crops, soil, condition, and characteristics supporting agri-system studies. Figure 1 shows a few use cases of remotely sensed data in agriculture.

Usually, in India, a practice to predict crop production for wheat, rice, jute, mustard, cotton, sugarcane, and sorghum is prevalent. With a strong background in satellite technology, spectral, temporal, and weather metrics are readily available for prediction. Therefore, a project known as "**FASAL**" (Forecasting Agricultural Output using Space, Agro-meteorology, and Land-based Observations) was taken

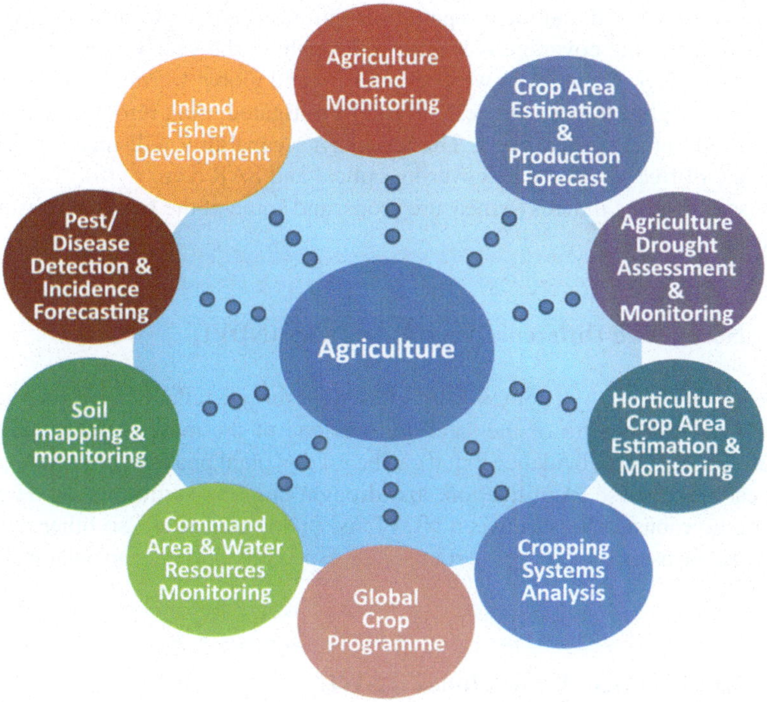

Fig. 1 Application of remotely sensed data in agriculture

up by **Mahalanobis National Crop Forecast Centre** (MNCFC) (MNCFC, 2022) to collect monsoon statistics and monitor crop cycle.

Under the project, "**KISAN**" by the MNCFC, multispectral and hyperspectral images from satellites and unmanned aerial vehicles (UAVs) were used to develop an optimum crop-cutting experiment (CCE) (Solomon, 2020b). Four states in India, Haryana, Karnataka, Maharashtra, and Madhya Pradesh, participated in the experiments, and 250 CCE locations were established in them. The locations were selected after considering parameters like date of sowing, indices like normalized difference vegetation index (NDVI), leaf area index (LAI), and the above surface biomass. As a result, the KISAN project was successful and enabled farmers to make crucial decisions before the harvest season.

2.2 *Remote Sensing Data in Agriculture*

It is vital to monitor the physical and biological growth of the crops closely. The advances in satellite technology and weather instruments allow close to real-time monitoring. The satellites are available in multispectral bands (visible, near infrared (NIR), short-wave infrared (SWIR)) as well as in the hyperspectral range. They are also available with submeter spatial resolution to moderate resolutions of 10–20 m. Moreover, the coverage is available throughout the year with a frequency of once in 5–12 days. Changes in weather patterns and global warming induce several risks, like droughts, floods, and pests. It can be mitigated using remotely sensed data by measuring crops in real time. The monitoring capability will increase with the abundance of free satellite data covering optical and SAR data.

The satellite uses indices to measure crops, and some of the famous measures are as follows:

2.2.1 Normalized Difference Vegetation Index (NDVI)

It depends on chlorophyll content and is based on surface reflectance captured using the red and NIR spectra of the satellite. It is one of the most popular and robust measures, with tons of evidence that it can be used in plant phenotyping, yield prediction, pest and disease identification, and drought conditions monitoring. However, there are no robust results between NDVI and yield predictions, so other measures should also be used to increase the robustness, especially during crop insurance claim settlement.

2.2.2 Land Surface Wetness Index (LSWI)

It is also vital to measure moisture in the soil and plants. While optical bands cannot sense water, SWIR—I or II bands—can easily do so. LSWI is an index developed to quantify wetness on the land. It is a good measure of the plant's stress and, when combined with NDVI and LSWI, becomes a very robust measure that can exemplify several deformities on the ground.

2.2.3 Radar Backscatter

Synthetic aperture radar (SAR)-based microwaves can also effectively measure soil and vegetation moisture. The backscatter has been successfully used to monitor crops (Yuzugullu et al., 2017). The factors like soil roughness and plant structure can impact the backscatter, and several papers have suggested the potential of backscatter in agri-measurements.

2.2.4 Fraction of Active Photosynthetic Radiation (FAPAR)

The FAPAR depends on the plant's chlorophyll condition and time of the day (Baret et al., 2007). In addition, the amount of CO_2 and the density of the biomass significantly impact FAPAR, a biophysical variable. Therefore, it provides more holistic measurements than vegetation indices (Meroni et al., 2013). The Copernicus Global Land Service (CGLS) provides FAPAR with a spatial resolution of 300 m 10 days interval.

2.3 Scaling Remote Sensing Solutions

There are challenges with the indices-based study—indices value for satellite image pixel range from -1 to + 1. First, we need to find the threshold point to separate classes. Ex. NDVI indicates the presence of vegetation on the ground. However, we need to know which values of NDVI represent no vegetation or high-level vegetation. An automated way of pixel value thresholding, like the Otsu method, has been ineffective (Open CV, 2022; Python, 2019). Secondly, the indices vary vastly with geographies and plant species.

We need to train the model to use satellite images, which can automatically identify the object of interest. Ground truth data must be collected continuously to scale the remote sensing solution and to improve the model's accuracy. It is used to annotate satellite images and used to train a model. The traditional ground truth collection approach by field officers uses the mobile app or trap to capture insects. This process is prone to error and lacks timeliness. A better system would be to use all three modes of precision farming data gathering; ground sensors, drones, and satellite imaging together and work together. The IoT-based intelligent farming system can monitor agricultural land with the help of sensors (soil moisture, humidity, light, and temperature) (Solomon, 2020c). CCTV cameras can continuously monitor crops, weeds, or insects. IoT sensors and CCTV cameras can be deployed in strategic locations all over the country. They constantly send crop and soil data to the backend system to continuously train remote sensing imaging models.

The challenge for these kinds of infrastructure is that AgriTech is a fragmented space where satellites, drones, and ground sensors are used in isolation and are operated by independent agencies. So, we need an orchestration to make all these disparate and isolated systems work collaboratively.

3 Need for Orchestration Platform

Orchestration is the automated configuration, management, and coordination of computer systems, applications, and services (Speck, 2020). The proposed orchestration platform aims to take data from multiple sensors, drones, and high-resolution imagery and combine and organize them. For example, suppose we have a nationwide network of such infra that can provide continuous ground-truthing data. In that case, agricultural data can be made available for the nation by leveraging remote sensing imagery. Moreover, this service can be made cost-effective on a subscription basis. So how is it going to help?

Let us illustrate a few use cases (Quantum, 2022). Plant stress detection has a growing interest in precision agriculture applications. There are numerous stressors to which crops can be subjected. Plant stresses are usually divided into biotic (stress resulting from exposure and interactions with other living organisms) and abiotic (environmental conditions and characteristics). Accurate laboratory tests provide valuable information regarding plants and soil's status and resilience to harsh conditions. However, such data is scarce and hard to collect for early detection. Figure 2 shows the impact of plant stress on the tobacco plant. The harsh conditions that plants face during their growth result in stunted growth.

A scalable approach is to use supervised machine learning modeling. Large amounts of data were gathered using satellite imagery, field measurements, disease records, and chemical and geological test results as labels to prepare training datasets.

A widely used procedure for the application of various indices is described in the paper (Veysi et al., 2020). The soil moisture content was determined for five points per each of the eight fields using the gravimetric method, resulting in forty ground truth data points. As input data, several spectral indices, including NDVI and temperature vegetation dryness index (TVDI), were constructed from the Landsat 8 satellite imagery. Numerous studies in the detection and forecast of plant stresses have been performed on small scales, that is, for fields of sizes a few square kilometers or less. Unfortunately, only some of such datasets are publicized and freely available. Having the ground sensor collect field data and make it available for modeling will help with early detection and save crops in the existing field.

Crop-type classification is a common task in agriculture. The main goal of the crop-type classification is to create a map/image where each point on it will have a class of what is growing there. There are many ways to solve this problem. However, one of the most explored methods is by using remote sensing data. A widely used

Fig. 2 Deterioration in plant growth due to stressors

method is multitemporal images, meaning taking a sequence of images and creating a time series-based digital profile of the crop (Quantum, 2022). A date range that covers the growing season is selected. The crop growth pattern has to be captured from the field during the growing season. Ground truth data is preprocessed, and a classifier is used to classify different crops and create a crop map of the study area.

4 Orchestration Platform: Unlocking the Remote Sensing Data for Agriculture

We propose three-tier architectures to unlock the remote sensing data for the agriculture sector, as shown in Fig. 3.

1. The physical layer consists of on-ground sensors, drones, CCTV, and other on-field data collection mechanisms. It also provides access to open-source remote sensing satellite data. Different agencies may operate these systems. Satellite data, drones, and CCTV can be in TBs. The best practice is to publish these data on the cloud platform to make it available for large-scale data processing.
2. The data processing layer consists of different pipelines to process the data collected. Satellite data processing platforms preprocess satellite data and return spectral data and indices like NDVI, EVI, IDBI, etc. IoT platforms ingest sensor data, especially soil sensors. Drones and CCTV cameras are the sources of images. These images need to be annotated with meta information. The human-in-the-loop annotation platform helps the operator interact with the automated system to ensure high-quality data. In general, a human-in-the-loop machine

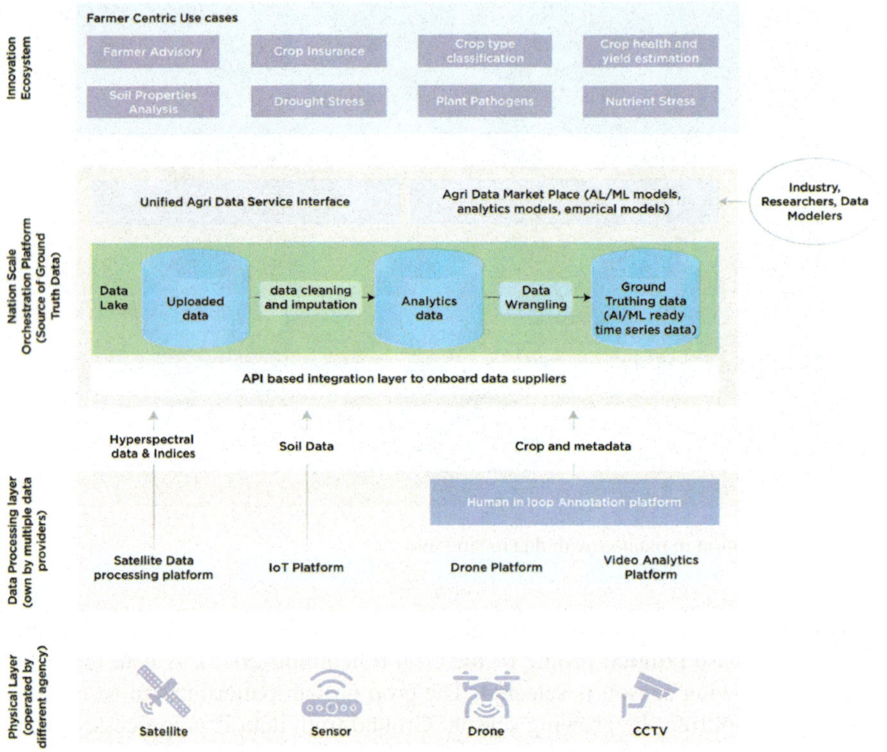

Fig. 3 Proposed architecture for the orchestration platform

learning process involves sampling good data for humans to label (annotation), using that data to train a model, and using that model to test more data for annotation (Heller, 2022). The data processing layer can provide a restful API to fetch data from a data store.

3. The orchestration layer integrates and provides access to annotated imagery and analytics-ready data. We also propose to create a marketplace for agriculture databases, AI/ML, and other data models. It is similar to how Apple and Android provide a platform for innovators to develop mobile apps. It will encourage innovators to develop innovative solutions and monetize them. All curated agriculture data will be available through unified Agri API or as a GIS layer. The Agri API layer is the restful API, and the GIS layer is exposed through OGC services. OGC defines several types of services for serving different kinds of data and maps. Typical OGC service types: Web Map Service (WMS) for helping collections of layers as map images and Web Map Tile Service (WMTS) for serving map layers as cached map tiles.
4. The innovation ecosystem layer consists of industries, researchers, and academicians who develop farmer-centric applications and use cases. Unified Agri API will provide the digital backbone on which farmer-centric applications can be

built. This platform will bring together data scientists, AI/ML engineers, and application developers. It will leverage ground truth data to build AI/ML-based prediction models to estimate data for larger geographical areas.

5 Case Study—Paddy Crop Insurance Using a Satellite-Based Composite Index of Crop Performance

5.1 Motivation

It is vital to protect the farmer against multiple risks throughout the crop season (Murthy et al., 2022). One such insurance scheme, Pradhan Mantri Fasal Bima Yojana (PMFBY), is based on yield. It provides a yield guarantee scheme over the mentioned geographic location. A group of villages (Known as Gram Panchayat) defines the Insurance Unit (IU) area. The past and the current yield measures form the basis of determining the loss and the payout to the affected farmers. A threshold yield (TY), the average best of five out of the past seven crops' yields at IU, is determined to benefit the farmers. A certain percentage between 70 and 90% of the average TY is guaranteed for the year in which loss is to be computed. Also, the insurance premium is charged based on crop and risks at IUs.

Under CCE, typically, yield is manually estimated by randomly picking crop-wise fields at each IU. Here, the number of measurements is minimal and full of manual errors. The measures have significant variations due to limitations in obtaining reliable global crop estimates. In turn, the estimated yield is way off the target, causing heart burns and disputes during claims settlements. Thus, it seems that the solution to the problem is to get a robust and accurate yield estimate at the IU level.

5.2 Problem Statement

The mathematical modeling of the yield is an extremely tough problem. It has many dynamic features derived from weather, crop type, soil quality, and post-harvest crop management. Therefore, data-driven machine learning models need suitable datasets, and variability is a big challenge in obtaining good data, especially at the local IU level. Furthermore, the limitations in scale, reliability, and uncontrolled errors at various levels prevent the development of robust yield estimation algorithms for insurance purposes. Nevertheless, the researchers from the Agriculture Insurance Corporation (AIC) and National Remote Sensing Centre (NRSC) proposed an interesting way to measure crop loss (Murthy et al., 2022). The paper also presents crop health factors (CHF), quantifying crop health. It is a complex index using physical and biophysical parameters of the crop to indicate its health. In this chapter, we will summarize the methodology and result of their study and suggest how it could have improved the automated mechanism of the ground-truthing dataset.

5.3 Study Area

The project implementation is done in Eastern India in West Bengal. It is mainly an agricultural-based state and dramatically relies on the southwest monsoon from the Bay of Bengal. It specifically cultivates rice, potato, wheat, mustard, and jute. Like many other places, agri-production is impacted by stressors and natural calamities. The cropped area is 9.4 million hectares, with irrigation available to 65% of the net cropped area.

5.4 Field Data Collection

The data was collected from the field using a mobile phone using set protocols. All the IUs are covered multiple times during the season. The field data has two points, reference, and random points, and the former were checked many times to monitor crops. In contrast, random points were checked only once during the crop season. During data collection, a group of IUs (block) had 10–20 points with 1–2 samples per IU. More than 100,000 samples were collected for 2020 with 15 factors from the field. It provided ample ground truth; samples were also linked to satellite indices.

5.5 Results

The result is a crop insurance method—The technology-based Bangla Shasya Bima Scheme which was implemented for the Aman season (July–December). Large farmers, about 86% (6.2 million out of 7.2 million), participated in the scheme. The ground truth data helped in robust verification, which used CHF and correlation. The deviations from CHF support mapping the crop conditions observed through a mobile application. The farmers also participated in the crop inference and set up a good agreement in actual field measures and CHF measured using remote sensing.

5.6 Analysis

It has been observed in West Bengal that the pre-harvest period is very vulnerable to floods, cyclones, and other natural calamities. They cause significant damage to the crop due to excessive water. Satellite images could effectively detect changes in yield and wetness. Combining their indices would produce a more objective assessment of the calamity-prone area. With robust analysis and data availability, the settlement process became fair, resulting in the timely disbursement of the claims. Such measures can significantly make the premium computation more objective. In addition, the data and methods are very systematic, with complete control at each step. Therefore, it will enable the use of ML algorithms quickly in the framework.

It is vital to evolve and reduce dependence on the current inefficient methods of yield estimation in India, especially with increased awareness and use of crop insurance. The paper (Murthy et al., 2022) addresses and showcases positive changes for crop insurance in the country. The newer methods of crop loss measurements are more objective and reduce the risk for the insurance companies while benefiting the farmers. Furthermore, the adoption of remote sensing for yield prediction is robust and overcomes the limitation of manual processes. But the model's accuracy dramatically relies on the features used in its construction. Therefore, scaling solutions for operational implementation beyond the algorithmic stage are very important as in the case where the CHF model is not operationally ready (Murthy et al., 2022).

Furthermore, the robust CHF measurement must account for undetected risks during the crop season. It can be done using a multimodal approach, e.g., using images and videos during data gathering. In addition, several challenges must be addressed for robust yield prediction using ML and DL algorithms (Klompenburg et al., 2020). The existing models use a limited dataset using attributes derived from weather and soil and apply classical ML algorithms. Therefore, their performance needs vast improvement, especially with the dynamic nature of agriculture (Filippi et al., 2019).

6 Conclusion

The value of remote sensing data in digital agriculture is well known. However, we need to leverage it more because of the unavailability and affordability of remote sensing data. NASA and ESA satellite data are primarily used. However, available reference data and indices are only sometimes valid in some geographies. Many innovations are happening in AgriTech, but it is a fragmented space. The proposed orchestration platform can provide a technology collaboration ecosystem to make ground-truthing data available from large geography through API continuously. It can help build a solution to take advantage of the economy of scale. It can help democratize the remote sensing data, and many application and service developers can leverage its benefits.

Additionally, a massive capacity-building exercise must be organized for field functionaries to effectively implement ground truth data collection. The availability of a nationwide orchestration platform for continuous ground truth data and picture collection across the seasons and multiple crops can significantly improve the model performance and make it operationally ready.

References

Agritech. (2022). Investment in agri tech startups jumps 2-fold to $4.6 billion in FY22. Business Standard. Available online. https://www.business-standard.com/article/companies/investment-in-agri-tech-startups-jumps-2-fold-to-4-6-billion-in-fy22-122113000675_1.html. Accessed 13th December 2022.

Baret, F., Hagolle, O., Geiger, B., Bicheron, P., Miras, B., Huc, M., Berthelot, B., Niño, F., Weiss, M., Samain, O., & Roujean, J. L. (2007). LAI, fAPAR and fCover CYCLOPES global products derived from VEGETATION: Part 1: Principles of the algorithm. *Remote Sensing of Environment, 110*(3), 275–286.

DAC&FW. (2021). Annual Report 2020–21. Department of Agriculture, Cooperation and Farmers Welfare, Ministry of Agriculture and Farmers' Welfare, Government of India. Available online. https://agricoop.nic.in/sites/default/files/Web%20copy%20of%20AR%20%28Eng%29_7.pdf. Accessed 13th December 2022.

Filippi, P., Jones, E. J., Wimalathunge, N. S., Somarathna, P. D., Pozza, L. E., Ugbaje, S. U., Jephcott, T. G., Paterson, S. E., Whelan, B. M., & Bishop, T. F. (2019). An approach to forecast grain crop yield using multi-layered, multi-farm data sets and machine learning. *Precision Agriculture, 20*(5), 1015–1029.

Heller, M. (2022). What is human-in-the-loop machine learning? Better data, better models. Available online. https://www.infoworld.com/article/3648456/what-is-human-in-the-loop-machine-learning-better-data-better-models.html. Accessed 13th December 2022.

Meroni, M., Fasbender, D., Kayitakire, F., Pini, G., Rembold, F., Urbano, F., & Verstraete, M. (2013, August). Regional drought monitoring using phenologicallytuned biomass production estimates from SPOTVEGETATION FAPAR. In *2013 Second International Conference on Agro Geoinformatics (Agro-Geoinformatics)* (pp. 495–499). IEEE.

MNCFC. (2022). Mahalanobis National Crop Forecast Centre (MNCFC). Available online https://www.ncfc.gov.in/about-us.html. Accessed 13th December 2022.

Murthy, C. S., Poddar, M. K., Choudhary, K. K., Pandey, V., Srikanth, P., Ramasubramanian, S., & Senthil Kumar, G. (2022). Paddy crop insurance using satellite-based composite index of crop performance. *Geomatics, Natural Hazards and Risk, 13*(1), 310–336.

Open CV. (2022). Image Thresholding. Available online https://docs.opencv.org/4.x/d7/d4d/tutorial_py_thresholding.html. Accessed 13th December 2022.

Python. (2019). Python: Thresholding techniques using OpenCV, Set-3 (Otsu Thresholding). Available online. https://www.geeksforgeeks.org/python-thresholding-techniques-using-opencv-set-3-otsu-thresholding/. Accessed 13th December 2022.

Quantum. (2022). Plant stress: What is it and how to detect it. Available online https://medium.datadriveninvestor.com/plant-stress-what-is-it-and-how-to-detect-it-649e3f77160. Accessed 13th December 2022.

Speck, D. (2020). Automation versus orchestration: What's the Difference? Available online. https://www.burwood.com/blog-archive/automation-vs-orchestration-whats-the-difference. Accessed 13th December 2022.

Solomon, R. (2020a). Precision Agriculture in India: New Technologies Are Here, But Wide Scale Adoption Is Far Off. Available online. https://www.globalagtechinitiative.com/in-field-technologies/precision-agriculture-in-india-new-technologies-are-here-but-wide-scale-adoption-is-far-off/. Accessed 13th December 2022.

Solomon, R. (2020b). Remote Sensing Technology Continues to Expand in Indian Agriculture. Available online https://www.globalagtechinitiative.com/in-field-technologies/sensors/remote-sensing-technology-continues-to-expand-in-indian-agriculture/. Accessed 13th December 2022.

Solomon, R. (2020c). How IoT Solutions for Indian Agriculture Are Working Despite Unique Challenges. Available online. https://www.globalagtechinitiative.com/digital-farming/how-iot-solutions-for-indian-agriculture-are-working-despite-unique-challenges/. Accessed on 13th December 2022.

Van Klompenburg, T., Kassahun, A., & Catal, C. (2020). Crop yield prediction using machine learning: A systematic literature review. *Computers and Electronics in Agriculture, 177*, 105709.
Veysi, S., Naseri, A. A., & Hamzeh, S. (2020). Relationship between field measurement of soil moisture in the effective depth of sugarcane root zone and extracted indices from spectral reflectance of optical/thermal bands of multispectral satellite images. *Journal of the Indian Society of Remote Sensing, 48*(7), 1035–1044.
Yuzugullu, O., Marelli, S., Erten, E., Sudret, B., & Hajnsek, I. (2017). Determining rice growth stage with X-band SAR: A metamodel based inversion. *Remote Sensing, 9*(5), 460.

An Algorithmic Framework for Fusing Images from Satellites, Unmanned Aerial Vehicles (UAV), and Farm Internet of Things (IoT) Sensors

Srikrishnan Divakaran

Abstract Satellites provide time series data in the form of multispectral images depicting land surface characteristics spanning several km^2, while unmanned aerial vehicles (UAVs) provide multispectral farm data with very high resolution spanning a few hundred square meters. In contrast, low-cost sensors and IoT sensors provide accurate spatial and time series data of land and soil characteristics spanning a few meters. However, in practice, each of these data sources has been separately used even though there is scope for optimizing farm resources and improving the quality of satellite and UAV data by exploiting their complementarity. In this chapter, we present an algorithmic framework that exploits the synergies among the three data sources to construct a high-dimensional farm map. We present an outline of how this framework can help in the construction of farm map in the context of crop monitoring.

Keywords Digital agriculture · Data science · Machine learning · Data integration · Data fusion · Remote sensing · Satellite · UAV · IoT

1 Introduction

In Digital Agriculture, farm management systems employ a mix of Satellites, Unmanned Arial Vehicles (UAVs), and Internet of Things (IoT) devices to acquire farm data. Over the past decade the increased availability of sophisticated remote sensing satellite services, widespread commercial use of UAVs due to a better quality of data, flexibility in data acquisition, and easing of regulations of its use, the easy deployment of inexpensive quality IoT sensors, standardized interfaces and programmable frameworks for data acquisition has resulted in an exponential increase in the volume of farm data. The availability of tools and techniques in Big Data, Data Science, and Machine Learning for filtering, representation, storage, retrieval, processing, and analysis of data has motivated the need for algorithms (Ghamisi et al., 2019; Zhu et al., 2018; Simões et al., 2021; Alvarez et al., 2021)

S. Divakaran (✉)
School of Engineering and Applied Sciences, Ahmedabad University, Ahmedabad, India
e-mail: srikrishnan.divakaran@ahduni.edu.in

© The Author(s), under exclusive license to Springer Nature Singapore Pte Ltd. 2023
S. Chaudhary et al. (eds.), *Digital Ecosystem for Innovation in Agriculture*,
Studies in Big Data 121, https://doi.org/10.1007/978-981-99-0577-5_4

for the fusion of IoT farm data with satellite and UAV data, thereby facilitating the development of various niche services in Digital Agriculture.

Satellites provide time series data in the form of multispectral images capturing land surface characteristics spanning several km^2; UAVs provide multispectral and hyper-spectral farm data with very high resolution spanning a few hundred square meters. In contrast, the low-cost sensors and IoT devices provide accurate spatial and time series data of land and soil characteristics spanning a few meters that help understand the factors contributing to the resilience and productivity of farmlands. In practice, each data source has been used separately even though there is scope for optimizing farm resources and improving the quality of agricultural services/predictions. This optimization is done by integrating satellite and UAV data by exploiting their complementarity and calibrating/validating their accuracy using farm data from IoT devices (Figs. 1, 2 and 3).

Notice that the modality employed by IoT sensors to acquire and represent farm data is vastly different when compared to satellites and UAVs. So integrating this

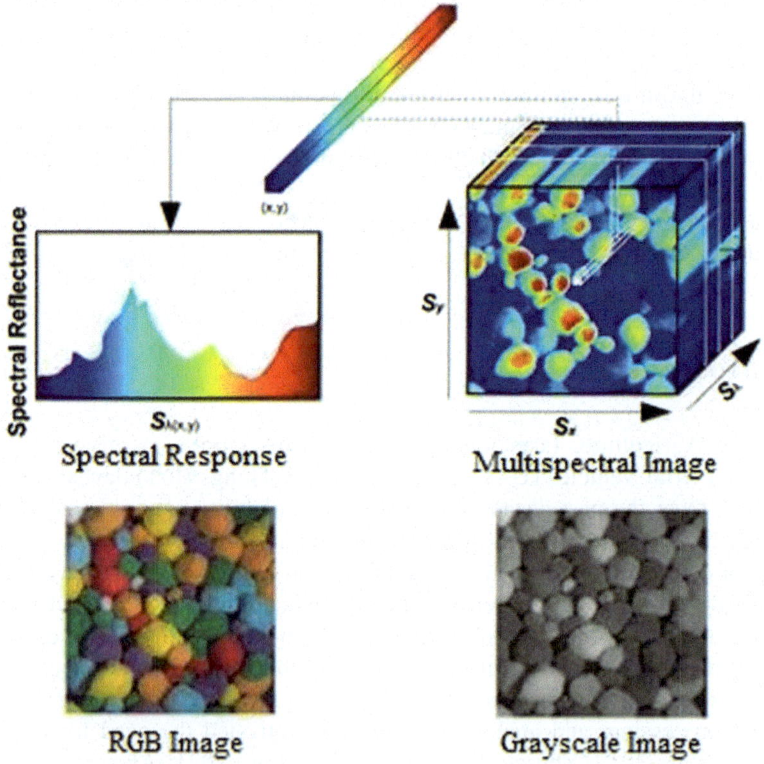

Fig. 1 Multispectral image stack, spectral response at the spatial location (*x*, *y*), an RGB image, and a grayscale image rendered from the image stack. (Khan et al., 2018, this figure was uploaded by Khurram Khurshid)

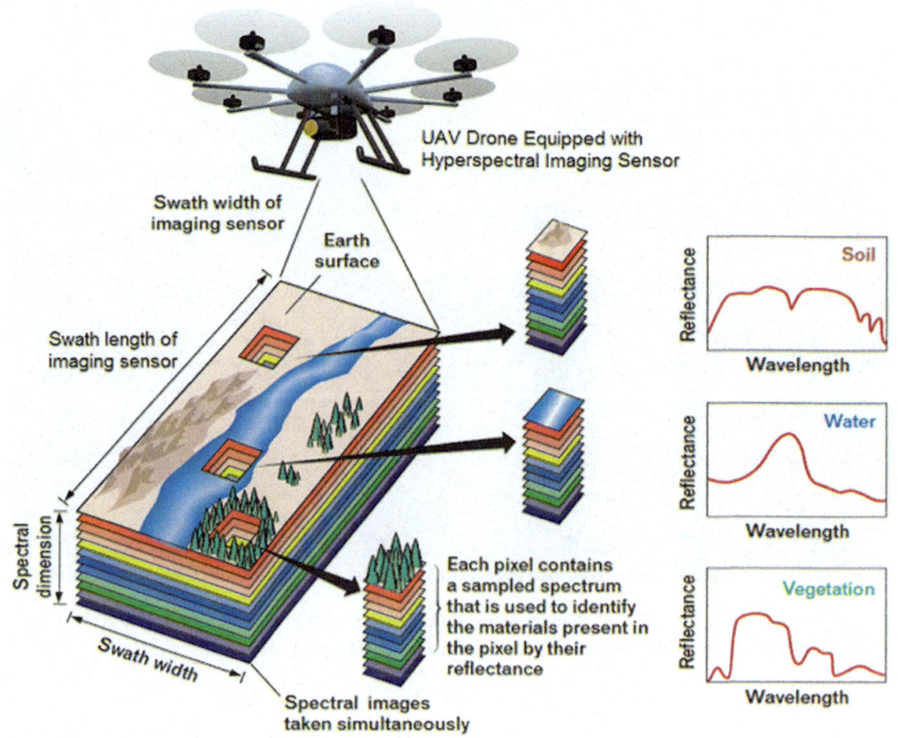

Fig. 2 Hyper-spectral imaging concept in remote sensing (Ozdemir et al., 2020)

data with satellite and UAV imaging data poses a serious challenge. However, this data reflects the ground truth accurately and can be used to calibrate/validate the data derived from the other two sources. This chapter presents an algorithmic framework for data fusion that employs standard Machine Learning algorithms to exploit synergies between sensing data from satellites, UAVs, satellites, and farm IoT sensors. However, existing approaches (Zhu et al., 2018; Alvarez et al., 2021) for data fusion are limited to combining images involving RGB, multispectral, and thermal sensors. For a detailed survey on various data fusion approaches, we refer the readers to (Ghamisi et al., 2019, Zhu et al., 2018 and Alvarez et al., 2021) (Fig. 4).

2 Overview of Data Fusion

In Digital Agriculture, most services use remote sensing data from satellites and UAV's, farm data from IoT sensors and other ancillary sources provided by relevant governmental agencies. Data fusion is one of the widely used strategies for integrating data for these services. The number of data sources and their heterogeneity

Fig. 3 Soil monitoring with IoT (Smart Agriculture, Manx tech group, February 24, 2021, https://www.manxtechgroup.com/soil-monitoring-with-iot-smart-agriculture)

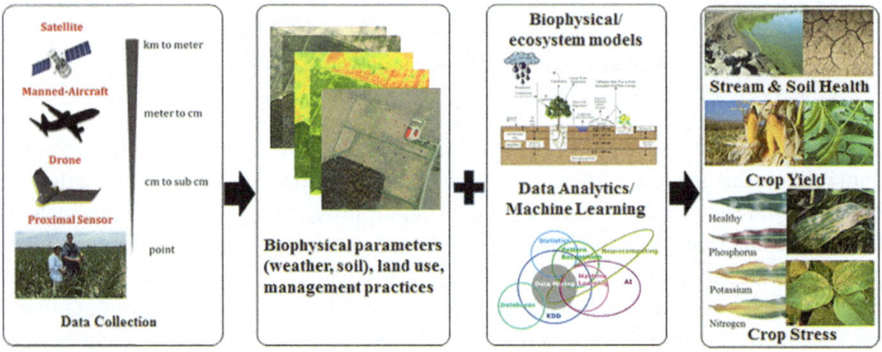

Fig. 4 Schematic illustrations of data collection and hypercube construction (Ag Sensing, Digital Ag, College of Food, Agriculture, and Environmental Sciences, The Ohio State University, 2022, https://digitalag.osu.edu/ag-sensing)

causes a significant challenge in developing a standard framework for data fusion. However, considerable progress in developing tools and techniques in Data Science, and Machine Learning for processing and analysis has helped drive the development of common algorithmic frameworks and metrics for data fusion.

Fusion approaches (Ghamisi et al., 2019; Zhu et al. 2018; Simões et al., 2021; Alvarez et al. 2021) are often concerned with the creation of a unified dataset of higher quality (i.e., higher resolution in the case of imaging data) by combining two or more datasets with different spatial, temporal and spectral resolutions. Figure 5 demonstrates schematically the multi-scale nature (different spatial resolutions) of diverse datasets captured by space-borne, airborne, and UAV sensors. In principle, there is a relation between spatial resolution and coverage/swathe, i.e., data with a coarser spatial resolution (space-borne date have broader coverage. In comparison, data with a finer spatial resolution (UAV data) have limited coverage. Figures 6 and 7 schematically illustrate how data fusion exploits local spectral and temporal attributes to create a unified dataset with better resolution.

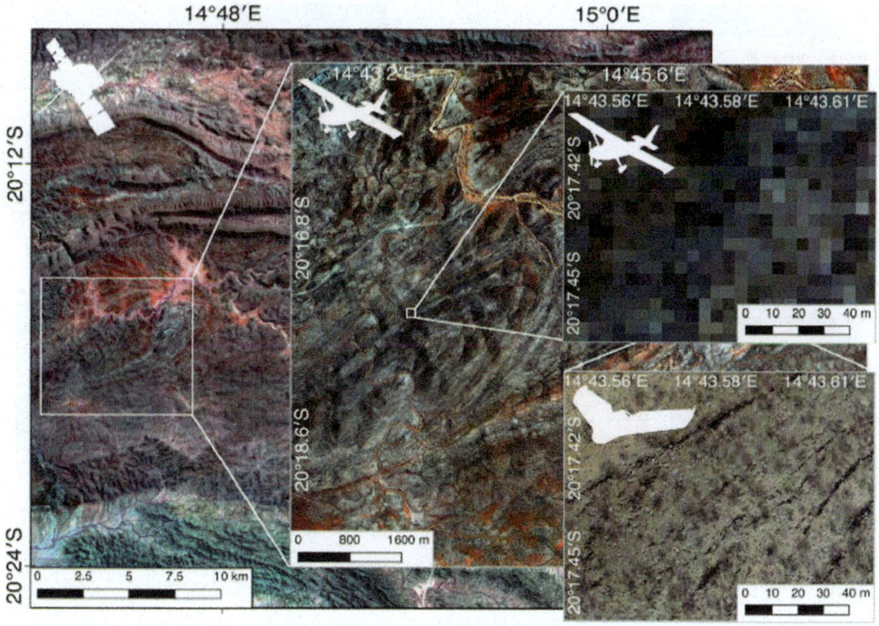

Fig. 5 Multi-scale nature of diverse datasets captured by multi-sensor data (Ghamisi et al., 2019 and Booysen et al., 2018)

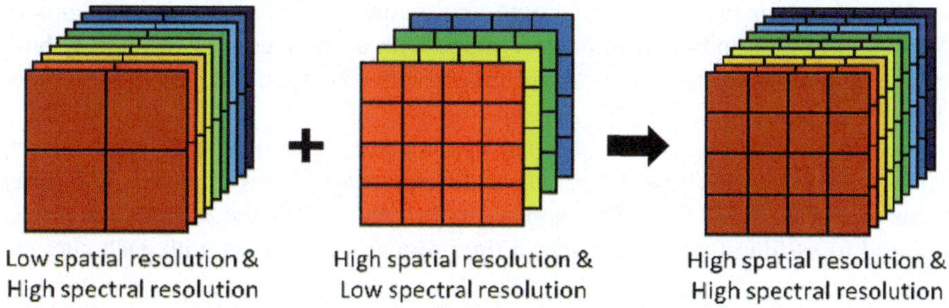

Fig. 6 Spatial–spectral and spatial–temporal data fusion (Ghamisi et al., 2019)

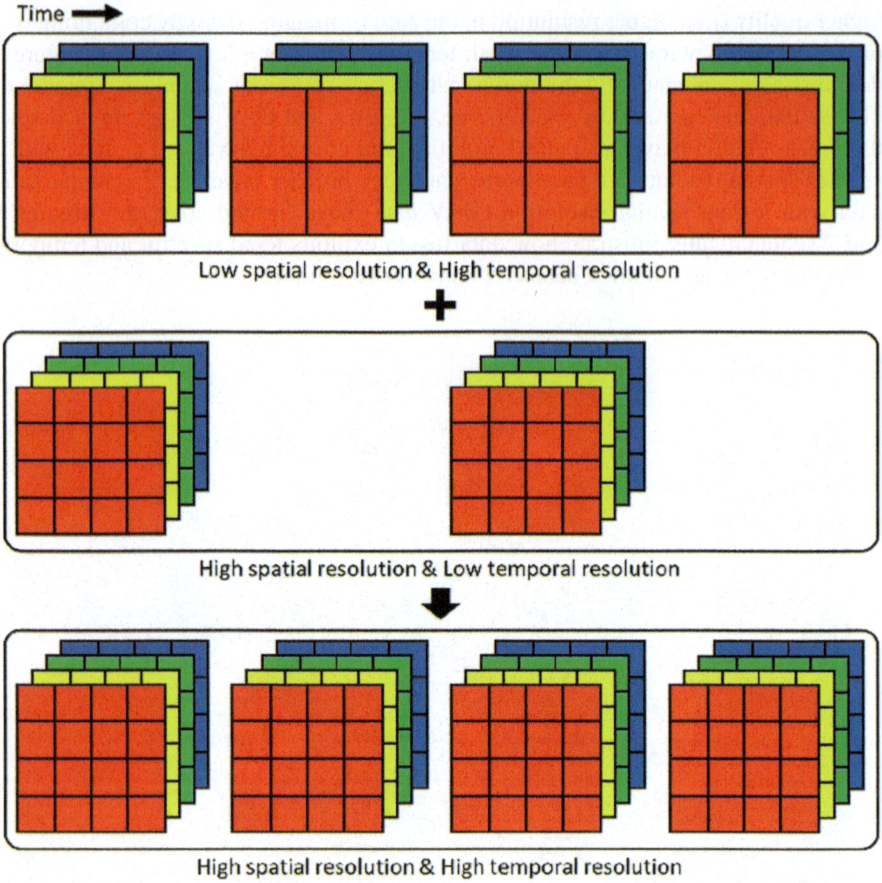

Fig. 7 Spatial–temporal data fusion (Ghamisi et al., 2019)

Spatial–spectral fusion is the fusion of an image with low spatial but high spectral resolution with an image with a high spatial but low spectral resolution to obtain an image with high spatial and spectral resolution. Similarly, spatial–temporal fusion is a technique for fusing an image with fine spatial resolution, but coarse temporal resolution (e.g., Landsat) data and an image with a fine temporal resolution, but a coarse spatial resolution to create an image with fine spatial–temporal resolution data.

In remote sensing, most traditional approaches perform data fusion involving images involving the same place but at different times. This approach is referred to as the space-first, time-later approach. In contrast, in the alternate approach referred to as time-first, space-later, a time series is associated with each site, and spatial post-processing is applied to incorporate neighborhood context. For agricultural/ecological applications that involve tracking changes in land surface characteristics, many researchers believe that the time-first, space-later methods are more effective when compared to space-first, time-later approaches.

All data fusion methods attempt to overcome the above measurement and sampling characteristics, which fundamentally limit the amount of information captured from a scene using an image obtained using a particular modality. Therefore, we need to understand the measurement and sampling characteristics of the data obtained and the relevant trade-offs involved in these modalities.

3 Farm Data Acquisition and Their Characteristics

Satellites can be broadly categorized as Geostationary (GEO), medium Earth orbit (MEO), and low Earth orbit (LEO) satellites. GEO satellites typically collect data at an hourly frequency; however, their spatial resolution is coarse (each pixel corresponds to a region whose length is greater than 100 m). MEO collects data on nearly ten spectral bands daily and at a spatial resolution ranging between 10 and 100 m. The LEO satellites collect data daily at a high spatial resolution (1–10 m) with a low spectral resolution with limited coverage. All remote sensing (RS) satellites have limited flexibility in acquiring data and are constrained by cloud cover, view angle, or acquisition time.

UAVs have been employed for remote sensing because of their lower weight, easily programmable interfaces, better spatial resolution, and the ability of end users to control: the type of sensor, angles of view, spatial resolution, time, and frequency of acquisition. The RS UAVs strength is its ability to acquire data at spatial resolution at centimetric to millimetric levels. It is due to its ability to acquire data close to the surface while not being affected by clouds. However, RS UAV has the disadvantage of having a smaller coverage of only a few square kilometers, often because of its battery power as well as due to rules for protecting air traffic and people's safety and privacy. Finally, UAVs complement the satellites in spatial and temporal resolution and swath and hence can operate synergistically with satellites.

IoT devices capture soil condition factors (soil carbon, soil pH, soil temperature, NPK, photosynthetic radiation, soil water potential, soil oxygen levels, wind, and water erosion) from individual farms that help determine the resilience (disease/pest prevalence) and productivity of farmlands (better accuracy in measuring factors contributing to crop yield and crop quality). Soil temperature and soil moisture content are monitored accurately using one or more probes buried in the soil. The soil temperature is an essential factor influencing root growth, respiration, decomposition, and mineralization of nitrogen. The moisture content of soil plays a vital role in soil chemistry, plant growth, and groundwater recharge. Soil moisture content and soil water potential influence crop yield by regulating soil temperature, serving as a crucial nutrient, transporting other nutrients, and being an essential component of photosynthesis. Solar radiation plays a vital role in photosynthesis and impacts plant growth and is measured using IoT sensors that essentially measure (i) photosynthetically active solar radiation, (ii) solar–UV, and (iii) solar–shortwave. Weather plays an essential role in plant growth and is determined by rainfall/precipitation, wind, humidity, and atmospheric pressure. Weather stations and soil sensors help measure the following: precipitation, temperature, humidity, air pressure, wind speed, and wind direction. NPK soil sensors are IoT sensors used to primarily critical soil nutrients nitrogen, phosphorus, potassium, and soil pH. In addition, existing IoT frameworks offer a wide range of wireless communication options that help integrate soil condition data with the data available through satellites and UAVs. Further, these solutions require low power and hence can be powered by battery, solar, or other renewable sources. It complements UAV and satellite data and can help recalibrate satellite data and identify trends. We refer the readers to (Ghamisi et al.,2019, Zhu et al., 2018, Alvarez et al., 2021, Adao et al., 2017, Arman et al., 2021, and Appel et al., 2018) for an excellent overview of relevant work on exploiting the synergy between UAV and satellite data.

Tables 1 and 2 summarizes the remote sensing data characteristics of satellites and UAVs.

Table 1 Remote sensing satellites and their data characteristics (Latchininsky et al., 2010)

Satellite and Country	Spectral bands	Pixel size (m)	Swath (Km)
Coarse resolution	–	–	–
NOAA-GOES (USA)	–	1000	> 2000
SPOT VEG (France)	–	1000	–
TERRA/MODIS (USA)	–	250	> 2000
—	–	500	–
—	–	1000	–
Moderate resolution	–	—	–
Landsat 5 (USA)	B, G, R, 3 IR	30	185
Spot 2 (France)	G, R, 2 IR	20	120
IRS 1C (India)	G, R, 2 IR	23	70, 142
Landsat 7 (USA)	B, G, R, 3 IR	30	185
TERRA/ASTER	G, R, IR	15	60
(Japan/USA)	4 IR 3 Thermal IR	30 90	—
EO-1 (USA)	—	30	37
Proba (ESA)	—	18, 36	14
SPOT-5	—	10	120

Table 2 List of hyper-spectral sensors (and respective characteristics) available for being coupled with UAVs (Adao et al., 2017)

Manuf	Sensor	Spectral range	No. bands	Spectral resolution (nm)	Spatial resolution (px)
BaySpec	OCI-UAV-1000	600–1000	100	< 5	2048
Brandywine Photonics	CHAI S-640	825–2125	260	5	640 × 512
Cubert GmbH	S 185-FIREFLEYE SE S485-FIREFLEYE XL Q-285 FIREFLEYE QE	450–950 355–750 450–950 550–1000 450–950	125 125 125	4 4.5 4	50 × 50 70 × 70 50 × 50
Headwell Photonics Inc	Nano-Hyperspec Micro-Hyperspec VNIR	400–1000 380–1000	270 775 837 923	6 2.5	640 1004 1600
Hyperspex	VNIR-1024 Mjol-nir-V-1240 Hypex Swir-384	400–1000 400–1000 1000–2500	108 200 288	5.4 3 5.45	1024 1240 384

4 Our Generic Algorithm for Data Cube Construction

In Digital Agriculture, data is typically obtained from remote sensing sensors (RGB, multispectral, hyper-spectral, and thermal) and farm IoT sensors (IoT devices that provide spatial and time series data of land and soil characteristics). For each region of interest (RoI) over a given period, each sensor constructs a 2D image consisting of pixels, where each pixel is associated with a 2D spatial coordinate and a 1D temporal coordinate. In addition, each pixel is associated with a feature vector that specifies the value sensed by that sensor. Suppose each value is specified by a one-dimensional vector, then the sensor measurement over the region of interest for a given time interval can be represented by a 3D cube. However, if each value is specified by a k-dimensional vector, then the sensor measurement over the region of interest for a given interval of time can be represented by a $k+3$ dimensional hypercube. Figure 8 schematically represents the measurement from a sensor by a 3D cube (hypercube for the general case), and Fig. 9 shows the measures from all sensed devices as a higher dimensional hypercube (we are showing them as a collection of cubes due to the inability to display a hypercube of dimension greater than 3). Notice that the spatial, temporal, and spectral characteristics of different sensors are different; hence, the sizes of data cubes are not necessarily the same.

Now, we need to construct a hypercube of low dimension by fusing the data cubes corresponding to each sensor. The resulting hypercube eliminates redundancy while capturing all the essential variations in the farm data. This fusion of data cubes depend on the relationship between the sensed data, the nature of the time

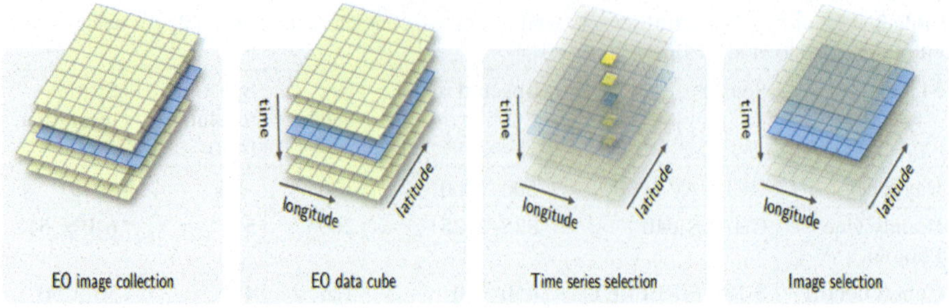

Fig. 8 Data cube corresponding each time series values from each IoT sensor (Simoes et al., 2021)

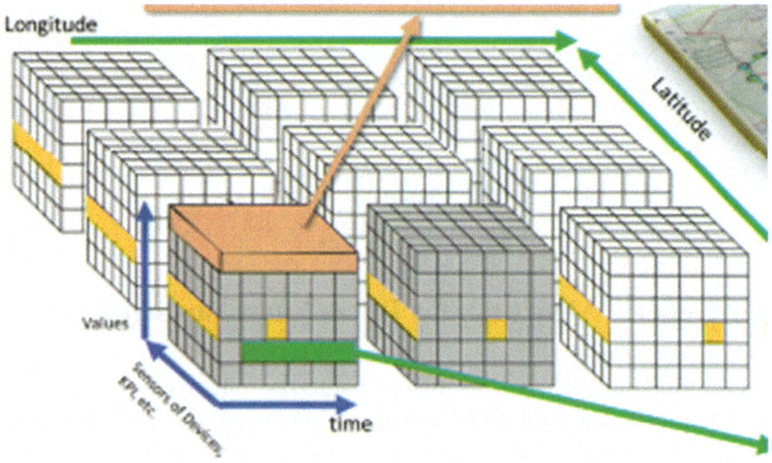

Fig. 9 Collection of data cubes of IoT time series and multispectral satellite and UAV data (Arman et al., 2021)

series extrapolation function, and the variation in the spatial, spectral, and temporal dimensions of data cubes corresponding to sensed data. So fusing the data cubes involves prioritizing the sequence in which the data cubes are combined, filling the gaps temporally, spatially, and spectrally so that all the data cubes are of the same resolution across every dimension.

For this, we describe our general algorithmic framework for data fusion consisting of the following steps:

1. *Preprocessing*:

   a. *Partitioning data cubes*: First, we partition the sensed attributes into *complementary* and *overlapping*. Complementary attributes correspond to data cubes of IoT farm sensors, and overlapping attributes correspond to data cubes of sensors in satellites and UAVs. Notice that the complementary attributes help in calibrating/validating satellite and UAV data. In contrast, the overlapping attributes provide data over the same RoI at different spatial, temporal, or spectral resolutions. Therefore, the data cubes corresponding to complementary and overlapping attributes are likely to be of different dimensions and resolutions depending on their sampling frequency, temporal, spectral, and spatial dimensions.

b. *Identifying an independent set of Informative overlapping attributes*: From each data point in RoI, we construct a raw feature vector for a given time instance by concatenating all the overlapping attributes. Then, we perform Principal Component Analysis (PCA) to identify the informative attributes. Notice these attributes are derived attributes (i.e., each attribute may be a linear combination of overlapping attributes). Next, we repeat this over the specified time interval and choose the overlapping attributes that frequently occur (determined by a frequency threshold) in the informative attributes. We refer to this set as the set of independent informative overlapping attributes.
c. Partition the independent set of informative overlapping attributes into layers such that the attribute with the highest resolution is at the bottom layer and the attribute with the lowest resolution is at the top of the hierarchy.
d. Define the non-informative overlapping attributes as a weighted function of an independent set of informative overlapping attributes using the coefficients of PCA in step 1b.

2. *Modeling time series data of complementary attributes through polynomial regression*: We employ polynomial regression for the time series data corresponding to a complementary attribute. In situations where the characteristics of the sensed attribute are well understood, we can fit the time series data using the user-specified function. We interpolate the time series data so that the data cubes of all complementary attributes have the same resolution along the temporal dimension.
3. Correct the values of this overlapping attribute at the lowest layer (i.e., highest resolution) based on the known models that characterize the relationship between the overlapping attribute and a subset of complementary attributes. At a given site, the farm IoT sensors provide the ground truth, and the UAV sensors provide the sensed values. We exploit the known relationship between the attribute measured by the farm IoT sensor to recalibrate the value of the related attribute sensed by a UAV/satellite sensor.
4. Now starting at a layer above the lowest layer repeatedly do the following until you reach the topmost layer: extract the key points and then perform point registration with the key points in the layer immediately below it to interpolate values to achieve the same resolution as the layer below.
5. We now construct the values of the non-informative overlapping attributes at the topmost layer by using the independent informative overlapping attribute values at the topmost layer and the weighted function defined in step 1d above.
6. Evaluating the values in the top layer using either a coherence function or predicting a property that is a function of the input data (e.g., land cover or vegetation biomass, through classification or regression-type approaches).

7. Return the data cube corresponding to the topmost layer with the corresponding coherence function value associated with that layer.

5 Data Fusion in the Context of Crop Monitoring

We now provide a very brief high-level description of how satellite images, multi-spectral drone images, and farm data from IoT sensors (humidity, pressure, wind, temperature) in the context of crop monitoring can be fused using our generic algorithmic framework presented in Sect. 4. For a broader understanding of the processes involved in data acquisition and exploitation of synergies of data fusion of satellite, UAV, and farm IoT sensor data, we refer the reader to a more recent excellent survey by Barrrile et al. (2022).

For monitoring of crops, we make use of remote sensing satellite and UAV data that involve the normalized difference vegetation index (NDVI), one of the frequently used indexes that relate the health of vegetation as a function of how different surfaces and objects absorb or reflection of light in the visible and infrared regions. From satellites, we acquire data on the surface reflectance of objects on the Earth's surface, and thereby the crops using their multispectral sensors. These multispectral images cover a wider area, but the coarseness of their resolution does not help in analyzing small parcels of land relevant to a large majority of small farmers. However, unmanned aerial vehicles (UAVs) address this issue by acquiring data over a narrow area but at a finer resolution using a variety of optical and airborne sensors involving multi-spectral and hyper-spectral imaging sensors. The NDVI is calculated by processing images that were taken by the multispectral and hyper-spectral sensors using the following formula: $\mathrm{NDVI} = (\mathrm{NIR} - \mathrm{Red})\ (\mathrm{NIR} + \mathrm{Red})$, where Red and NIR are spectral surface reflectance measurements obtained in the visible (red) and near-infrared regions, respectively. In contrast, the low-cost sensors and IoT devices provide accurate spatial and time series data of land and soil characteristics spanning a few meters that is used to accurately compute NDVI which then can be used as ground truth to validate/recalibrate the satellite and UAV data. This recalibration/validation of satellite data is done through data fusion in our algorithmic framework by employing standard Machine Learning algorithms.

Multispectral sensors used in satellites typically use from 3 to 10 bands, whereas hyper-spectral sensors compared to multispectral sensors cover hundreds of narrower bands providing greater resolution. This higher spatial resolution helps distinguish smaller objects and hence detect spectral surface reflectance more accurately. The farm IoT sensors that can be used in the accurate time series computation of NDVI are soil moisture, leaf wetness, pH, temperature, humidity, and pressure sensors.

6 Conclusions and Future Work

In this chapter, we present an algorithmic framework that exploits the synergies among the satellites, UAVs, and farm IoT sensors to construct a high-dimensional farm map. First, we describe our algorithmic framework in terms of some standard methods in Data Science and Machine Learning for creating a hypercube of low dimension by fusing the data cubes corresponding to each sensor so that the resulting hypercube eliminates redundancy while capturing all the essential variations in the farm data. This fusion of data cubes involves prioritizing the sequence in which the data cubes are fused, filling the gaps temporally, spatially, and spectrally so that all the data cubes have the exact resolution across every dimension. We are in the process of exploiting our algorithmic framework in the context of farm monitoring. One of the immediate challenges we are facing is (i) to obtain data for sufficient number of sites from the three sources from satellite, UAV, and IoT sensors and (ii) standardize the data acquisition parameters and data formats so that data values acquired across different sensors are comparable.

References

Adão, T., Hruška, J., Pádua, L., Bessa, J., Peres, E., Morais, R., & Sousa, J. J. (2017). Hyperspectral imaging: A review on UAV-based sensors, data processing and applications for agriculture and forestry. *Remote Sensing, 9*(11), 1110. https://doi.org/10.3390/rs9111110.

Arman, A., Bellini, P., Bologna, D., Nesi, P., Pantaleo, G., & Paolucci, M. (2021). Automating IoT data ingestion enabling visual representation. *Sensors, 21*(24), 8429. https://doi.org/10.3390/s21248429.

Appel, M., Lahn, F., Buytaert, W., & Pebesma, E. (2018). Open and scalable analytics of large earth observation datasets: From scenes to multidimensional arrays using scidb and gdal. *ISPRS Journal of Photogrammetry and Remote Sensing, 138*, 47–56.

Audebert, N., Saux, B. L., & Lefvre, S. (2017). Beyond rgb: Very high resolution urban remote sensing with multimodal deep networks. *ISPRS Journal of Photogrammetry and Remote Sensing, 140*, 20–32.

Ball, J. E., Anderson, D. T., & Chan, C. S. (2017). Comprehensive survey of deep learning in remote sensing: theories, tools, and challenges for the community. *Journal of Applied Remote Sensing, 11*, 042 609–1–042 609–54.

Barrile, V., Simonetti, S., Citroni, R., Fotia, A., & Bilotta, G. (2022). Experimenting agriculture 4.0 with sensors: A data fusion approach between remote sensing, UAVs and self-driving tractors. *Sensors, 22*(20), 7910. https://doi.org/10.3390/s22207910.

Booysen, R., Zimmermann, R., Lorenz, S., Gloaguen, R, Nex, P. A. M., Andreani, L., & Mockel, R. (2018). Towards multiscale and multisource ¨ remote sensing mineral exploration using rpas. A case study in the lofdal carbonatite-hosted ree deposit, namibia. Remote Sensing.

Alvarez-Vanhard, E., Corpetti, T., & Houet, T. (2021). UAV & satellite synergies for optical remote sensing applications: A literature review. *Science of Remote Sensing, 3*, 100019. https://doi.org/10.1016/j.srs.2021.100019

Ghamisi, P., et al. (2019). Multisource and multitemporal data fusion in remote sensing: A comprehensive review of the state of the art. In *IEEE Geoscience and Remote Sensing Magazine*, March 2019, vol. 7(1) (pp. 6–39). https://doi.org/10.1109/MGRS.2018.2890023.

Khan, M., Adeel, Y., Khurram, K., Faisal, S., & Asad, A. (2018). Automated forgery detection in multispectral document images using fuzzy clustering. https://doi.org/10.1109/DAS.2018.26.

Latchininsky, A.V., & Sivanpillai, R. (2010). Locust habitat monitoring and risk assessment using remote sensing and GIS technologies. In Ciancio, A., & Mukerji, K. (Eds.), Integrated management of arthropod pests and insect borne diseases. Integrated Management of Plant Pests and Diseases, vol 5. Springer, Dordrecht. https://doi.org/10.1007/978-90-481-8606-8_7

López, A., Jurado, J. M., Ogayar, C. J., & Feito., F. R. (2021). A framework for registering UAV-based imagery for crop-tracking in precision agriculture. *International Journal of Applied Earth Observation and Geoinformation, 97*, 102274.https://doi.org/10.1016/j.jag.2020.102274

Ozdemir, A., & Polat, K. (2020). Deep learning applications for hyperspectral imaging: a systematic review. *Journal of the Institute of Electronics and Computer, 2*, 39–56. https://doi.org/10.33969/JIEC.2020.21004.

Simões, R. E. O., Câmara, G., de Queiroz, G. R., de Souza, F. C., de Andrade Neto, P. R., Santos, L. A., Carvalho, A., & Ferreira, K. R. (2021). Satellite image time series analysis for big earth observation data. *ArXiv* abs/2204.11301.

Solberg, A. H. S. (2006). In Data fusion for remote-sensing applications.

Zhu, X., Cai, F., Tian, J., & Williams, T. K.-A. (2018). Spatiotemporal fusion of multisource remote sensing data: Literature survey, taxonomy, principles, applications, and future directions. *Remote Sensing, 10*(4), 527. https://doi.org/10.3390/rs10040527.

Globally Scalable and Locally Adaptable Solutions for Agriculture

Gogumalla Pranuthi and Rupavatharam Srikanth

Abstract Precision agriculture and smart farming will soon replace the conventional methods adopted in agriculture; these technologies are capable of increasing the productivity of crops and also optimizing the inputs, and at the same time sustaining the environmental resources. The ability of satellite data in precision farming has been made evident through several projects and research activities adopted across India. The past decade has seen a rapid increase in the use of satellite-based remote sensing data in crop mapping and crop monitoring. This growth can be attributed to at least three factors—the availability of open remote sensing data, advanced machine learning methods, and access to cloud computing platforms that can handle big data storage and processing. In this book chapter, we propose Globally Scalable and Locally Adaptable Solutions for Agriculture that can facilitate in crop monitoring at a larger scale. This chapter focuses on the use of open-source high-resolution (in terms of spectral, spatial, and temporal resolution) satellite data; open source cloud-based platforms, and big data algorithms that are reforming agriculture. This book chapter will detail the available open-source satellite data and platforms with use cases. We also discuss in detail the methodology of developing a crop monitoring system using Google Earth Engine that can be globally scaled and locally adaptable, using a case study of the wheat crop for Samastipur district, Bihar, India.

Keywords Artificial Intelligence · Big data platforms · Google Earth Engine · Open-source satellite data · Sentinel

1 Introduction

Crop monitoring involves determining several crop-based parameters from sowing to harvesting to be able to judge the status of the crop for appropriate decision-making. The scale of the crop monitoring system is very important as it is based

G. Pranuthi · R. Srikanth (✉)
International Crops Research Institute for the Semi-Arid Tropics, Patancheru, Hyderabad 502324, India
e-mail: r.srikanth@cgiar.org

© The Author(s), under exclusive license to Springer Nature Singapore Pte Ltd. 2023

S. Chaudhary et al. (eds.), *Digital Ecosystem for Innovation in Agriculture*, Studies in Big Data 121, https://doi.org/10.1007/978-981-99-0577-5_5

on the purpose of monitoring crops. For example, farmers require field-level crop monitoring throughout the season, whereas policymakers require crop monitoring at a larger scale (village/taluk/Mandal/district/state); for example, estimates on crop yield and crop losses due to calamities. An interactive crop monitoring system with high-resolution products with national or global coverage is desired as this system can be used by different stakeholders for estimating crop status. The first satellite-based crop monitoring program, Large Area Crop Inventory Experiment (LACIE), was launched by the United States Department of Agriculture (USDA) in the year 1973 (Erickson, 1984). The approach of the program was mainly to estimate wheat area and production on a larger scale using weather data and Landsat satellite data. In India, acreage estimation and production forecasting of important agricultural crops using satellite remote sensing data were initiated under the project "Crop Acreage and Production Estimation" by the Space Applications Centre (SAC) and carried out over a period of two decades at the behest of the Ministry of Agriculture, Govt. of India (MOA). Subsequently, a more inclusive model, FASAL (Forecasting agricultural output using Space, Agro-meteorology, and, land-based observations) was initiated in 2007–08, and SAC was entrusted with the responsibility of implementing space technology-based production forecast of crops and upgradation of the procedure with new data availability.

The use of remotely sensed satellite data powered by artificial intelligence and machine learning in developing crop-based solutions has proved to be effective (Reddy et al., 2021; Tawade & Virnodkar, 2022). In the recent past, many web-based or mobile-based crop monitoring systems have been launched that are helping farmers improve their yields. The availability of open-source satellite data; advanced data analytical tools that can handle big data; and access to open-source cloud computing platforms, which allow users to store big data and apply big data algorithms; also provides tools to visualize and present the products in a format that can be easily understood by the farmers. High-resolution (spatial, temporal, and radiometric) satellites like Landsat-8, Sentinel-1, Sentinel-2, and MODIS enable the measurement of crop parameters at the field level throughout the cropping season without interruption. These satellites regularly visit every part of the globe, allowing agricultural organizations/researchers to develop systems that farmers can use for farm-level monitoring as well as other stakeholders to estimate crop production on a national scale. However, the datasets corresponding to agriculture are huge, ranging from satellite datasets to crop, soil, and climate data. Processing of these datasets for reliable and meaningful products requires big data platforms and big data algorithms. Here in this book chapter, we will be reviewing the different open source satellite data, open source big data platforms, and big data algorithms that can be employed to develop solutions for agriculture that can be "Globally Scaled and Locally Adaptable." There are numerous satellite datasets, however we present the most popular open-source satellite datasets in this chapter. In the case of big data platforms, the internet provides many options, but the best known and suited applications for scaling up technologies have been briefed in this book chapter. In this chapter, we also discuss how several crop monitoring tools have used open-source resources in developing meaningful and reliable products that can be directly accessed by users. We will discuss how each open source dataset, platform, or big data algorithms has been leveraged with one example for each resource.

2 Open Source Satellite Data

With the launch of every new open satellite mission, the spatial, spectral, and temporal characteristics have been improving, which enables researchers to develop new tools for farmers and policymakers to increase their decision-making ability. The present open satellite missions are providing data in most regions of the electromagnetic spectrum, ranging from ultra violet to microwave regions (Fig. 1). Hyperspectral sensors (Li et al., 2022) and Light Detection and Ranging (LiDAR) (Bates et al., 2021) sensors are gaining attention in monitoring agriculture. They are typically used on unmanned aerial vehicles (UAVs) or low-orbit satellites due to their extremely high cost. The open-source satellite datasets are provided by organizations such as NASA/USGS National Aeronautics and Space Administration/United States Geological Survey and the European Space Agency (ESA) for land, climate, and agriculture monitoring. Aerial imagery is also available as an open source for the USA under NAIP (National Agricultural Imagery Program), which is updated annually (Prasai et al., 2021). These are high-resolution datasets that allow the stakeholders to assess the crop condition during the cropping period at a larger scale with accuracy.

Open satellite data has been available since the 1970s, but the increased spatial, spectral, and temporal capabilities of newly launched satellites are allowing for crop monitoring from the field to the national and global levels. Here we discuss the

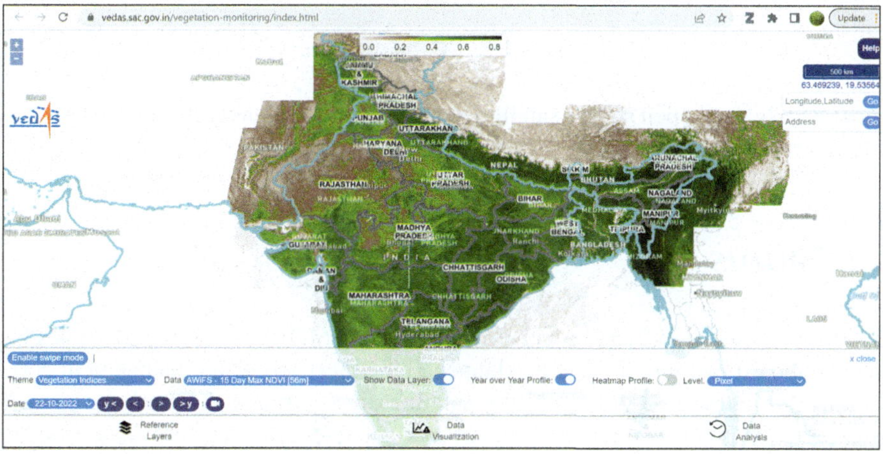

Fig. 1 Screenshot of a vegetation and crop monitoring system developed under VEDAS program by Space Application Centre India

Table 1 List of open-source satellite datasets, with their spectral, spatial and temporal resolution

Satellite	Bands	Spectral resolution (μm)	Spatial resolution (m)	Revisit time (days)	URL
MODIS	36 (10—Vis; 7—NIR; 2—SWIR; 16—TIR)	0.4–1.45	250–1000	16	https://www.modis.gsfc.nasa.gov/data/
Landsat	10	0.4–2.29	30	16	https://www.modis.gsfc.nasa.gov/data/
Sentinel-1	4 (VH, VV, HH and HV)	5.405 GHz	20	12	https://www.sentinels.copernicus.eu/web/sentinel/missions/sentinel-1
Sentinel-2	13 (4—Vis; 3—Red Edge; 4—NIR; 2—SWIR)	0.4–2.28	10–20	5	https://www.sentinel.esa.int/web/sentinel/missions/sentinel-2
PlanetScope	4	0.4–0.86	3	1	https://www.planet.com

most widely used open source satellite data and how these datasets are used for crop monitoring (Table 1).

2.1 MODIS

Moderate Resolution Imaging Spectro-radiometer (MODIS) (Didan, 2015) is a key sensor aboard two satellites, Terra (launched in 1999) and Aqua (launched in 2002) by NASA. These two polar-orbiting satellites orbit around the earth in opposite directions in order to photograph every part of the earth's surface once in every one or two days. It offers a unique combination of spectral, temporal, and spatial resolution compared to previous global sensors, making it a good candidate for large-scale crop growth monitoring. MODIS acquires data in 36 spectral bands, the spectral wavelengths ranging from 0.4 μm (Visible) to 14.4 μm (Thermal Infrared bands). The spatial resolution of MODIS satellites ranges from 250 to 1000 m with a revisit time of 8 to 16 days. MODIS data has played a key role in monitoring agriculture and climate on a national scale as the coverage of this satellite is large. There are many climate and crop-based products developed from MODIS data, which makes crop monitoring easy on a larger scale. The Foreign Agricultural Service (FAS) Crop Explorer (USDA, 2005, 2016), developed by the International Production Assessment Division (IPAD) under the US Department of Agriculture (USDA), gives global crop and climate assessments and also provides cropped area, progress of sowings, yield, and production on a national scale for several major crops. The primary objective of FAS Crop Explorer is to provide reliable crop estimates to increase the food security of all nations. The FAS Crop Explorer uses MODIS-derived products and makes crop-based assessments and also provides a crop monitoring system that is updated at an 8 to 16-day interval with a spatial resolution of 250 m.

Within 2–4 h after satellite captures, the MODIS rapid response system gives FAS researchers access to georeferenced, calibrated, and mosaicked daily global MODIS images for FAS regions of interest. FAS experts can study and assess the impact of disaster occurrences on crops in real-time because of the quick access to data. For monitoring crop growth, remotely sensed vegetation indices data is used to track how the growing season has changed in comparison to reference long-term mean conditions when crop conditions are being monitored for a particular location. In order to associate vegetation indices from multiple sensors, their temporal and spatial behavior (means, ranges, and anomalies) at target FAS agricultural locations around the world is examined. These statistical patterns are then interpreted in light of the various crop producing seasons. All this processed data is made available for different parts of the world through a web application (https://ipad.fas.usda.gov/cropexplorer/), which has a great repository of all the products that can help in monitoring crops up to district and sub-district level. The FAS Crop Explorer is the only web application that has been providing crop and climate-based estimates consistently and regularly since 2002.

Mahalanobis Crop Forecast Centre (MNCFC) is also providing district level NDVI, NDWI, and drought indices derived from MODIS datasets through a web platform (https://www.ncfc.gov.in/maps.html). The Visualization of Earth Observation Data and Archival System (VEDAS) platform developed by the SAC of India is providing applications for a large number of disciplines, including agriculture, land use, photogrammetry, forestry and environment, coastal and marine resources, hydrological studies, climate change studies, urban and infrastructure, etc. This system also provides a "Vegetation and Crop Monitoring" web application which uses MODIS data along with AWiFS, OCM, and PROBA to monitor crop condition through NDVI measurements. Free and open source tools including GDAL, Python, Flask (REST API), GeoServer, PostgreSQL, and JavaScript APIs were used to construct the entire application. A method for searching, tracking, analyzing, and displaying water level data for a chosen river or reservoir site in India is provided by VEDAS (Mishra et al., 2020) (Fig. 1).

2.2 Landsat

Landsat earth-observing satellites were launched jointly by NASA (National Aeronautical Space Agency) and United States Geological Survey (USGS) for the first time in the year 1972 (Alfoldi & Ryerson, 1973). The Landsat satellites that are currently operating are Landsat-8 and Landsat-9 (Vermote et al., 2016). These Landsat satellites have 10 spectral bands ranging from visible, NIR, and thermal infrared regions with a spatial resolution of 30–60 m and a revisit time of 16 days. Landsat is one of the oldest open-source satellite data sources, with a massive archive of satellite images dating back to 1972. This long-term data helps to monitor climate change and its impact on different land use classes, vegetation monitoring, trends in agricultural dynamics, etc. Landsat data is extensively used for land use and land cover mapping. Landsat satellites have the optimal ground resolution and spectral bands to efficiently track land use and document land change due to climate change, urbanization, drought, wildfire, biomass changes (carbon assessments), and a host of other natural and human-caused changes. A significant commercial consumer of Landsat imagery is Geospatial Data Analysis Corporation (GDA Corp.) which compiles and evaluates a significant volume of global Landsat imagery (https://www.globaldata.com/). For crop mapping, monitoring, evaluation, and forecasting, Landsat's combination of intermediate spatial resolution (60-m and 30-m), multispectral and thermal bands, and its extensive historical record all make for a suitable database. For its global crop analysis, GDA relies on its time series of surface reflectance calibrated by Landsat mosaics and provides field-to-country assessments. The correlations between time series spectral features and crop types that have been established during the previous 10–15 years provide the foundation for GDA's current season crop assessments. These connections make it possible to perform highly automated in-season field level crop mapping, which forms the basis for regional crop monitoring, frequent assessments of the state of the crops' areas, and forecasts of their end-of-season yields and output.

As Landsat receives spectral information in the spectral region of thermal infrared, it also facilitates crop water studies. FAO's Water Productivity Open Access Portal (WaPOR) (Geshnigani et al., 2021) (Fig. 2) monitors crop water productivity and crop evapotranspiration for the African continent using Landsat data. This application helps farmers achieve considerable yields with optimum use of water resources. This application performs several computations to estimate crop evapotranspiration, crop water requirement, and crop water productivity using the Google Earth platform and provides products at one-to ten-day intervals with 30–100 m resolution. The products that the WaPOR application provides to its users are Gross Biomass Water Productivity; Net Biomass Water Productivity; Actual Evapotranspiration; Above Ground Biomass Production; Transpiration; Net Primary Production; Precipitation (https://wapor.apps.fao.org/home/WAPOR_2/1). This application provides subnational estimates of crop productivity which can be locally adaptable and also provides continental estimates. The WaPOR model uses a remote sensing data-based Penman–Monteith (PM) model to estimate actual evapotranspiration, which is also

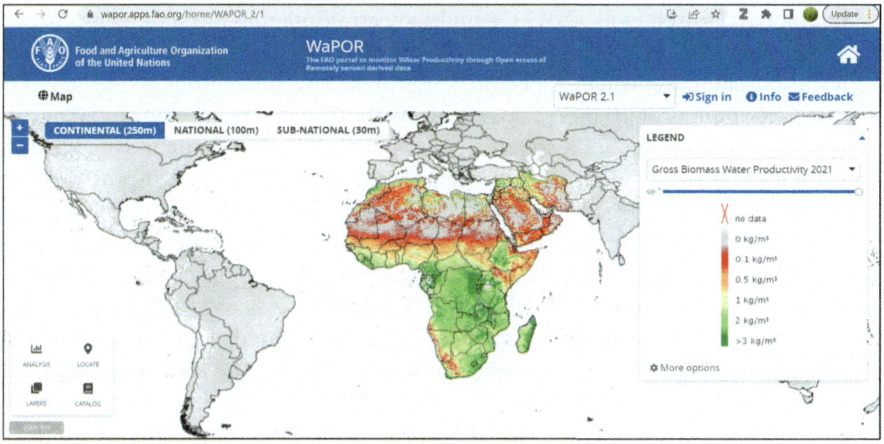

Fig. 2 Screenshot of a WaPOR a crop water productivity portal developed by Food and Agriculture Organization of the United Nations

the approach commonly used by FAO and described in the FAO-56 Irrigation and Drainage series paper (Pereira & Allen, 1999).

2.3 *Sentinel Missions*

The European Space Agency (ESA) has launched next-generation satellites Sentinel 1–6 to replace the older generation satellites to ensure continuity in monitoring the earth's surface. However, Sentinel-3 and 6 are intended for marine observations to monitor sea surface topography and temperature, ocean color, and other ocean-related observations. The Sentinel-4 and 5 missions' primary focus is air quality monitoring, which measures air quality, climatic forcing, ozone, and UV radiation, with great spatiotemporal resolution. Only the Sentinel-1 and Sentinel-2 satellite missions provide land-based observations, which can help in vegetation and crop monitoring studies.

The Sentinel-1 mission comprises a constellation of two polar-orbiting satellites operating day and night and performing C-band synthetic aperture radar imaging, enabling them to acquire imagery regardless of the weather. The data is provided at an outstanding time interval of 10 days and a spatial resolution of 10 m (Filipponi, 2019). The Sentinel-1 data is not affected by clouds; SAR images have a great advantage in the phenological monitoring of crops. The Sentinel-1 data has four dual polarized bands: VH (Vertically, Horizontally polarized band), VV (Vertically, Vertically polarized band), HH (Horizontally, Horizontally polarized band), and HV (Horizontally, vertically polarized bands). However, VH and VV bands are used for monitoring land-based products, while HH and HV are used for monitoring water-based or ocean-based products.

Sentinel-2 (Drusch et al., 2012), a multi-spectral imaging mission, comprises of two polar-orbiting satellites (Sentinel-2A and Sentinel-2B) positioned at different angles; each satellite carries 13 spectral bands in the spectral regions visible and NIR at spatial resolutions of 10 m; red edge and SWIR at spatial resolutions of 20 m; and atmospheric bands at spatial resolutions of 60 m, with each satellite having bands at almost the same bandwidth makes the temporal resolution 5 days. It provides data suitable for assessing the state and change of vegetation, soil, and water cover.

Sentinel-1 data is mostly used in India for rice crop monitoring, which is grown as a major Kharif crop. During Kharif, the presence of the monsoon trough hinders multispectral crop monitoring. The state of Tamil Nadu has used Sentinel-1 data to estimate the crop damage in rice paddies due to extreme drought conditions that occurred in the year 2017 (European Space Agency, 2017). The analysis of Sentinel-1 data provided the location and extent of crop damage, and these estimates were used by the state government of Tamil Nadu to disburse compensation to almost 2 lakh farmers as quickly as possible. The timely disbursement of crop insurance to farmers whose farms have been affected by natural calamity can alleviate a little of the suffering of the farmers and, in particular, the smallholder farmers. This methodology can be employed at national scale and global scale to estimate loss of crop and yields. and for disbursement of insurance to farmers.

There have been several studies which have tested Sentinel-2 for crop monitoring and have proved it to be reliable as it has better spatial, temporal, and radiometric resolutions. The Agrio mobile-based application for precision agriculture (https://play.google.com/store/apps/details?id=com.agrio&hl=en_IN&gl=US) uses Sentinel-2 level II processed images for monitoring crop chlorophyll content, which is considered a good indicator for crop health status. The users can monitor the chlorophyll content and locate if there are any changes in chlorophyll content in any part of the field, and can detect crop damage, and take appropriate measures. Chlorophyll content of the crop is estimated from Sentinel-2 data using the radiative transfer (RT) model inversion method. Many researchers have used Sentinel-2 for monitoring crop growth and development because of its good spatial resolution (10–20 m) and bands in the VIS, NIR, Red Edge, and SWIR regions. VIS and NIR regions are used to develop indices that are used to monitor the greenness and health of the crop. While red edge bands are used to measure processes such as photosynthesis and to calculate a crop's chlorophyll content. Similarly, SWIR bands are used to measure the crop water content, which can indicate the moisture stress and need for irrigation.

2.4 Planet Scope

PlanetScope (Planet, 2016) is a constellation of 180 low-orbiting high-resolution satellites known as Doves. These satellites travel across the whole surface of the earth at least once every day, and they measure the spectral reflectance in four spectral zones (Red, Green, Blue, and NIR). The spatial resolution of these satellite pictures ranges from 1 to 3 m. Data from PlanetScope is a great resource for monitoring vegetation.

Planet data provide improved spatial resolution and better temporal coverage, which is crucial in overcast places where there is a higher probability of capturing a cloudless image. Although planet data has a charge, it can be used for free for research and educational purposes.

The Manna Irrigation Intelligence solution (Manna Irrigation, 2022) is a sensor-free, software-based solution that provides site-specific irrigation recommendations at the touch of a button, without the hassle of soil or plant sensors. It uses PlanetScope data along with IoT data to deliver actionable information to farmers so they can make better-informed, more confident irrigation decisions to help combat the global water and food crisis. The PlanetScope data makes the tool more efficient and effective in locating the anomalies within the field, providing better prediction services for irrigation of agricultural fields. The use of PlanetScope data helps in lowering the cost of irrigation monitoring and also increases the accuracy of monitoring.

An agro-technical technique has been created and proven by Manna Irrigation, which generates irrigation recommendations based on a mix of PlanetScope satellite data, precise weather data, and crop-specific models. The Manna model is primarily based on the evapotranspiration (ET) approach; the model generates the observed ET of each field pixel, or the precise quantity of crop water consumption. The irrigation recommendations are then adjusted to account for variables including soil properties and regional growing methods. The Manna agro database has hundreds of local crop protocols for dozens of crops in different countries, which growers and advisors can choose from when planning stress measures.

2.5 *Ecostress*

ECOsystem Spaceborne Thermal Radiometer Experiment on Space Station (ECOSTRESS) (Fisher et al., 2020) was launched to the International Space Station in the year 2018 to measure the temperature of plants and to better understand how much water plants need and how they respond to stress. ECOSTRESS collects data at a spatial resolution of about 70 m using a multispectral thermal infrared radiometer to measure the radiance of the Earth's surface in five spectral bands and one shortwave infrared band.

3 Open Source Cloud Computing Platforms

Large quantities of geospatial data produced by earth observation satellites in recent years are freely accessible to the public and researchers. Traditional spatial data processing software falls short in efficiently storing, processing, disseminating, and analyzing these large data sets. Array database systems, MapReduce systems, and web services have been suggested and developed as novel technologies to access and analyze large amounts of earth observation data in order to satisfy these demands.

Here we discuss web-based services like Google Earth Engine, Python, Google Colab, Sentinel Hub, and Open EO that provide a solution to store, process, analyze, and present big earth observation data in a meaningful way.

3.1 Google Earth Engine (GEE)

GEE (Gorelick et al., 2017) has been widely used to process data for crop management, crop monitoring, climate monitoring, and environmental protection. GEE is a cloud-based platform for planetary-scale geospatial analysis that brings together Google's vast computational resources. The platform has completely changed the area of remote sensing by enabling researchers to perform studies that would have otherwise required an enormous amount of resources to obtain (download), store, and perform. The Google Earth Engine Data Catalogue (https://developers.google.com/earth-engine/datasets/catalog) is a list of the datasets that are available in GEE. Scientists, researchers, and developers can leverage this data and GEE's cloud computing capabilities to create products from satellite images at a lower cost and in less time, as well as to create use-case-specific applications that make data analysis and visualization simpler.

GIS and remote sensing tools on Google Earth Engine allow farmers and policymakers to determine the crop and land suitability of any location on the globe. Crop Suitability Mapping Tool (Peter et al., 2020) provides crop suitability for any location on the earth for 18 different crops based on soil and climate (https://cartoscience.users.earthengine.app/view/crop-niche) (Fig. 3). The geovisualization program was created as a user-friendly method of defining agricultural climate adaptability and enabling practitioners and researchers from many fields to interact with geographic data. The findings outline the application's user interface, inputs, and outputs before evaluating the crop's critical climate niche in Africa (i.e., the potential geographic range based on temperature and rainfall suitability). It provides maps that illustrate the spatial expansion of a particular variety of crop that can survive higher temperatures. The temporal adaptability of the model application is further illustrated by a panel comparison of the crop niche in study area over moving four-month growing seasons and a comparison of two distinct years to demonstrate crop. The process for producing the maps involved the collection and aggregation of temperature, precipitation, and NDVI data for the date range and season range selections (first aggregating by season, then aggregating across selected years), geographic delineation, and suitability characterization based on the user-selected crop phenology parameters for temperature and rainfall.

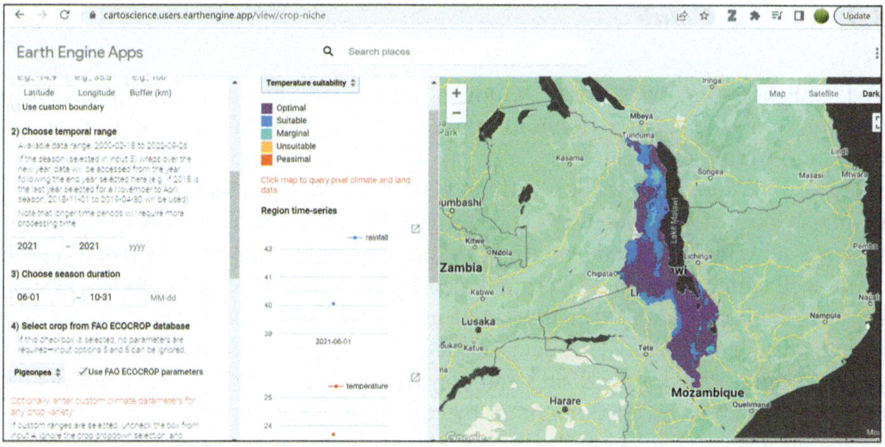

Fig. 3 Screenshot of a crop and climatic suitability tool developed for 18 crops for African countries

3.2 *Python*

Python is an open-source, multi-tasking, popular, and easy-to-use programming language. It is used for various purposes, ranging from data analysis and data visualization to the development of web and mobile applications. Python is also used as an open-source cloud computing platform for GIS and remote sensing analysis.

The big advantage of Python is that it can access a large number of "packages". This includes not only ArcGIS, Open Source GIS, and Google Earth Engine but can also hold various data types such as satellite images, UAV images, climate data, crop data, and much more. Python is highly efficient and highly flexible for big data analysis and helps to reduce the repetitive work done on graphic user interface platforms like ArcGIS, ERDAS, and QGIS. It is highly flexible with different data types and formats. It also enables third-party software to be operated on its platform.

Wageningen University, the Netherlands, developed PCSE (Python Crop Simulation Environment) as a Python package for building crop simulation models. It provides the environment to run crop simulation models by accessing the crop, weather, and soil data from its repository. It also gives provision for the use of user-specific data and simulates crop-based parameters such as phenology, biophysical parameters, respiration, photosynthesis, and evapotranspiration. PCSE also includes the packages for implementing WOFOST and LINTUL3. The WOFOST crop simulation model in PCSE is used in monitoring crops and predicting crop yields under the MARS project that is implemented for most crops in Europe and beyond.

3.3 Google Colab

The Collaboratory, or "Colab" (Google, 2016), is a cloud computing platform that works on the Jupyter (Python) application developed as an output of Google Research. Colab allows all users to write and execute complex algorithms free of charge on the Jupyter notebook without needing to install it separately. However, it requires Google login credentials and Google cloud space to use Colab. Colab allows anybody to write and execute arbitrary Python code through the browser and is especially well suited to machine learning, data analysis, and education. Users can share Jupyter notebooks with other users and user groups without having to download, install, or run anything. All the data, scripts, and outputs can be stored on the Google cloud/Google drive for accessing and viewing for later use. Even GEE scripts can be run on Colab using GEE plugin or by calling the GEE libraries.

Using Google Cloud with TensorFlow, the Agrology application (Agrology, 2019) runs machine learning models on sensor data to provide predictive insights for growers. Agrology is a comprehensive crop monitoring system that deploys IoT sensors above and below ground to measure temperature, rainfall, and soil conditions. The IoT data in combination with AI/ML supports predictive agriculture that helps farms identify the stresses early and take the right and timely actions that can save the crop from damage. Agrology works on Google Colab that helps store data from different IoT sensors and also runs big data algorithms such as Tensor Flow to make reliable crop estimates from the IoT data. It provides service to several specialty farms across the USA, from California to Virginia.

3.4 Sentinel Hub

A platform created by Sinergise called Sentinel Hub offers Sentinel data access and visualization capabilities. Public users can access this private platform at (https://www.Sentinel-Hub.Com/). Most of the crop monitoring studies that used Sentinel data have used Sentinel Hub to download data and process them on other platforms, as Sentinel Hub places restrictions on feature access under various price tiers, and allows only viewing, picking, and downloading of raw data under free plan. Paid access provides higher resource access restrictions, data processing, mobile application data access, Open Geospatial Consortium (OGC) protocol data access, and technical assistance.

3.5 Open EO

In order to address the need for technology consolidation for storage, processing, and analysis of huge volumes of earth observation data, the OpenEO project was launched in the year 2017. OpenEO (https://openeo.org/)creates an open application programming interface (API) that unifies and connects large Earth observation cloud back-ends to clients like R, Python, and JavaScript. OpenEO has a data repository with raw satellite data and also Analysis Ready Data (ARD) that can be used through the APIs provided in the Open EO library. The JavaScript, Python, and R APIs can be modified according to the user specifications and requirements.

4 Big Data Analytics

Using big data analytical tools helps to increase the accuracy in predicting, estimating, or monitoring agriculture. Agricultural monitoring requires data from various sources such as crops, soil, climate, satellite images, UAV images, etc. However, the processing of these images requires platforms with strong computational abilities. Leveraging artificial intelligence and machine learning (AI/ML) tools for monitoring agriculture has proved to be effective. AI sensors can also detect and target pests, diseases, and weeds while applying herbicides in precision agriculture, which optimizes the use of chemical control. The different AI/ML algorithms that have been used in agricultural research studies are presented in the Table 2. which has been taken as an excerpt from the article (Bhat & Huang, 2021), which compares and discusses the complexity, characteristics, limitations, and application of each AI/ML tool for agricultural studies and monitoring.

Smart Farm (Smart Farm, 2019) is a mobile-based crop monitoring tool that uses machine learning algorithms to predict crop yields for past and future using measured and simulated climatic data. The random forest machine learning tool has improved the accuracy of yield predictions of crops with five climatic parameters substantially in comparison with other AI/ML algorithms. The tool claims to give up to 90% accuracy in predicting crop yields with a random forest algorithm with ten-fold cross-validation. The random forest model is rained on uses 20 decision trees, which gave better accuracy in predicting crop yields.

Plantix is an android mobile-based application that uses the latest AI algorithm to detect plant, and crop pests, diseases, and other stresses with outstanding accuracy (Srikanth et al., 2020). It is a crop advisory app for farmers, extension workers, and gardeners around the world, developed by PEAT, an Indo-German start-up in 2015 (Rupavatharam et al., 2018). The Plantix application can detect more than 500 plant pests and diseases, and give the possible measures that can be taken up to control that pest or disease. With every new image, the detection accuracy of the application increases. It provides a platform for the users to interact and exchange knowledge about crop cultivation practices, best control practices, management practices, etc. Plantix provides a detailed description of biological and chemical control methods for hundreds of diseases through the library. Based on GPS location, the application also provides weather forecasts and current weather conditions that would help the farmer in planning the control applications. A newly introduced crop guide helps farmers to increase their crop production by giving detailed information from sowing

Table 2 Comparison of different AI/ML techniques used for agricultural monitoring (Bhat & Huang, 2021)

Technique	Complexity	And constraints	Applications
ANN (Artificial neural networks)	High	High power consumption; susceptible to overfitting; requires large datasets; parameter tuning is time-consuming	Classification of crop patterns; crop estimation; soil properties; and analysis
Decision trees	Low	Variation sensitivity; inconsistency in prediction accuracy; susceptibility to overfitting	Crop classification; estimation of crop yield based on soil variables;
Support vector machines	Low	Fast and accurate; simple to implement; few training samples required; less prone to overfitting	Classification of disease, soil mapping; retrieval of vegetation attributes
Random forests	High	Problem of overfitting; high prediction accuracy; easy parametrization	Attribute mapping; classification, and regression
Deep learning	High	It is highly promising for agricultural applications; it has high accuracy, high operational costs	Estimating yield by measuring vegetation characteristics

to harvesting and valuable hints for their specific crop. As the images are geotagged, their real-time monitoring of pests and diseases can be monitored.

5 Case Study: Developing a Globally Scalable and Locally Adaptable Crop Monitoring System for Wheat Crop

The crop monitoring system development involved the following steps (Fig. 4):

- Visualizing time series of vegetation indices derived from open source satellite data
- Deriving biophysical parameters from open source satellite data
- Developing crop prediction models on GEE and developing a graphic user interface to view the predicted yields of the study area.

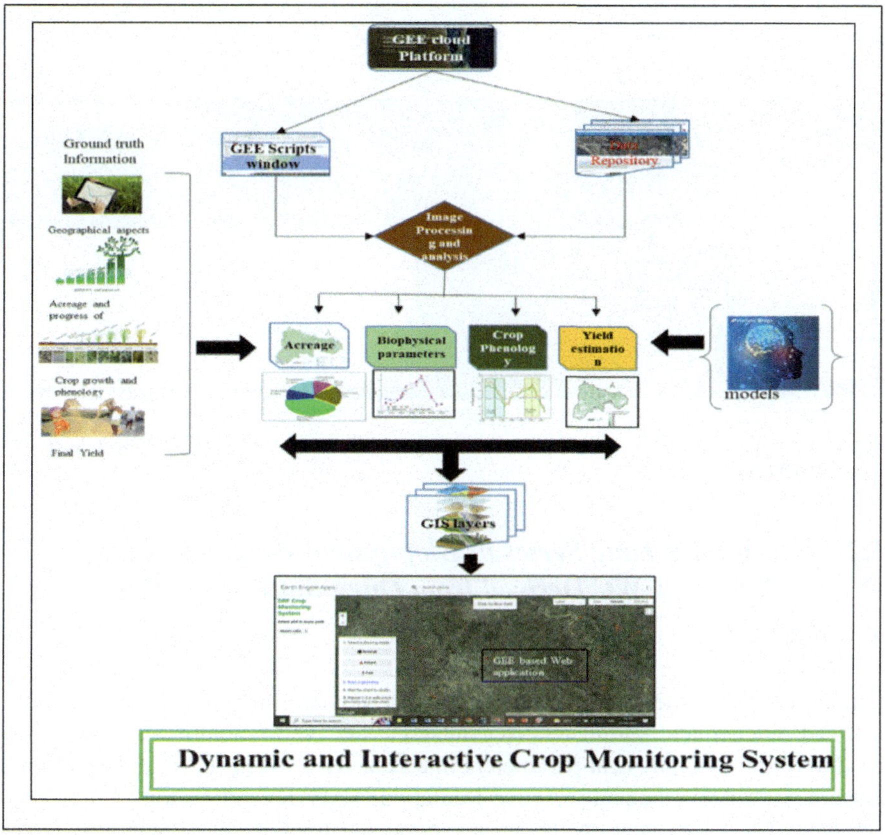

Fig. 4 Flowchart to develop a dynamic and interactive crop monitoring system using open source satellite data on GEE

5.1 Visualizing Time Series Vegetation Indices Derived from Open Source Satellite Data

The crop growth is progressive, and to study the crop growth one needs time series of the vegetation indices to be able to detect any deviations from the normal or the past. Therefore, the first thing that we derive from the remote sensing data is the time series of vegetation indices. On GEE platform, the required images are filtered using date and location filters and the vegetation index for those images is computed and the mean or maximum of the vegetation index is visualized on the mapper of the GEE. This is then developed into an interactive map using the UI tools in GEE. User can select the location using the pointer or provide a shape file to get the time series of vegetation index. In the Fig. 5, the most commonly used vegetation index Normalized differential Vegetation Index (NDVI) time series is presented for better understanding.

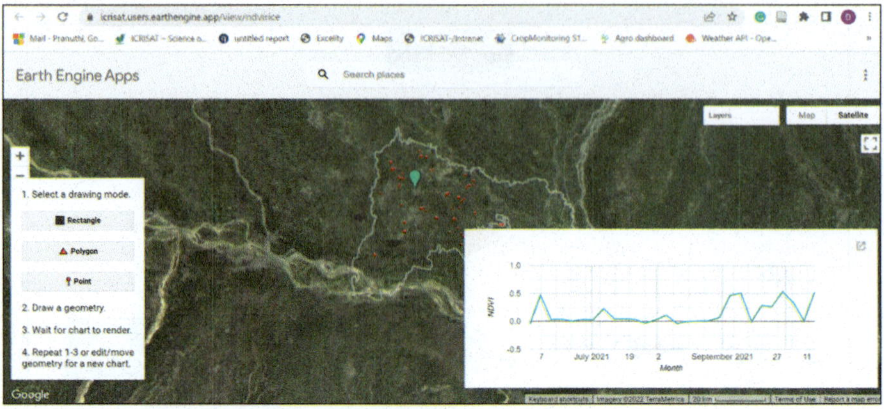

Fig. 5 Screen shot of GEE web application for deriving time series of vegetation indices for a specific location

5.2 Developing Time Series of Biophysical Parameters like LAI, CCC, CWC Derived from Open Source Satellite Data

Satellites measure the varying reflectance properties determined by canopy biophysical and biochemical parameters (Kganyago et al., 2021). Sentinel-2 images were used to retrieve biophysical parameters like Leaf Area Index (LAI) and Fraction of Absorbed Photosynthetically Active Radiation (FAPAR) and Canopy Chlorophyll content (CCC). These parameters were estimated using 8 bands of Sentinel-2 image in GEE using the same algorithm that the biophysical algorithm in SNAP toolbox uses (https://step.esa.int/docs/extra/ATBD_S2ToolBox_L2B_V1.1.pdf).

These parameters indicate the health and status of the crop as any deviation to these parameters is linked to crop status. Time series of LAI estimated for wheat crop using Sentinel-2 images is presented in Fig. 6. It is observed that LAI for wheat crop in Samastipur district is maximum during the 3rd week of Feb and declined thereafter (Fig. 6). The observed flowering period coincided with the peak LAI stage estimated from satellite data.

5.3 Predicting Crop Yields Using Sentinel-1 and Sentinel-2 Derived Indices

Crop yield prediction from satellite data is generally done in two different stages that is the flowering stage (F1) and the harvesting stage (F2). For the model development, ground truth data collected for wheat crop is used. The vegetation indices developed from Sentinel-1 and Sentinel-2 images are used as proxies to yield. This methodology

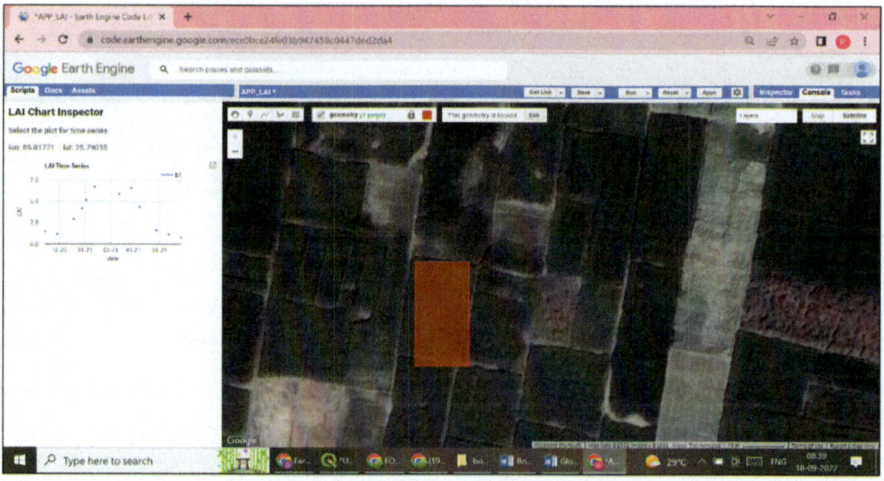

Fig. 6 Time series of LAI derived from Sentinel-2 images visualized using GEE interactive interface for regions of interest

enables us to estimate crop yields with minimum or no ground truth data. Steps in developing prediction models:

1. The time series of vegetation indices are developed for all the demo and control plots and the images with cloud cover are excluded.
2. Indices of each image are correlated with the yield; indices with a coefficient of correlation of more than 0.2 were selected to be used in the model. Then step-wise regression is performed on the selected variables.
3. The step-wise regression (SWR) again eliminates variables that are collinear and less responsive to the dependent variable.
4. After performing SWR, deep learning models like neural networks (ANN) and machine learning models like random forest (RF) are applied to the data.

For F1 yield estimation of wheat crop, satellite data till from November 1 to February 28 is used for model development; for F2 satellite data till April 30 is considered for model development. From SWR analysis, the vegetation indices that influenced the yield are Atmospherically Resistant Vegetation Index (ARVI), and Normalized Photochemical Reflectance Index (PRIN) derived from Sentinel-2 and Vertical Horizontal Polarized Band (VH) and Normalized Vertical Vertical Polarized Band (NVV) from Sentinel-1 data. Only those variables that are more correlated with the yield and are non-collinear are selected using the step-wise regression method. The random forest performed better for F1 (with R^2 of 0.84 and 0.78 and RMSE of 0.55 t/ha and 0.65 t/ha for calibration and validation datasets respectively) and F2 stages (with R^2 of 0.90 and 0.88 and RMSE of 0.48 t/ha and 0.40 t/ha for calibration and validation datasets respectively) (Fig. 7). All the above scripts and analysis can be published as a web application using GEE. The web application can be accessed using the URL: https://ICRISAT.users.earthengine.app/view/drf-crop-monitoring-system.

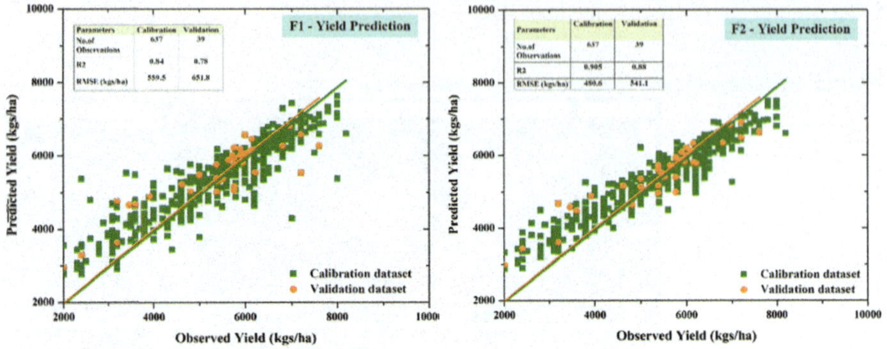

Fig. 7 Scatterplots of observed and predicted wheat yields for calibration (blue color) and validation (orange color) datasets at F1 and F2 stages

6 Challenges in Monitoring Crops Using Open-Source Satellite Data

The first step in the crop monitoring is crop classification which is the most difficult task. If the study area cropping pattern is diverse, it can lead to reduced accuracy of the crop monitoring, while in monocrop cropping pattern crop monitoring becomes easy. The quality and quantity of the training and validation datasets used for developing a crop monitoring system also influence the accuracy of the crop monitoring systems.

Most of the data offered as open source is multispectral data which has better spatial and temporal resolutions. When these datasets are used in developing crop monitoring system, clouds are the major challenges that are encountered in the process.

Integration of datasets from different satellites can provide solutions to many challenges such as cloud, low spatial, and temporal resolutions. However, due to different spatial and temporal resolutions the integration of data becomes complex and hinders in providing a comprehensive crop monitoring system.

7 Conclusion

Open-source satellite datasets, big data platforms, and big data analytical tools have substantially increased the number of crop monitoring systems that have impacted the livelihoods of farmers and also helped policymakers make appropriate and timely decisions. These satellite datasets, used singly or in combination with big data platforms and big data analysis tools provide solutions that can be locally adapted and can be globally scaled for the use of different stakeholders with different needs.

As Sentinel data is a high spatial resolution data, it can be adapted locally, and due to GEE computational capabilities, this tool can be scaled globally. Combining high-resolution Sentinel-1 and Sentinel-2 data, GEE, and big data analytics like artificial intelligence and machine learning tools for developing a crop monitoring system for wheat crop has made this study exceptional and remarkable.

References

Agrology. (2019). *The Climate is Changing Invest in a platform that keeps up*. https://www.agrology.ag/.

Alfoldi, T., & Ryerson, R. (1973). ERTS imagery interpretation package: Suggestions for laboratory design. *NASA STI/Recon Technical Report N, 75*, 22720.

Bates, J., Montzka, C., Schmidt, M., & Jonard, F. (2021). *UAS mounted LiDAR for estimating LAI type metrics for winter wheat*. EGU21-5982.

Bhat, S. A., & Huang, N.-F. (2021). Big data and ai revolution in precision agriculture: Survey and challenges. *IEEE Access, 9*, 110209–110222.

Didan, K. (2015). MOD13Q1 MODIS/Terra vegetation indices 16-day L3 global 250m SIN grid V006. *NASA EOSDIS Land Processes DAAC, 10*, 415.

Drusch, M., Del Bello, U., Carlier, S., Colin, O., Fernandez, V., Gascon, F., Hoersch, B., Isola, C., Laberinti, P., & Martimort, P. (2012). Sentinel-2: ESA's optical high-resolution mission for GMES operational services. *Remote Sensing of Environment, 120*, 25–36.

Erickson, J. (1984). The LACIE experiment in satellite aided monitoring of global crop production. *The Role of Terrestrial Vegetation in the Global Carbon Cycle: Measurement by Remote Sensing, 23*, 191–217.

European Space Agency. (2017, August 17). *Sentinel-1 speeds up crop insurance payouts*. https://www.esa.int/Applications/Observing_the_Earth/Copernicus/Sentinel-1/Sentinel-1_speeds_up_crop_insurance_payouts.

Filipponi, F. (2019). Sentinel-1 GRD preprocessing workflow. *18*(1), 11.

Fisher, J. B., Lee, B., Purdy, A. J., Halverson, G. H., Dohlen, M. B., Cawse-Nicholson, K., Wang, A., Anderson, R. G., Aragon, B., & Arain, M. A. (2020). ECOSTRESS: NASA's next generation mission to measure evapotranspiration from the international space station. *Water Resources Research, 56*(4), Article 4.

Geshnigani, F. S., Mirabbasi, R., & Golabi, M. R. (2021). Evaluation of FAO's WaPOR product in estimating the reference evapotranspiration for stream flow modeling. *Theoretical and Applied Climatology, 144*(1), Article 1.

Google. (2016). *Welcome to colaboratory*. https://colab.research.google.com/.

Gorelick, N., Hancher, M., Dixon, M., Ilyushchenko, S., Thau, D., & Moore, R. (2017). Google Earth Engine: Planetary-scale geospatial analysis for everyone. *Remote Sensing of Environment, 202*, 18–27.

Kganyago, M., Mhangara, P., & Adjorlolo, C. (2021). Estimating crop biophysical parameters using machine learning algorithms and Sentinel-2 imagery. *Remote Sensing, 13*(21), 4314.

Li, Q., Gao, M., & Li, Z.-L. (2022). Ground hyper-spectral remote-sensing monitoring of wheat water stress during different growing stages. *Agronomy, 12*(10), 2267.

Manna Irrigation, T. (2022, September 20). Irrigation intelligence from outer space. *Manna Irrigation Intelligence*. https://manna-irrigation.com/.

Mishra, S., Chander, S., Pradhan, R., Dubey, A. K., Oza, M. P., & Sharma, S. A. (2020). Webgis for water level monitoring and flood forecasting using open source technology. *Journal of Geomatics, 14*(1).

Pereira, L., & Allen, R. (1999). 5.1 crop water requirements. *CIGR Handbook of Agricultural Engineering*, 213.

Peter, B. G., Messina, J. P., Lin, Z., & Snapp, S. S. (2020). Crop climate suitability mapping on the cloud: A geovisualization application for sustainable agriculture. *Scientific Reports, 10*(1), Article 1.

Planet. (2016). *Planet imagery product specification: PlanetScope & RapidEye.*

Prasai, R., Schwertner, T. W., Mainali, K., Mathewson, H., Kafley, H., Thapa, S., Adhikari, D., Medley, P., & Drake, J. (2021). Application of Google earth engine python API and NAIP imagery for land use and land cover classification: A case study in Florida, USA. *Ecological Informatics, 66*, 101474.

Reddy, G. O., Dwivedi, B., & Chary, G. R. (2021). Big data in smart farming: Challenges and opportunities. *Indian Farming, 71*(11).

Rupavatharam, S., Kennepohl, A., Kummer, B., & Parimi, V. (2018). Automated plant disease diagnosis using innovative android App (Plantix) for farmers in Indian state of Andhra Pradesh. *Phytopathology (TSI), 108*(10), Article 10.

Smart Farm, S. (2019). *We deliver the information you need to grow.* https://www.smartfarm.ag/.

Srikanth, R., Rekha, K. S., Kiran, D. R., & Raju, P. V. (2020). *Use of artificial intelligence in ipm.*

Tawade, A., & Virnodkar, S. (2022). *Remote sensing image fusion using machine learning and deep learning: A systematic review.*

USDA, F. (2005). The FAS crop explorer: A web success story. http:%www.Fas.Usda.Gov/Info/Fasworldwide/2005/062005.

USDA, F. (2016). The crop explorer. *USDA-FAS, Washington, DC*, http://www.Pecad.Fas.Usda.Gov/Cropexplorer/Mpa_maps.Aspx. Last Accessed: December, 26, 2016.

Vermote, E., Justice, C., Claverie, M., & Franch, B. (2016). Preliminary analysis of the performance of the Landsat 8/OLI land surface reflectance product. *Remote Sensing of Environment, 185*, 46–56.

A Theoretical Framework of Agricultural Knowledge Management Process in the Indian Agriculture Context

Ram Naresh Kumar Vangala and Gaurav Mishra

Abstract A practical knowledge management (KM) process can trigger continuous innovations in the overall development of agriculture and its allied fields. It can also help improve the livelihood of rural communities in developing countries like India. However, little information on agriculture knowledge management processes and the ecosystem required for their implementation in the literature is available. This chapter attempts to derive a theoretical framework to understand the KM process using an Indian milk cooperative as a case. In addition, the chapter discusses the systemic factors which may drive the KM process in any agriculture-based organization. The framework proposed in this chapter could be used by any agricultural organization irrespective of size, location, and economic background. This structure could direct agricultural organizations in their knowledge management initiatives by exploring various ecosystem components for better operation and usage.

Keywords Agricultural knowledge management (AKM) · Information and communication technologies (ICTs) · Knowledge management (KM) · Milk cooperatives

1 Introduction

In the current era of globalization, knowledge has developed into a crucial organizational asset (Paul, 2004), a necessary component for economic advantage, and successful organizational performance (Rajnish, 2011). Knowledge management (KM), also known as the management of knowledge, is one of the most effective strategies to succeed in the information era. It is widely used by many enterprises, research

R. N. K. Vangala (✉)
Food and Agribusiness School (FABS), Hyderabad, Telangana, India
e-mail: ramnaresh@sgih.ac.in

G. Mishra
Development Management Institute (DMI), Patna, Bihar, India
e-mail: gmishra@dmi.ac.in

© The Author(s), under exclusive license to Springer Nature Singapore Pte Ltd. 2023
S. Chaudhary et al. (eds.), *Digital Ecosystem for Innovation in Agriculture*,
Studies in Big Data 121, https://doi.org/10.1007/978-981-99-0577-5_6

institutes, and academics. KM is a dynamic and ongoing activity that enables the business to improve innovation processes (Karadsheh et al., 2009) and makes sure that employees have the relevant information at the right time and place (Carla Jackson, 2001).

Since ancient times, India has been practicing agriculture. Therefore, knowledge in this sector is huge. Handling of agricultural knowledge/information happens at different positions: individuals, communities, organizations, or institutions and networks of them (Engel, 1990). It is found that various actors like the government, non-profit organizations, farmers' producers organizations, and private entities are working in this sector with various aims like production improvement, toward the improvement of farmer's income, development of farm communities, and creating opportunities for farmers in developing as entrepreneur (Vangala 2018) (Gummagolmath et al., 2011).

It is stated that most agricultural knowledge is not generated, articled, or shared by a single entity. A large part of knowledge in organization is tacit, which is more utilized than explicit (Ahmed, 2009). Different organizations like government and private agencies and research and extension institutions often build diverse knowledge. The lack of cooperation or partnership between these institutions is one of the causes for the ineffective transfer of knowledge (Saravanan, 2012). Therefore, there is an enormous opportunity for KM in agriculture.

1.1 Agricultural Knowledge Management (AKM)

Knowledge is one of the critical factors for a healthy agriculture system. It will help to generate relevant information, increase productivity, and increase profitability and reliability. Lack of relevant knowledge may result in poor production and a decrease in stability and may cause harm to natural resources. Therefore, proper knowledge from the right people at the right time may improve agricultural performance. Knowledge consists of experience, skills, values, attitudes, and relative information that enables it to function more effectively and systematically. When knowledge is formal, it is called explicit knowledge; when it is intangible and implicit, it is called tacit knowledge. It has been said that the creation of knowledge materializes in four ways inside an organization: (i) socialization (tacit to tacit); (ii) externalization (tacit to explicit); (iii) combination (explicit to explicit), and (iv) internalization (explicit to tacit) (Nonaka & Takeuchi 2007).

In order to strengthen themselves and achieve sustainable growth, many business organizations/firms are investing and implementing knowledge management. The knowledge management process—from knowledge generation through knowledge transfer—has been explained using a variety of frameworks. Table 1 gives a selection of pertinent articles that outline the theoretical underpinnings of knowledge management across various industries.

Practically speaking, very few studies have examined the significance of KM use in agricultural enterprises. A better agriculture knowledge management (AKM) system will assist stakeholders including farmers, agricultural organizations or institutions, subject matter experts, policymakers, and others in addressing issues like climate change, reliance on rainwater for agriculture, and environmental sustainability, among others. It is assumed that subject specialists or scientists are responsible for producing new information in the fields of agriculture and extension services. With the assistance of extension and field personnel, the information is subsequently shared with the farmers. Consequently, neglecting local knowledge production and sharing, as well as the importance of articulating farmer demands and fostering their empowerment (Vangala 2018).

According to the National Knowledge Commission (NKC) of India, the effective use of KM in agriculture will strengthen the agricultural economy and offer farmers a competitive advantage on the international market (Vangala 2014). Enhancing food security, generating ongoing advances, and increasing agriculture can all be made possible with effective AKM. To establish an effective AKM, it is necessary to often and continuously interact with a variety of stakeholders, including research institutions, universities, governments, the private sector, non-profit organization, topic specialists, knowledge workers, and farm communities.

From Table 1, it can be summarized that most of the knowledge management frameworks have focused on education and information technology. Moreover, the proposed framework of agriculture knowledge management in other countries like Tanzania, Colombia, and Africa explained different types of knowledge generation and transfer in agricultural systems at community level. Not many studies have been carried out on the AKM process in India's agricultural sector. Furthermore, the given frameworks predominantly focused on specific targets, which do not necessarily encompass interactions among various stakeholders.

The authors Venkatasubramanian and Mahalakshmi, claims that AKM in India is still in its infancy (Venkatasubramanian and Mahalakshmi, 2012). It is a relatively new idea; thus, its potential has not been fully realized. There is a need for Indian agriculture to build an AKM process framework that takes into account the many stakeholders and gives farmers and experts simple access to all forms of traditional and modern knowledge, providing access to the different levels of expertise in agriculture. This is necessary for Indian agriculture to participate in this new era of the economic world and to overcome issues like scarcity of natural resources, address the issue of climate change, and enhance productivity.

The rest of the paper is structured as follows: Sect. 2 presents the methodology for developing the proposed framework. Section 3 presents the conceptual agricultural knowledge management framework. The conclusion is discussed in the last section of the chapter.

Table 1 Agricultural knowledge management frameworks

Framework	Sector	Objective	Type of knowledge	Beneficiaries
Rahaman (2000)	Rural development	To harness traditional knowledge	Traditional knowledge, scientific knowledge	Local communities
Bhusry et al. (2013)	Educational institute	To harness explicit knowledge	Tacit knowledge, explicit knowledge	Students, faculty
Singh and Soltani (2010)	IT industries	To manage and transfer knowledge in organization	Tacit knowledge, explicit knowledge	Software, engineers
Lwoge (2010)	Agriculture	To manage and integrate indigenous and exogenous knowledge	Indigenous knowledge, exogenous knowledge	Rural communities
Meijer et al. (2015)	Agriculture	To combine extrinsic and intrinsic factors in farmers decision in adopting new technology	Tacit knowledge, explicit knowledge	Small hold farmers
Howland et al. (2015)	Agriculture	To understand the attitudes skills, and practices of fruit- growing farmers and to develop effective knowledge-sharing system	Tacit knowledge, explicit knowledge	Farmers
Alemu et al. (2018)	Agriculture	Integrating scientific and indigenous knowledge through agricultural knowledge management and developing a knowledge-sharing system	Indigenous knowledge, scientific knowledge	Farmers and value chain

2 Methodology

Based on the inquiry and observation from the case studies, the framework was created. This kind of research might be regarded as theory construction. KM is not a new topic, but there is little information available on it in Indian agriculture. This exploratory inquiry first examines the history before looking at the current situation.

Information systems (IS) employ both quantitative and qualitative research methods. Through formal, survey, and numerical approaches, the quantitative method is used in social research. While the qualitative method is developed in social science to enable researchers to understand people, social, and cultural contexts. In this approach, action research, case studies, and ethnography are included. According to Creswell (2016), each methodology has unique ontological, epistemological, axiological, rhetorical, and methodological qualities (John and Cheryl, 2016). One of the most important and common strategies employed in the qualitative method in the information system is the case study approach.

A case study is a type of research methodology that aims to provide comprehensive and multifaceted knowledge of a complicated problem in its actual setting. It is used to examine situations in which (1) various factors and relationships are present, (2) there are no fundamental laws defining which factors and relationships are significant, and (3) the factors and relationships can be observed directly (Fidel, 1984). It enables us to gather a variety of data, including participant observations both direct and indirect, participant interviews, documentary reviews, and historical documents.

The study's goals are to define the AKM flow in agricultural organizations and create a conceptual/theoretical framework. The case study approach can be useful in this topic. In this regard, case selection is crucial. Three criteria were used to pick the organizations: First, they needed to demonstrate a strong desire to engage farmers and share knowledge with them; second, they needed to be interested in knowledge projects; and third, they needed to be active participants in KM programs. After carefully choosing the organizations, we approached the appropriate individuals for authorization by the standard procedure for entering an organization. Email and phone calls were used for all first communications.

2.1 *Profile Case A: Mulukanoor Women's Cooperative Dairy (MWCD)*

MWCD is located in the Muluknoor village, Karimnagar district of Telangana state. 89 percent of the families in the area are small-scale, mixed-crop farmers, with less than 2 hectares of land under cultivation. This is the first women's cooperative dairy in India. The dairy came into operation through the support of Cooperative Development Foundation (CDF), Mulukanoor Cooperative Rural Bank and Marketing Society, and National Dairy Development Board (NDDB). At present, it has more than 192 primary women cooperative dairies. These dairies were spread around Karimnagar, Warangal, Siddipet, and Janagoan districts.

2.2 *Profile of Case B: Mehsana District Cooperative Milk Producer's Union Ltd. (Dudhsagar Dairy)*

The primary goal is to assist consumers by offering high-quality and secure milk products while also giving milk producers fair profits. The organization hopes that the dairy industry will contribute more to the rural economy by creating jobs for many local farmers and boosting their knowledge of how to run their businesses through cooperatives. The Dudhsagar Dairy runs leadership development programs, cleans milk production campaigns, and offers animal health, veterinary, and feed and fodder services.

2.3 *Findings*

It is clear that the flow of information is strong between the stakeholders in the organizations. It was found that both organizations benefit from networking partners like research institutes, state agriculture universities, state veterinary universities, and farmers for acquiring knowledge on clean milk production techniques, livestock management, and other relevant information. Both organizations use different forms of repositories like annual reports, user manuals, books, compact disks (CDs), and Web portals for storing knowledge.

Knowledge dissemination occurs through field workers, extension officers, cluster heads, para-veterinary professionals, and veterinary doctors. Both organizations provide capacity-building programs, community development programs, and training programs. The organizations arrange weekly and monthly meetings with farmers to share knowledge on the latest technology in livestock management, feed, fodder development, etc. In addition to checking cattle, the field supervisors make personal visits to farmers' homes and impart knowledge and information.

3 The Theoretical Framework of Agricultural Knowledge Management Process

Based on the findings of case studies (involving two organizations) and secondary source of information (literature review of various knowledge management frameworks proposed by different researchers), the proposed theoretical framework was developed. Figure 1 is the proposed theoretical agricultural knowledge management framework, integrated with four process elements along with core idea of information and communication technologies (ICTs) required to facilitate innovation in the organization.

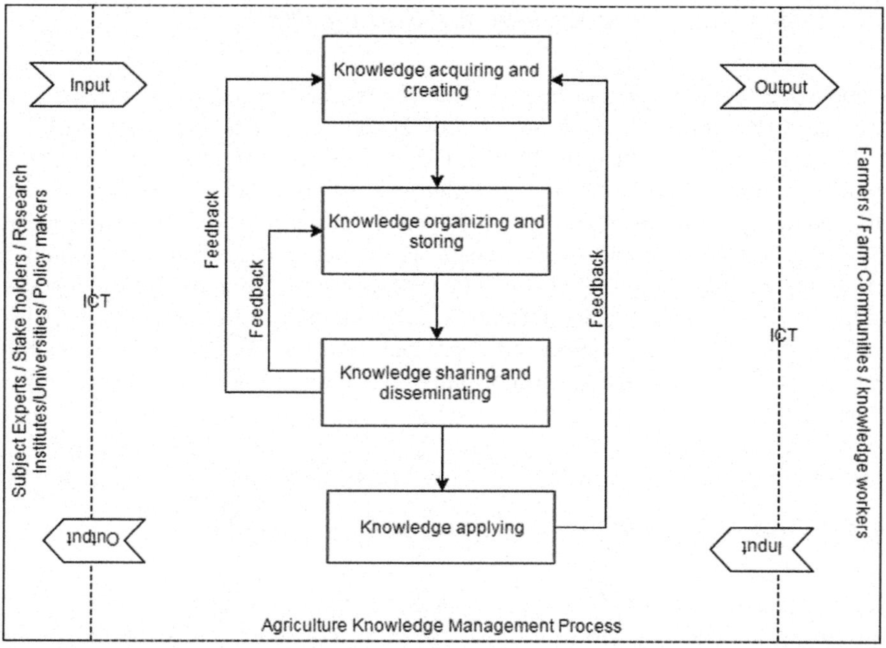

Fig. 1 Theoretical framework of AKM process

3.1 Knowledge Acquisition and Creation (KAC)

It is the first stage of the framework. In this stage, the organization's members will gain, collect, and create the required and valuable knowledge to perform their work. As part of the KAC, new content will be developed, and existing knowledge will be updated with the with the tacit and explicit knowledge of the organization (Brian, 1995). The process of creating new ideas and best practices is known as knowledge creation, which obtains from both internal and external sources that reside with the people working in the organization (Charles, 2001). It is essential to incorporate tacit knowledge used by the farmers into the AKM process and convert it into explicit knowledge by providing a feedback system. That is where socialization and externalization happen when a farmer's tacit knowledge is shared verbally with experts through a participatory process. It is then tested in local research platforms and converted into explicit knowledge. There is a continuous interplay between tacit and explicit knowledge when knowledge moves through individuals and groups at different organizational levels (Maryam & Dorothy 2001). The interaction between different stakeholders/actors is fundamental for creating and acquiring knowledge, as it allows for exchanging research findings, questions, and other knowledge-sharing activities. For the process of creating and acquiring knowledge, organizations will conduct training programs, workshops, or seminars (Shizhong et al., 2006). Web technologies enable employees or members of organizations to gather knowledge for work purposes (Maria and Yi-Chen, 2011).

3.2 *Knowledge Organization and Storage (KOS)*

Knowledge organization and storage (KOS) is the second stage of the process. At this stage, the knowledge is evaluated, refined, indexed, and stored in the organization's repository for further use. Prior to use, knowledge is validated or evaluated to make sure it is correct and beneficial. After evaluation, explicit knowledge is categorized in a structure for indexing so that it can be easily retrieved from repositories. By updating the stored knowledge, organizations can reduce redundancy, minimize obsolescence, and improve quality (Thomas et al., 1998). The archive enables access to a wide range of knowledge formats which including codified knowledge, electronic databases, videos, photographs, and other formats. The knowledge repository can serve as storage for organized knowledge and facilitate sharing and dissemination (Vangala 2018).

3.3 *Knowledge Sharing and Dissemination (KSD)*

Knowledge sharing and dissemination (KSD) is the fourth stage of the model. In this stage, knowledge sharing happens between individuals and/or groups in the organization. Through this process, employees learn new things, and the creation of new knowledge happens among them. Knowledge can be shared in horizontal and/or vertical directions. Knowledge is transferred between coworkers in a horizontal manner and between organizations in a vertical direction. Knowledge dissemination can enhance the employee's skill, attitude, and development in the organization, which can lead to achieving the organizational goals (Christopher and Kevin, 2003). KSD is considered one of the primary practices of the framework because the main objective of AKM is to encourage knowledge sharing among individuals, groups, and organizations.

Knowledge dissemination can be sparked by official, informal, and private strategies, including gatherings, conversations, social media, teamwork, in-person interactions, etc. Knowledge is transferred inside institutions through social networks, teamwork, interaction, and conversation. In one of the studies, it was discovered that an absence of horizontal and vertical fragmentation inside the company is a major threat to collaboration and social learning (Hermans et al., 2015). Providing incentives, rewards, and appreciation to employees can encourage the dissemination of knowledge in the organization (Varintorn & Nazrul 2006). As a result, the workforce will become a friendly place for knowledge creation and sharing.

3.4 *Knowledge Application (KA)*

The last stage of the framework is knowledge application. It is to apply best practices in their job and make use of knowledge from the various stakeholders. In this process, knowledge is kept in practice, and practitioners apply lessons learned from their previous experience. It also includes a decision-making process and problem-solving, leading to knowledge creation (Sayyed et al., 2011).

3.5 *Information and Communication Technologies (ICTs)*

Information and communication technologies (ICTs) are used as both singular and plural nouns. ICT advancements can enable the continuous exchange of knowledge generation processes from the lab to the field and vice versa. It helps place data systematically for the decision-making process (Hilbert, 2016). Data mining techniques are used to extract farmers' embedded knowledge from large data sets to identify best practices (Sylvain et al., 2016). Some of the emerging technologies that are enabling an agricultural knowledge revolution include big data, machine learning, artificial intelligence, and technology-driven local markets. Various communication channels are being used for disseminating agriculture knowledge to farm communities. Facilitators, knowledge brokers, and ICTs tools will play an essential role as intermediary roles in the agriculture knowledge management process.

The framework (Fig. 1) suggests that there is potential for conceptualizing the use of information for agricultural development within the targeted community. It can help the organization to manage knowledge (both tacit and explicit) at different levels. The four main stages of the AKM process included in the framework are acquisition and creation, organization and storage, sharing and dissemination, and application. Each element of knowledge passes through a cyclical process from acquisition to application. In this process, new knowledge is produced, and the process is continuous. There is a provision to include both tacit and explicit knowledge at each level.

As part of the process of AKM, all stakeholders-from organizational leadership to agricultural workers, regulatory authorities to farm communities-contribute tacit (local knowledge) and explicit (scientific knowledge that exists at various levels) knowledge. Hence, information and knowledge would flow more readily from lab to land and vice versa. All the phases of KM were included in the four stages of the developed framework, and this will aid in blending the multiple view points and involvement with extensive knowledge. A method for analyzing and updating knowledge at different levels is enabled by the framework's feedback system. If generated knowledge is not communicated correctly, and there is no mechanism for feedback, then the AKM process cycle is likely to decay. Various actors like farmers, non-profit organizations, governments, researchers, policymakers, industrialization, and commercialization are included in the framework and will contribute to the knowledge repository.

4 Role of Digital Technologies in Theoretical Framework

Digital technology can play a crucial role in transforming organizations' knowledge management processes. With the help of digital technologies, knowledge can be easily acquired, organized, stored, and disseminated. The rapid development of new ICTs is not only doing jobs easy but also creating new opportunities for knowledge workers and organizations to organize resources in digital format. It enables the AKM approach in gathering, storing, and organizing a large set of data at affordable price and can benefit in the process of producing explicit knowledge that can be incorporated into innovation (Mahr and Lievens, 2012).

Digital technologies like the Internet and artificial intelligence (AI) are some of the most powerful tools that help acquire knowledge and make better decisions faster by understanding data and extracting it. For example, natural language processing (NLP) can help read and understand the text more effectively. It will help the farmers to develop localized content that can be stored in the organizational repository. In addition, machine learning (MI) can help predict future outcomes based on the pattern behavior of past data. Online visual tools can help collaboration between farm communities and subject expert groups to create knowledge. The network of collaboration will help generate and acquire knowledge among the various stakeholders in agriculture.

Cloud computing, data mining, big data, and other related technologies are new technologies available to store, organize, and retrieve data from the organizational repository. ICTs tools, like mobile phone, laptops, desktops, and others, are essential to store, classify, retrieve, and disseminate both tacit and explicit knowledge.

Creating digital networks can help coordinate and cohesion between farmers and subject experts and farmers to market. The development of mobile technology allows the farmers (owners of mobile phones) to access information and the possibility of engaging and participating in many social interactions through mobile phones. Mobile-based applications can be used to share agriculture knowledge among peer groups.

5 Conclusion

In the literature, it is found that KM frameworks were developed in the context of the organizational environment (service sectors, industry sector, and educational institutes). Consequently, they fall short of meeting the demands of rural communities, where both tacit and explicit knowledge are gained and disseminated in unique ways. Moreover, it was observed that tacit knowledge was gathered and disseminated within local network of communities. Therefore, there is a predominant loss of tacit (local) knowledge. On other hand, farmers are not considered producers of knowledge. This is because agricultural research and extension think that scientists generate knowledge, and it is shared by the field workers and followed by farmers

(Ram Naresh et al., 2015). Since many farmers lack the expertise to address their farming issues, they are willing to share their experiences with professional groups to strengthen their knowledge base. Therefore, there is an urgent need of AKM process that can manage and integrate both tacit and explicit knowledge.

Any agricultural body, whatever its size, geography, or economic status, can use the proposed AKM framework. It will help the organization take initiatives for implementing AKM and also help in analyzing the factors that need to be addressed. Organizations can use the AKM framework for establishing teams for implementing AKM to achieve the organization's objectives. With the combination of ICTs, people, and the AKM process, the proposed AKM framework may be implemented as a comprehensive strategy.

6 Future Direction

There is a need to see agricultural organizations' constraints in implementing the knowledge management framework. Moreover, metrics need to be developed for measuring each phase in the framework to help the organization develop its knowledge management performance.

References

Alavi, M., & Leidner, D. E. (2001). Knowledge management and knowledge management systems: Conceptual foundations and research issues. *MIS Q*. 107–136.

Alemu, D., Jennex, M., & Assefa, T. (2018). Agricultural knowledge management system development for knowledge integration. In *Proceedings of the 51st Hawaii International Conference on System Sciences*.

Allameh, S. M., Zare, S. M., et al. (2011). Examining the impact of km enablers on knowledge management processes. *Procedia Computer Science, 3*, 1211–1223.

Bhusry, M., Nagar, R., & Ranjan, J. (2013). Implementing knowledge management in higher educational institutions in India: A conceptual framework. *Liceo Journal Higher Education Research,7*(1), 155.

Chen, S., Duan, Y., Edwards, J. S., & Lehaney, B. (2006). Toward understanding inter-organizational knowledge transfer needs in smes: insight from a UK investigation. *Journal of knowledge management*.

Collins, C. J., & Clark, K. D. (2003). Strategic human resource practices, top management team social networks, and firm performance: The role of human resource practices in creating organizational competitive advantage. *Academy of Management Journal, 46*(6), 740–751.

Creswell, J. W., & Poth, C. N. (2016). *Qualitative inquiry and research design: Choosing among five approaches*. Sage publications.

Davenport, T. H., Prusak, L., et al. (1998). *Working knowledge: How organizations manage what they know*. Harvard Business Press.

Delerce, S., Dorado, H., Grillon, A., Rebolledo, M. C., Prager, S. D., Patiño, V. H., Garcés Varón, G., & Jiménez, D. (2016). Assessing weather-yield relationships in rice at local scale using data mining approaches. *PloS One,11*(8), e0161620.

Engel, P. G. H. (1990). Knowledge management in agriculture: Building upon diversity. *Knowledge, Technology and Policy,3*(3), 28–35.
Fidel, R. (1984). The case study method: A case study. *Library and Information Science Research,6*(3), 273–288.
Gummagolmath, K. C., & Purushottam, S., et al. (2011). ICT initiatives in Indian agriculture-an overview. *Indian Journal of Agricultural Economics, 66*(902-2016-67322).
Hermans, F., Klerkx, L., & Roep, D. (2015). Structural conditions for collaboration and learning in innovation networks: using an innovation system performance lens to analyse agricultural knowledge systems. *The Journal of Agricultural Education and Extension,21*(1), 35–54.
Hilbert, M. (2016). Big data for development: A review of promises and challenges. *Development Policy Review,34*(1), 135–174.
Howland, F. C., Muñoz, L. A., Staiger, S., Cock, J., & Alvarez, S. (2015). Data sharing and use of ICTs in agriculture: Working with small farmer groups in Colombia. *Knowledge Management for Development Journal, 11*(2), 44–63.
Jackson, C. (2001). Process to product: Creating tools for knowledge management. In *Knowledge management and business model innovation* (pp. 402–413). IGI Global.
James, P. (2004, October). Strategic management meets knowledge management: A literature review and theoretical framework. In *5th KM Conference, Australia, Canberra.*
Karadsheh, L., Mansour, E., Alhawari, S., Azar, G., & El-Bathy, N. (2009). A theoretical framework for knowledge management process: Towards improving knowledge performance. *Communications of the IBIMA,7*, 67–79.
Lee, M. R., & Lan, Y.-C. (2011). Toward a unified knowledge management model for SMEs. *Expert systems with applications, 38*(1), 729–735.
Lwoga, E. T., Ngulube, P., & Stilwell, C. (2010). Managing indigenous knowledge for sustainable agricultural development in developing countries: Knowledge management approaches in the social context. *The International Information & Library Review, 42*(3), 174–185.
Mahr, D., & Lievens, A. (2012). Virtual lead user communities: Drivers of knowledge creation for innovation. *Research Policy,41*(1), 167–177.
Meijer, S.S., Catacutan, D., Ajayi, O. C., Sileshi, G. W., & Nieuwenhuis, M. (2015). The role of knowledge, attitudes and perceptions in the uptake of agricultural and agroforestry innovations among smallholder farmers in sub-Saharan Africa. *International journal of agricultural sustainability, 13*(1), 40–54.
Nonaka, I., & Takeuchi, H. (2007). The knowledge-creating company. *Harvard Business Review, 85*(7/8), 162.
O'dell, C., & Grayson, C. J. (1998). If only we knew what we know: Identification and transfer of internal best practices. *California Management Review, 40*(3), 154–174.
Pentland, B. T. (1995). Information systems and organizational learning: The social epistemology of organizational knowledge systems. *Accounting, Management and Information Technologies, 5*(1), 1–21.
Rafea, A. (2009). Managing agriculture knowledge: Role of information and communication technology. In *Science Forum 2009 Workshop: ICTs Transforming Agricultural Science, Research and Technology Generation.*
Rahman, A. (2000). *Development of an integrated traditional and scientific knowledge base: A mechanism for accessing, benefit-sharing and documenting traditional knowledge for sustainable socio-economic development and poverty alleviation*. Citeseer.
Rai, R. K. (2011). Knowledge management and organizational culture: A theoretical integrative framework. *Journal of knowledge management.*
Saravanan. R. (2012). ICTs for agricultural extension in India: Policy implications for developing countries. In *Proceedings of 8th Asian Conference for Information Technology in Agriculture, AFITA* (pp. 1–11).
Singh, A., & Soltani, E. (2010). Knowledge management practices in Indian information technology companies. *Total Quality Management,21*(2), 145–157.

Supyuenyong, V., & Islam, N. (2006). Knowledge management architecture: Building blocks and their relationships. In *2006 Technology Management for the Global Future-PICMET 2006 Conference* (Vol. 3, pp. 1210–1219). IEEE.
Vangala, R. N. K. (2018). *Developing a framework of agriculture knowledge management process in Indian context*. Ph.D. Thesis, Dhirubhai Ambani Institute of Information and Communication Technology.
Vangala, R. N. K., Hiremath, B. N., & Banerjee, A. (2014). A theoretical framework for knowledge management in indian agricultural organizations. In *Proceedings of the 2014 International Conference on Information and Communication Technology for Competitive Strategies* (pp. 1–7).
Vangala, R. N. K., Mukerji, M., Hiremath, B. N. (2015). ICTs for agriculture knowledge management: insights from Dhruva, India. In *Proceedings of the Seventh International Conference on Information and Communication Technologies and Development* (pp. 1–4).
Venkatasubramanian, V., & Mahalakshmi, P. (2012). Innovative institutional approaches for agricultural knowledge system management in India.

Problems and Applications of Digital Agricultural Transformations

Simple and Innovative Methods to Estimate Gross Primary Production and Transpiration of Crops: A Review

Jorge Celis, Xiangming Xiao, Jeffrey Basara, Pradeep Wagle, and Heather McCarthy

Abstract Sustainable agricultural production to support food security is one of the main targets of the 2030 agenda for global sustainable development goals (SDGs). New technology advancements and sources of information play a critical role in supporting agriculture to achieve the SDGs goals and increase production capabilities to meet rising food demands. Gross primary production (GPP) and transpiration (T) of crops are the largest carbon and water fluxes in agroecosystems. GPP data products are used to estimate crop yield, an important metric for agricultural resiliency and food security. Over land surfaces, T is the largest component of evapotranspiration and represents more than 60% of the precipitated water returned to the atmosphere. T is used to represent water use and improve irrigation in croplands, thereby helping to reduce production costs and support sustainable crop production. In this chapter, we review the state-of-the-science approaches for estimating GPP and T, including in situ and remote sensing methods, while focusing on the biophysical foundation behind the major available techniques. Furthermore, given the linkage between GPP and T through the behavior of stomates, it is possible to simplify the calculation of both variables by estimating common biophysical parameters (i.e., minimum stomatal conductance) that influence water loss and vegetation carbon uptake. We highlight innovative approaches that enhance the calculation of both variables based on this relationship and review the extent and limitations of the algorithms used in some of the most popular satellite products like Moderate Resolution Imaging Spectroradiometer (MODIS) global evapotranspiration product (MOD16) and EcoStress. Furthermore,

J. Celis · X. Xiao (✉) · H. McCarthy
Department of Microbiology and Plant Biology, University of Oklahoma, Norman, OK 73019, USA
e-mail: xiangming.xiao@ou.edu

J. Celis
e-mail: jcelis@ou.edu

J. Basara
Department of Meteorology, University of Oklahoma, Norman, OK 73019, USA

P. Wagle
USDA-ARS, Oklahoma and Central Plains Agricultural Research Center, El Reno, OK 73036, USA

© The Author(s), under exclusive license to Springer Nature Singapore Pte Ltd. 2023
S. Chaudhary et al. (eds.), *Digital Ecosystem for Innovation in Agriculture*, Studies in Big Data 121, https://doi.org/10.1007/978-981-99-0577-5_7

this chapter explores and describes the potential for using solar-induced fluorescence (SIF) data to calculate agricultural GPP and T. Finally, given the importance of certain variables in agricultural and forest meteorology, this chapter also includes a section highlighting the main conversion factors useful in calculating energy, carbon, and water fluxes.

Keywords Gross primary production · Transpiration · Solar-induced chlorophyll fluorescence · Agriculture · Crop monitoring · Remote sensing

Acronyms and Abbreviations

APAR	Absorbed photosynthetically active radiation
$APAR_{chl}$	Absorbed photosynthetically active radiation by the chlorophyll
CO_2	Carbon dioxide
CP	Change point detection
c_p	Specific heat
E	Evaporation
E_{bs}	Bare soil evaporation
EC	Eddy covariance
ECOSTRESS	The ecosystem spaceborne Thermal radiometer experiment on space station
ECT	Eddy covariance tower
ER	Ecosystem respiration
ET	Evapotranspiration
EVI	Enhanced vegetation index
E_{wc}	Evaporation from wet canopy
fPAR	Fraction of photosynthetically active radiation
FVS	Flux variance similarity
F_{wet}	Water cover fraction
G	Ground soil heat flux
GMAO	Global modeling and assimilation office
GOSAT	Greenhouse gases observing satellite
GPP	Gross primary production
H	Sensible heat flux
HSR	High spatial resolution
IR	Infrared
LAI	Leaf area index
LE	Latente heat flux
LST	Land surface temperature
LSWI	Land surface water index
LUE(ε_g)	Light use efficiency
LUE0(ε_0)	Maximum light use efficiency
MIP	Model intercomparison project

MODIS	Moderate resolution imaging spectroradiometer
MP	Moving-point-transition
MSR	Moderate spatial resolution
NBP	Net ecosystem carbon balance
NDVI	Normalized differentiate vegetation index
NEE	Net ecosystem exchange
NPP	Net primary production
NPV	Non-photosynthetically active vegetation
OCO-2	Orbiting carbon observatory-2
P-M	Penman–Monteith
PAR	Photosynthetically active radiation
PAV	Photosynthetically active vegetation (chloroplast)
PPFD	Photosynthetic photon flux density
QC	Quality controls
r_a	Aerodynamic resistance
r_c	Canopy stomatal resistance
REF	Reference
RH	Relative humidity
Rn	Net radiation
SE	Standard error
SEB	Surface energy balance
SIF	Solar-induced chlorophyll fluorescence
SMAP	Soil moisture active passive (SMAP)
SVAT	Surface-vegetation-atmosphere transfer
SW	Total incoming shortwave solar radiation
SWIR	Shortwave infrared
T	Transpiration
TEM	Terrestrial ecosystem model
TIR	Thermal infrared
T_{opt}	Optimal air temperature for photosynthesis
u*	Friction velocity
VIs	Vegetation indices
VPD	Vapor pressure deficit
VPM	Vegetation photosynthesis model
VTM	Vegetation transpiration model
WUE	Water use efficiency
γ	Psychrometric constant
Δ	Vapor pressure–air temperature curve
ρ	Air density

1 Introduction

Gross primary production (GPP) and transpiration (T) are the largest terrestrial carbon dioxide (CO_2) and water vapor (H_2O) fluxes, respectively, and are crucial variables in agriculture (Bondeau et al., 2007; Foley et al., 1996; Sitch et al., 2003; Zhao & Running, 2010). In situ measurements (Aubinet et al., 2012) and remote sensing (Running et al., 2004; Xiao et al., 2004a) have been extensively utilized to study terrestrial carbon and water fluxes. The accessibility of remote sensing observations and the development of global observation networks (e.g., Fluxnet) (Hulley et al., 2017; Pastorello et al., 2020) have made it viable to assess and quantify the variability of carbon and water dynamics at multiple spatial scales. Seasonal and interannual variabilities in GPP are a strong marker on the role of vegetation in sequestering atmospheric CO_2 (Amthor, 1995; Sitch et al., 2003). GPP estimates are used to improve crop yield assessments (Doughty et al., 2018; He, et al., 2018; Wu et al., 2020) and to enhance our understanding of key practices in the livestock industry like rotational grazing and pasture renewal (Liáng et al., 2020; Zhou, et al., 2019). T, the major component of evapotranspiration (ET) in agroecosystems, is important for water resource availability at the land surface given that 57% of the precipitated water in the surface is returned to the atmosphere and precipitates back into the continental area (Ent et al., 2010), and making the information helpful to improve efficiency of irrigation in croplands, which can reduce production cost and support food security (White et al., 2000). However, despite the advances in instrumentation, algorithms, and data coverage, there are still several limitations that must be considered when using GPP and T estimates for agricultural applications.

GPP and T are challenging to estimate and scale up at the field level in agroecosystems (Chen & Coughenour, 2004; Wei et al., 2017a). GPP varies over different fields, agricultural practices (field disturbances), and crop varieties. T varies widely in space and time in agroecosystems, making its contribution to ET crucial to complete the water balance. T is challenging to estimate at individual crop fields due the variability in the vegetation response to environmental conditions and management practices.

The eddy covariance (EC) method (Montgomery, 1948; Obukhov, 1951; Swinbank, 1951) measures heat, mass, and momentum under horizontal homogeneity conditions. The EC technique can be applied to calculate the density of the flux between the vertical column of wind and the turbulent fluctuations of the variable of interest (e.g., CO_2, H_2O) (Aubinet et al., 2012; Longdoz & Granier, 2012). The use of the EC theory has increased since the late 1990's, thanks to the technological advancements in sonic anemometers and gas analyzers, which improved measurement of land surface CO_2 exchange and motivated the installation of the early EC towers (Aubinet et al., 2012). EC data are considered the golden standard of in situ CO_2 exchange information, it can be used for regional studies and cross-site comparisons, thanks to networks like Fluxnet, which was created as a network to standardize existing EC information (Pastorello et al., 2020). EC towers provide CO_2 and H_2O fluxes, meteorological and environmental data including soil heat fluxes, soil temperature, and soil moisture. The CO_2 flux data are measured at 10 Hz frequency (10 measurements per second), and typically reported at a 30-min timestep. The EC flux data are processed through a standard data processing pipeline (Fig. 1) to allow regional comparisons and global analyses. However, within the suggested data processing pipeline, each site can select different friction velocity (u*) thresholds for filtering nighttime CO_2 fluxes that can introduce uncertainties in cross-comparison and regional studies (Pastorello et al., 2020).

Furthermore, many studies use EC data. For example, EC data are used in the ecosystem level validation to parametrize ET and T in the surface-vegetation-atmosphere transfer (SVAT) model (Boulet et al., 2000). However, frequently there is a scale disparity in the validation of bare soil evaporation (E) and T (Williams et al., 2004). To address this issue and to enable the partition of the ET flux components, measuring the stable isotope composition of the flux between surface and atmosphere have proven highly accurate when combined with EC data (Harwood et al., 1999; Yepez et al., 2003). The success and accuracy of stable isotopes is linked to the fact that E and T from vegetation produce unique isotopic signals in the water vapor that are easily distinguished (Williams et al., 2004).

Contemporary approaches utilize different types of models (e.g., statistic, deterministic, stochastic, and physical) to estimate GPP and T, but some of the most successful models (physically-based and light use efficiency)—based on their accuracy and spatiotemporal data consistency—are driven by climate data fused with dynamic vegetation parameters from satellite images (Celis, et al., 2021; Zhang et al., 2017). A major difference among these popular models is the approach used to include the effects of environmental conditions on GPP and T estimates. For example, the photosynthetically active radiation absorbed by the chlorophyll ($APAR_{chl}$) provides better insights than models that don't distinguish between the non-photosynthetically active and photosynthetically active vegetation to estimate GPP. On the other hand, soil moisture has a strong impact on ET, but T is typically

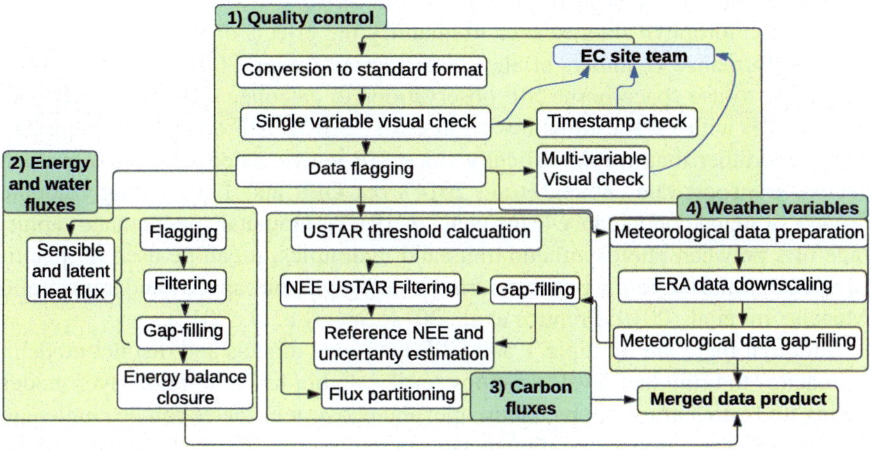

Fig. 1 Fluxnet eddy covariance data processing pipeline [adapted from Pastorello et al. (2020)]

included as a function of non-changing vegetation and soil moisture parameters, introducing high uncertainties into ET and T models (Mahfouf et al., 1996; Matsui et al., 2005; Vivoni et al., 2007a). Furthermore, given the complexity of the techniques to estimate ET and T and the difficultly in scaling them, there's a need for a simpler approach to provide consistent time series T data that can be applied over different latitudes and regions with limited ground data.

The information in this chapter aims to support the work of scientists and researchers in government agencies, academia, as well as the private sector, as the variables included in our discussion are critical for agricultural production and management, and the T component is a fundamental element to fully represent the water balance in the future climate change scenarios and weather risk management. To address these components, this chapter contributes to (1) better understanding of the physiological and biophysical components behind GPP and T, (2) providing a clear framework of the current methods and limitations for in situ and remote sensing-based approaches to calculate GPP and T, and (3) highlighting the main physiological and biophysical components needed to improve the estimation of the CO_2 and H_2O fluxes, and the advantages of some innovative approaches.

2 Biophysical Link Between GPP, T, and SIF

Stomatal opening and closure control the exchange of CO_2 and H_2O between plants and the atmosphere, influencing photosynthesis directly and coupling GPP and T (Campbell & Norman, 1998; Dorman & Sellers, 1989; Nobel, 1999). Photosynthesis is a widely studied process, where solar radiation is absorbed by chlorophyll to provide the energy needed to transform CO_2 into sugars for plant use. The role of chlorophyll in the photosynthesis process and the fact that the chlorophyll *a* associated with the photosystem II emits fluorescence (long-wave radiation) allows for the use of chlorophyll fluorescence to quantify the effects of stress and photosynthetic performance (Lambers et al., 2008; Taiz & Zeiger, 2002). Consequently, it is possible to use space-borne SIF observations to calculate crop productivity and connect SIF with GPP estimations (Yan et al., 2015). SIF and T are empirically connected rather than mechanistically since SIF is associated with photosynthetic electron transport rate (Zhang et al., 2014), but GPP and T are strongly coupled through stomatal conductance (Stoy et al., 2019). As stomatal conductance captures trade-offs between photosynthetic traits and hydraulics, it can be used to quantify carbon gains and hydraulic risk to determine carboxylation capacity and transpiration (Mencuccini et al., 2019; Prentice et al., 2014; Sperry et al., 2017).

Although there are multiple T and GPP leaf level models and distinct modeling approaches to bring leaf level to canopy level, i.e., big leaf and multi-layer models, scaling up from leaf to canopy, landscape, and ecosystem scales can be challenging given the extensive parametrization that the models require and the large uncertainties introduced given the number of variables involved in the model configuration (Xiao, 2006). Further improvements in the T and GPP calculations can lead to a more accurate representation of ecosystem productivity, water use, and water use efficiency (WUE; GPP/T) in crops (Stanhill, 1986). The strong linkage between GPP and T has led to suggestions that it may be possible to use SIF as a WUE predictor at the ecosystem level (Lu et al., 2018).

3 Main Methods and Limitations to Calculate GPP

Gross primary production at the ecosystem scale is inferred using direct measurements of NEE between ecosystems and the atmosphere (Reichstein et al., 2005) because it is not possible to directly measure the carbon flux components (GPP and ER) at ecosystem, regional or global scales (Ma et al., 2015); data-driven modeling approaches can be used to estimate GPP (Beer et al., 2010; Jung et al., 2009; Xiao et al., 2004a). On the other hand, there are alternative approaches for net photosynthesis measurements, but they can only capture small spatiotemporal samples and are very time consuming (Anav et al., 2015; Baldocchi, 2003). Due to the environmental link between stomatal conductance and photosynthesis, many efforts have been made to use leaf level information to predict canopy conductance and canopy photosynthesis rate (Cox et al., 1998). Canopy level photosynthesis models, either ignore the dependence of the fluxes in leaf canopy structure (leaf angle, LAI, etc.) to directly fit the leaf level data into the canopy (Stewart & Verma, 1992); consider only the LAI and assume that all leaves contribute equally to the canopy CO_2 and H_2O fluxes; or others more complex include functions for the canopy assimilation with some of the factors that limit photosynthesis throughout the canopy like the light absorption relationships, where toward the base of the canopy rubisco capacity will reduce (Sellers et al., 1992); recent algorithms enable the input of canopy architecture parameters to improve the accuracy of the canopy light environment representation in crops (Song et al., 2013). However, despite the potential of some canopy level photosynthesis models, these models are heavy limited in scaling up across heterogeneous landscapes, multiple assumptions are needed causing structural uncertainties, the dependence and challenges of spatiotemporal leaf level data constraint their usage, limitations for cross-comparison at larger scales, and complexity of some of the models make us consider focusing on other approaches within this chapter (Chen et al., 1999; Cox et al., 1998; Song et al., 2017). For these reasons, along with the importance of scalability, this chapter focuses on GPP (in situ) fetch footprint-level estimation and some of the major data-driven GPP estimation methods.

3.1 In Situ GPP Estimates

The EC method is a micrometeorological nondestructive technique resultant from the conservation equation (Monteith & Szeicz, 1960) that measures NEE based on the change in CO_2 storage in the canopy air space under conditions of horizontal homogeneity and above-canopy turbulent flow exchange. The carbon uptake in the ecosystem is calculated as the difference between the measured NEE and ER; daytime and nighttime ER calculated based on fitted exponential equations (Reichstein et al., 2002, 2005).

There are no observation techniques available to directly quantify vegetation carbon uptake (Ma et al., 2015), and for that reason GPP can only be inferred from direct measurements of NEE (Anav et al., 2015; Reichstein et al., 2002, 2005). GPP derived from EC-measured NEE (GPP_{EC}) is available worldwide for selected locations via networks like Ameriflux, Fluxnet, Euroflux, and Chinaflux. Networks like Fluxnet provide biological and meteorological information, CO_2 and H_2O flux data, and surface energy balance at the ecosystem level. The standard data processing pipeline (Fig. 1) adopted in these networks can be divided in four main blocks (Pastorello et al., 2020). **(1)** Quality control (QC) that combines manual and automatic checks for each variable; **(2)** energy and H_2O fluxes processing block that includes latent heat (LE) and sensible heat (H) flux variables and the correction factor for energy fluxes that uses H_2O to estimate the divergence from the energy balance closure of the EC site; **(3)** CO_2 fluxes processing block that calculates u* range to exclude periods of low turbulence, which enables filtering and the calculation of random uncertainties, then uses three different approaches (daytime, nighttime, and sundown) to partition net ecosystem CO_2 exchange (NEE) into ecosystem respiration (ER) and gross primary production (GPP); **(4)** this block focuses on processing of weather variables and gap filling using downscaled reanalysis data, and using some of these meteorological variables to process some flux variables. The final step of the pipeline focuses on creating the final dataset by combining the data products produced in the previous steps with the related quality flags at the different temporal aggregations (Pastorello et al., 2020). Additional information can be found at (Pastorello et al., 2020), and this implementation pipeline is available at the open network-enabled flux processing pipeline (ONEFlux).[1]

The partitioning methods might differ within each network, given the influence of atmospheric conditions during the NEE processing leading to different USTAR thresholds and filtering methods. The filtering process is important to avoid underestimation of the fluxes that can occur as a result of low turbulence conditions that prevent the EC system from recording this information (Pastorello et al., 2020).

The dataset provided from any of the flux networks might include multiple GPP_{EC} estimates depending on the partitioning (daytime, nighttime, and sundown reference respiration) and USTAR threshold estimation and filtering methods. The most recent Fluxnet processing pipeline suggests two methods to estimate USTAR using a complete year of nighttime information, then grouping by season and temperature, and finally partitioning into 20 USTAR categories (Pastorello et al., 2020). Then, the mean NEE is calculated, and the USTAR computation approach is the difference between both of the following methods. **(1)** Change point detection (CP) that takes two linear regressions between NEE and USTAR with an imposed 0 slope in the second regression, where the two lines intersect is the change point and this is

[1] https://github.com/AmeriFlux/ONEFlux/.

frequently utilized to identify anomalies in climate data. **(2)** Moving-point-transition (MP) compares the average NEE in the 10 higher USTAR categories and the average NEE value in each of the 20 USTAR categories. Nonetheless, for most sites and networks, users will encounter GPP_{EC} estimates using mainly the variable USTAR threshold (VUT) and the constant USTAR threshold (CUT) that are the constant variants of the USTAR calculation methods and are used to calculate the NEE uncertainty due to the selected USTAR threshold (Pastorello et al., 2020). The VUT approach uses the joint population data from the previous step to obtain USTAR thresholds from multiple selected years, including those previous and subsequent of the target year, enabling to have year-to-year variability within the 40 thresholds extracted, and it allows influence from conditions in neighboring years. The CUT method focuses on across years information and uses the information of all years available to filter and extract the 40 USTAR values. Finally, the GPP_{EC} variable is provided according to the NEE USTAR filtering ($GPP_{EC\text{-}VUT}$ and $GPP_{EC\text{-}CUT}$) and gap filling with the corresponding suffixes. The 40 USTAR values are then used to estimate uncertainty from the threshold estimation (percentiles) and the values in the dataset. Finally, the processed information is typically organized into 30-min intervals, where the information can be presented using the average (mean) which have a smoothing effect and reference (REF) that uses the most representative values obtained in the previous steps. The standard error (SE) based in the standard deviation is often included. Figure 2 displays the comparison between daily GPP_{EC} estimates calculated under different partitions (daytime and nighttime) and different USTAR filtering (VUT and CUT) for an agroecosystem in the USA over a 10-year period (2002–2012) (Fig. 3).

The GPP_{EC} dataset downloaded from one of the flux networks should be carefully considered since the selection could potentially impact the outcome of the user needs. For example, the time series displayed on Fig. 2 of the mean monthly average daily GPP values of a rotational agroecosystem that cultivated wheat, corn, soybeans, and canola with different management practices (fertilizer, tillage, herbicide, and burn) during the 2002–2012 period indicates that there are consistent GPP_{EC} values for both partition methods (daytime and nighttime), but GPP_{EC} from nighttime partition presents negative GPP_{EC} values and in average can be up to 2 umol CO_2 M^{-2} S^{-1} larger than GPP_{EC} estimates from daytime partition which will affect the monthly and annual total GPP estimates. For both daytime and nighttime partition methods, the GPP_{EC} estimates were constantly similar whether the USTAR filtering was CUT or VUT. However, for the daytime partition, the largest variance between CUT and VUT (Fig. 2a and b) occurred between the mean and reference (REF) datasets, where REF was larger in the GPP_{EC_VUT} estimates.

3.2 *Model-Based GPP Estimates*

There are many techniques to estimate GPP that include machine learning based on EC data, enzyme kinetic (process-based) models, statistical models that use SIF or vegetation indices and light use efficiency (LUE) models. The LUE models are the

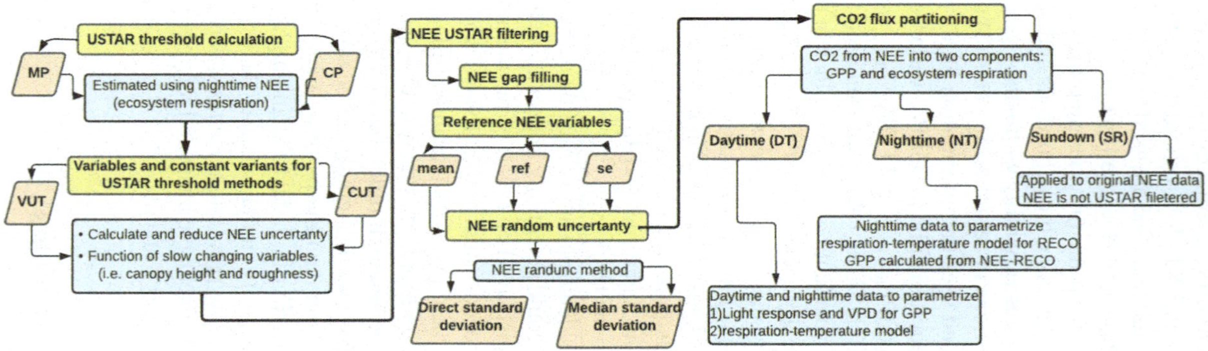

Fig. 2 Fluxnet carbon flux partitioning pipeline

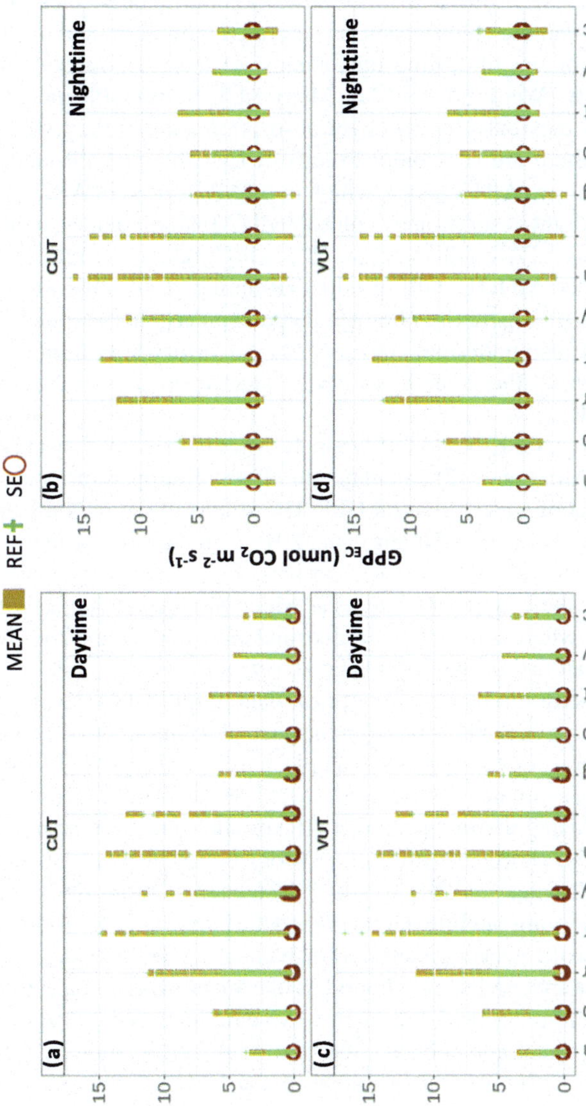

Fig. 3 Comparison between daily daytime and daily nighttime partition methods for GPP_{EC} estimates (2002–2012) in rotational crops using CUT and VUT USTAR filtering

most widely used because of their lower complexity and greater data availability (Zhang et al., 2017). The LUE-based GPP models are based on the portion of solar energy that is absorbed by vegetation, and their accuracy is affected by environmental conditions (i.e., temperature and water) (Chang et al., 2021; Huang et al., 2019a). The photosynthesis curve at the leaf level varies among plant species, because it becomes nonlinear when the chloroplasts are saturated with light, and this becomes a limitation at the moment of scaling up for most of the LUE models in heterogeneous environments (Wei et al., 2017a). Many LUE-based GPP models can be grouped by the ways they calculate (1) light absorption and (2) the LUE. In terms of light absorption, there are two major groups of LUE-based GPP models, light absorption by the canopy ($APAR_{canopy}$), which is $APAR_{canopy} = FPAR_{canopy} * PAR$. The other group uses light absorption by chlorophyll ($APAR_{chl}$) in the canopy, which is $APAR_{chl} = FPAR_{chl} * PAR$. (Zhang et al., 2017) highlighted the limitation of the $APAR_{canopy}$ LUE-based MOD17 GPP product (Running et al., 2004) to represent the seasonal changes of vegetation photosynthetic capacity that resulted in global underestimations and also underlined the superior performance in multiple biomes of the vegetation photosynthesis model (VPM 2.0) (Xiao et al., 2004a; Zhang et al., 2017) that uses $APAR_{chl}$.

In terms of LUE (ε_g), LUE-based GPP models differ in the way they (1) calculate the maximum LUE parameter, (2) calculate the effect of temperature on GPP, and (3) calculate the effect of water stress on GPP. The effects of temperature and water stress on GPP (Chang et al., 2021; Huang et al., 2019a) are introduced to improve the accuracy of the estimates, since these variables have an impact on vegetation carbon uptake. For example, there are temperatures at which photosynthetic activity is at a maximum and others where it would be limited, and atmospheric and soil conditions which would induce stomatal to prevent excessive water loss.

The approaches using remote sensing optical data to quantify ecosystem carbon fluxes have been gaining in popularity given the pace of technological advancements and their spatiotemporal consistency. For example, red and red edge (670–750 nm) are used in SIF models to estimate GPP (Li et al., 2018; Pacheco-Labrador et al., 2019), ER can be estimated using thermal infrared (TIR) or land surface temperature(LST) (Rahman et al., 2005), shortwave infrared (SWIR) is linked with the columns of CO_2 concentrations(Rayner & O'Brien, 2001). The later relationship is used in the greenhouse gases observing satellite (GOSAT) and the Orbiting Carbon Observatory-2 (OCO-2) to estimate the net ecosystem carbon balance (NBP) (Kondo et al., 2015), and vegetation indices are crucial in LUE models like VPM (Xiao et al., 2004a; Zhang et al., 2017). Furthermore, optical and microwave data have been used to estimate biomass through empirical approaches (Blackard et al., 2008; Imhoff, 1995; Toan et al., 1992). Figure 4 provides an overview of the carbon components that can be quantified using different regions of the electromagnetic spectrum. For further details, the readers are referred to Xiao et al. (2019).

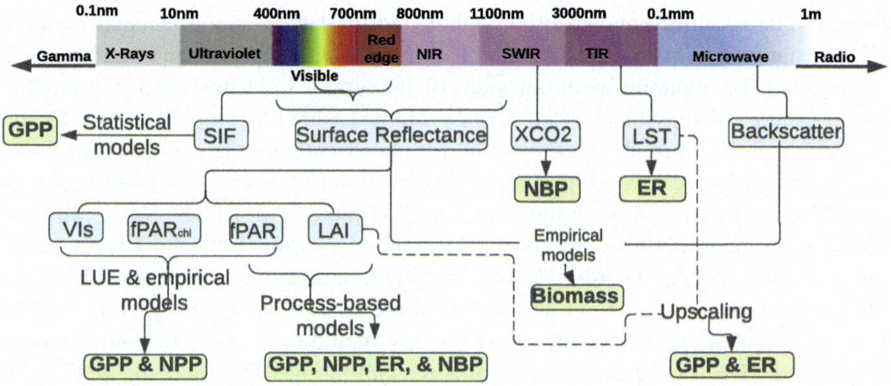

Fig. 4 Main approaches with remote sensing used to estimate the terrestrial carbon flux components [adapted from Xiao et al. (2019)]

3.3 Current Work and Limitations (In situ, Remote Sensing, and Modeling-Based)

The ecosystem exchange measured with the EC technique is subject to errors (random and systematic) related to turbulent flux transport and the calibration, location, and accuracy of the instrument (Loescher et al., 2006). Errors linked to the environmental drivers on seasonal phenology, ecosystem disturbances, and carbon balance are different from the uncertainties regarding the understanding of the biological response to weather variables. Nonetheless, the biggest limitation of the technique is in forest and tall canopies, where defining the environments below the canopy is more challenging (Loescher et al., 2006), because of changes in stability due to variation in the characteristic roughness, zero-plane displacement, and field disturbances in agroecosystems (Loescher et al., 2003). Low turbulence conditions are another limitation of the EC technique, and these conditions can occur frequently at nighttime due to cooler conditions, lack of surface heating induce thermal stratification, and increased atmospheric stability. The fluxes can be sensitive to the calculation method because of partial decoupling of above-canopy measurements, increasing the need for USTAR thresholds to determine adequate conditions to calculate the turbulent exchanges over a given period (Loescher et al., 2006). Finally, one of the biggest challenges is the upscaling of EC data. Ameriflux and Fluxnet have led the efforts in storing, providing, and standardizing EC data worldwide and every year the number of EC sites increases. The data serve as the gold standard for validation despite the sources of uncertainty. Although there is a growing network of sites, it is still challenging to determine the carbon and water dynamics at larger scales like the regional level.

Early light use efficiency (LUE) models assumed that all canopy leaves had the same photosynthesis curve and CO_2 limiting the scale-up with leaf area index (LAI) (Wei et al., 2017a). Modern LUE models assume a nonlinear relationship between the leaf photosynthesis and absorbed light (Medlyn, 1998), including environmental factors that represent changes in the temperature, water response, and nutrient availability, and the apparent quantum yield or maximum light use efficiency (LUE0) (Xiao et al., 2004a; Zhang et al., 2017). Many LUE-GPP models (i.e., MOD17) presume that the optimal air temperature for photosynthesis (T_{opt}) is constant. Since T_{opt} actually changes over time and biome, this assumption can lead to uncertainties in GPP of up to 10 Pg C/yr at the global scale (Chang et al., 2021). Some improvements have been made to the newest MOD17 algorithm (C55 and C6 products), but limitations persist regarding biome-specific lookup tables and weather drivers in the parameterization (Zhao et al., 2005). In the newest MOD17 algorithm, APAR is the amount of energy for the entire canopy including non-photosynthetic vegetation, and LUE is considered constant across all biomes (Zhang et al., 2017), [76, p. 200]. Multiple studies have demonstrated that leaf quantity (density) and quality (chlorophyll and nitrogen contents) directly impact the photosynthetic capacity of vegetation (Wu et al., 2016); however, the MOD17 algorithm doesn't consider leaf quality. Given the spatial variability of leaf quality within each biome (Kattge et al., 2009), the MOD17 data products are limited in their spatial and temporal representation of GPP; this is confirmed when the products are compared with GPP_{EC} estimates.

The VPM model is a LUE-GPP model that has been widely tested at local, regional, and global scales (Chang et al., 2021; Doughty et al., 2018; Wu et al., 2018, 2020; Xiao et al., 2004a). In VPM 2.0, general equation ($GPP = APAR_{chl} * LUE_g$) lies in the partitioning of the photosynthetically active vegetation (PAV, chloroplast) and non-photosynthetically active vegetation (NPV, stems, senescent vegetation) in leaf and canopies. The LUE component incorporates the fraction of each individual biome with the LUE0, temperature and water stress, and phenology. The temperature stress is included using the equation based on the Terrestrial Ecosystem Model (TEM) (Lutz et al., 2019; Melton & Arora, 2016) and the plant's response to the water stress condition is based on the land surface water index (LSWI) (Xiao et al., 2005).

On the other hand, other approaches like satellite-derived vegetation indices as proxies of GPP and net primary production (NPP) have been widely used for carbon climate studies worldwide (Dash & Curran, 2004; Saleska et al., 2016; Tucker et al., 2001). This approach combines different spectral bands to represent canopy greenness, while considering canopy structure, chlorophyll content, and plant phenology (Xiao et al., 2019). Multiple studies have found a strong link between VIs and GPP (Rahman et al., 2005) and aboveground NPP (Ponce-Campos et al., 2013). For this reason, there has been a push to estimate GPP exclusively based on satellite-derived vegetation indices (Huang et al., 2019b). However, significant differences have been reported between GPP_{EC} and VIs in biologically driven and weather-driven agroecosystems (Restrepo-Coupe et al., 2016). Finally, process-based terrestrial models (diagnostic (Farquhar et al., 1980) and prognostic (Hazarika et al., 2005)) can use remote sensing data (fPAR and LAI) to quantify carbon fluxes (Xiao et al., 2019). However, limitations exist due to the data availability during the models spin up

Table 1 Summary of relevant recent approaches for estimating GPP at the ecosystem scale

Approach	Advantages	Disadvantages	Cite
EC method	High quality and frequency, widely available, standardized data. Gold standard for GPP estimates	Subject to systematic and random errors. Nighttime conditions limitations	(Loescher et al., 2006)
VI proxy	Simple and highly scalable	High uncertainty in biologically and weather driven ecosystems	(Restrepo-Coupe et al., 2016)
MOD17 (LUE)	Consistent global and temporal records	LUE remains constant for individual biomes. Biome-specific lookup table limitations. APAR used is absorption by the entire canopy	(Zhao et al., 2005)
VPM (LUE)	Based on the partition of photosynthetically active (PAV) and non-photosynthetically active vegetation (NPV)	Weather forcing introduces uncertainty into the model. Land-cover type spatial scale limits the precision of the C3/C4 partition	Xiao et al., 2004a; Beer et al. 2010; Wu et al., 2020)
Process-based	Supports understanding of the terrestrial ecosystem status and mechanistic linkages	Large uncertainties at site and regional scales	(Farquhar et al., 1980; Hazarika et al., 2005)
SIF (statistical)	Information available at ecosystem and global scales	Coarse spatial scale limits its use for site and field level analyses. Systematic estimation of GPP requires consideration of clear-sky bias, canopy structure and leaf morphology, and SIF yield sensitivity under stress conditions	(Pacheco-Labrador et al., 2019)

process, and several studies have shown large uncertainties of these models at the site-level and regional scales (Huntzinger et al., 2012; Liu et al., 2018) (Table 1).

3.4 *Innovative and Simple Approaches to Estimate GPP*

The VPM model can substantially improve GPP estimates in croplands and grasslands by improving the calculations that account for temperature and water stress. This will enhance the VPM model's in-season estimates which already has proven a superior performance and accuracy compared to other high scalable models like the MOD17 algorithm. Such improvements include simplification in the water stress calculation and use of the 2.11–2.29 μm spectral region of the shortwave infrared to estimate land surface water index (LSWI) (Loescher et al., 2003). Furthermore, the VPM model uses EVI rather than NDVI because multi-year data analyses have yielded stronger linear relationships between EVI and GPP (Xiao et al., 2004b). This improvement makes it easier for in-season calculation of GPP in near real time and improves the sensitivity to the moisture content of soil and vegetation (Blum, 2009; Myneni, et al., 1999). In addition, considering the uncertainties in GPP estimations (Cox et al., 1998), the new VPM version will use a well-suited approach based on the EVI to estimate site-specific optimum temperature (Cox et al., 1998; Wei et al., 2017b)).

Moreover, some contemporary approaches have successfully estimated global CO_2 fluxes combining OCO-2 data and a CO_2 (0.1 degrees) inverse modeling system (Maksyutov et al., 2021) fed with ERA5 reanalysis meteorological data and refining the interhemispheric transport rate and vertical profiles in the troposphere of the model. Other approaches have used observation-based methods with fluxes from terrestrial vegetation and ocean derived machine learning (Bakker et al., 2016; Quéré et al., 2018). The orbiting carbon observatory-2 (OCO-2) v10 provides information of the CO_2 column average mixing ratios, and it has the goal of providing global coverage. The model intercomparison project (MIP) enables users to obtain 10-day fluxes data and land data aggregated to 1.33 s averages, ocean data to 5 s averages to reduce problem size. Flexpart lagrangian transport model has been used coupled with an atmospheric tracer transport model to simulate surface CO_2 flux footprints on 0.2 degrees grid for land and on 0.4 degrees grid for ocean (Maksyutov et al., 2021).

The SIF is the energy emitted by the canopy after chlorophyll absorbs light (Guanter et al., 2014). Recent works have assimilated space-borne SIF to estimate specific biophysical parameters to include in the calculation of GPP, as it links the vegetation responses to some stressors (Pacheco-Labrador et al., 2019). Furthermore, there is a strong consistency between GPP and SIF that depends on the C3/C4 photosynthetic pathway (Guanter et al., 2014). SIF could serve to support the GPP estimates and use as a proxy APARchl (Ma et al., 2018). After acknowledging the link between GPP and T with the stomatal conductance and common meteorological drivers, some recent studies propose to use SIF for T monitoring (Li et al., 2018; Lu et al., 2018; Ma et al., 2018).

4 Main Methods to Calculate ET Components (T and E)

The quantification and estimation of terrestrial H_2O fluxes in agroecosystems is challenging due to the complex dynamics of water storage in the ecosystem, transport, and hydrometeorological conditions (Mencuccini et al., 2019; Stoy et al., 2019). Water transport is crucial for the water particle exchange between the surface and atmosphere, and it is determined by the soil properties (e.g., porosity and conductivity), vegetation structure (e.g., conductance, photosynthetic capacity, and xylem–phloem transport properties), and vegetation surfaces, and it can exhibit nonlinear responses to hydrometeorological drivers (Celis et al., 2021; Mencuccini et al., 2019; Stoy et al., 2019; Vivoni et al., 2008).

There are diverse approaches to measure and estimate the components of ET (T and E) such as plant-level sap flow, leaf gas exchange, soil lysimeters, canopy chambers, photometers, soil heat pulse methods, and stable and radioisotope techniques (Stoy et al., 2019), but they have specific scalability challenges. SAPFLUXNET (Poyatos et al., 2016) addresses some of the challenges by building a global network to facilitate the access to this information. However, given the need for this information in agroecosystems, the challenges of using these measuring techniques in large agricultural fields, and the extensive literature reviews regarding these methods (Abtew & Melesse, 2012; Kool et al., 2014), this section focuses on existing and emerging approaches that can be easily applied to unmeasured ecosystems due their high scalability potential for estimating T at the site and ecosystem scale with low complexity for industrial agroecosystems.

4.1 In situ and Model-Based Approaches to Calculate T and ET

Transpiration is the largest component of ET in agroecosystems, and ET can be measured using EC at the ecosystem scale (Wilson et al., 2001). Typically estimated from the latent heat of vaporization (LE) or actual evapotranspiration, the EC technique provides a valuable source of information at the site scale, and the information is available globally through Fluxnet.

4.1.1 In situ Eddy Covariance Measurements

The ET partitioning technique using EC data must acknowledge the uncertainty in GPP_{EC} estimates (Sect. 3.3) and consider two main important aspects: (1) canopy and atmosphere decoupling (Jocher et al., 2017) and (2) the energy balance closure (Stoy et al., 2013). The current approaches using EC data to partition ET complements the calculation with other techniques (i.e., in situ) that measure or predict T and E components at resolutions that align with the EC fetch footprint and the common half-hourly or daily period (Stoy et al., 2019). For example, (Scanlon & Kustas, 2010, 2012) used the flux variance similarity (FVS) theory, which is based on separating non-stomatal processes (bare soil evaporation and respiration) and stomatal driven processes linked with the EC-measured high frequency (10 or 20 Hz) CO_2 and H_2O fluxes. The Monin–Obukhov FVS theory (Campbell & Norman, 1998; Wagle et al., 2023) uses a leaf level water use efficiency (WUE) equation that establishes variations through the canopy due to environmental conditions, proposing a direct link between the changes of CO_2 and stomatal uptake. Under this WUE theory, there are deviations from the average CO_2 mixing ratio (c) and water vapor mixing ratio (q) due to respiration and E from non-stomatal pathways, resulting in a change in correlation between the stomatal and non-stomatal carbon exchange processes, which allows for partitioning of ET into T and E (Campbell & Norman, 1998; Stoy et al., 2019). The flux-variance approach has been demonstrated to accurately represent the ET partition in agroecosystems like corn during the growing season (Maksyutov et al., 2021).

4.1.2 Remote Sensing Model-Based Approaches

The main remote sensing approaches to calculate ET and T are based on (1) statistical models linking ET with vegetation indices (Glenn et al., 2008), (2) surface energy balance (SEB) that estimates ET as the residual of the surface energy budget and depend on thermal-IR data (Kustas & Anderson, 2009), and (3) physically-based models based on the Penman–Monteith (P–M) equation (Ivanov et al., 2004; Mu et al., 2011). The Ecosystem Spaceborne Thermal Radiometer Experiment on Space Station (ECOSTRESS) ECO3ETALEXIU product uses a SEB model that depends on their level 2 land surface temperature and emissivity to estimate ET (Mencuccini et al., 2019; Quéré et al., 2018). The model provides a high-resolution ET output (70 m), but with some limitations in the ET partitioning (Anderson 2021). Remote sensing-based SEB models can produce ET estimates that exceed the incoming net radiation due to advection of the sensible heat flux (H) (Glenn et al., 2007). This is a consequence of the complexity to accurately estimate H using the aerodynamic resistance (Glenn et al., 2007; Gowda et al., 2008). The Moderate Resolution Imaging Spectroradiometer (MODIS) MOD16 improved ET algorithm (Mu et al., 2011) builds on (Mu et al., 2007a) and has its origins in a modified version of the (Cleugh et al., 2007) model that uses the P–M approach driven by remote sensing data. Some of the improvements of the new MOD16 ET algorithm include a simplified approach to calculating vegetation cover, sum of daytime and nighttime ET calculations, the estimation of the ground heat flux, and improvements in the stomatal conductance and ET partitioning (Mu et al., 2011). The forcing data comes from the daily meteorological reanalysis dataset from NASA's Global Modeling and Assimilation Office (GMAO) and half-hourly observations from Ameriflux EC systems. The model vegetation parameters (1 Km^2) MODIS land cover type (MOD12Q1), FPAR/LAI (MOD15A2) (Myneni et al., 1999), and white-sky MODIS albedo (MOD43C1, 0.05-degree) are used in the input and then calibrated following (Zhao et al., 2005) method.

Transpiration is a key indicator of crop productivity (Blum, 2009) and the largest component of the ET in crops (Celis, et al., 2021; Wei et al., 2017b). Most ET and T modeling approaches are based on the Penman–Monteith (P-M) equation (Monteith, 1965; Penman, 1948) that requires complex parametrizations. In the P–M equation, actual ET is defined as λE (Eq. 1), where the slope of the saturation vapor pressure–air temperature curve (Δ), net radiation *(R_n)*, soil heat flux (G, frequently considered 0), aerodynamic resistance(r_a), canopy stomatal resistance (r_c), vapor pressure deficit(D), specific heat(c_p), psychrometric constant(γ), and air density(ρ) are used to calculate λE.

$$\lambda E = \frac{\Delta(Rn - G) + \gamma \frac{\rho m \lambda \upsilon \rho q a}{\gamma a}}{\Delta + \gamma\left(1 + \frac{\gamma a}{\gamma s}\right)} \tag{1}$$

Although the P–M (Eq. 1) is one of the most reliable ways to estimate ET and T (Mu et al., 2007b), since it is a combination of (1) estimating the stomatal conductance; (2) estimating evaporation from the ground surface, and (3) estimating transpiration from the wet canopy; the total control of stomata and the assumptions used to estimate the leaf transpiration and potential ET is one of the biggest limitations especially in remote sensing models (Jarvis & McNaughton, 1986).

4.2 *Limitations of Current Methods and Techniques*

The importance of stomata controls in the P–M (Monteith, 1965; Penman, 1948) approach limits the precision of the MOD16 improved algorithm, because of its assumption that stomata close during nighttime. Furthermore, the role of soil moisture is crucial for ET calculations (Vivoni, et al., 2008), because soil moisture conditions affect the water stress in vegetation and stomatal closure as a response to prevent water loss. MOD16 (Mu et al., 2011) accounts for the canopy water interception and calculates the total daily ET (ET = E_{wc}+E_t + E_{bs}) as from evaporation from wet canopy (E_{wc}), transpiration from dry canopy (E_t), and bare soil evaporation (E_{bs}). However, the soil moisture component is represented using the water cover fraction (F_{wet}) proposed by Fisher (2008) that is based on relative humidity (RH). Moreover, water-limited ecosystems adopt different responses to avoid water loss, and the relation between soil moisture and ET is stronger. In this chapter, we highlight the need to improve the MOD16 algorithm by (1) including surface and root zone soil moisture from the soil moisture active passive (SMAP) level 4 and (2) estimating continuous stomatal resistance (r_s) using Schulze energy-limited relation (Schulze et al., 1994, 1995) and Irmak's upscaling method based on the photosynthetic photon flux density (PPFD) (Irmak et al., 2008) (see Table 2). Some of the equations to consider for including the soil moisture component in some of these complex models are displayed in Eqs. 2 and 3 and are fully described in Ivanov's (2004) modeling framework.

$$\beta e = \min\left(1, \frac{\boldsymbol{\theta}_{100}}{0.75\,\boldsymbol{\theta}_s}\right) \tag{2}$$

$\boldsymbol{\theta}_s$ = saturation soil moisture; $\boldsymbol{\theta}_{100}$ = soil moisture top 0.1 m

Table 2 Summary of important contemporary techniques to estimate the components of ET at the ecosystem scale

Approach	Advantages	Disadvantages	Cite
EC method	Uses high quality and high frequency widely available data	Calculation terms rarely shared. Relies and sensitive to WUE data assumptions. Scale can be limited to site footprint	(Campbell & Norman, 1998; Scanlon & Kustas, 2010)
ECOSTRESS (SEB-based)	High spatial resolution information (70 m), global availability	High uncertainty in the partitioning of ET Limitations calculating H using aerodynamic resistance	(Hulley et al., 2017)
MOD16 (Physical-based)	Long consistent temporal and spatial global records	Moderate spatial resolution (500 m), nighttime stomata closure assumptions, and water cover fraction is based on RH	(Glenn et al., 2008)
VTM	Simple and accurate approach to estimate transpiration. Easy to scale and low parametrization	Despite having a strong foundational structure due to the coupling with GPP, further testing in agroecosystems is needed	(Alfieri et al., 2009)
tRIBS (Physical-based)	High precision and highly scalable at very high spatial resolutions. Input can be at multiple spatial scales	High complexity and extensive parametrization. Data continuity and scale will depend on the forcing	(Celis, et al., 2021; Irmak et al., 2008; Ivanov et al., 2004)
SIF	Information available at ecosystem and global scales	Empirical link with TR, uncertainty in SIF retrieval, low understanding of their mechanistic relationship	(Pacheco-Labrador et al., 2019)

$$\beta t = \min\left(1, \frac{\theta_{top} - \theta_r}{0.75\theta_s - \theta_r}\right) \tag{3}$$

θ_r = saturation soil moisture; θ_{top} = soil moisture top 1 m.

4.3 *Innovative and Simple Approaches to Estimate Water Fluxes*

Stable isotopes can be employed to partition the components of ET and have proven highly accurate when combined with EC data (Stoy et al., 2019; Williams et al.,

2004). Isotope-based ET partitioning outputs are used to validate the SVAT model parametrizations, contributing to the predictability of flux responses in the SVAT models (Calvet et al., 1998). The isotope approach has a larger sample area footprint than EC, despite collecting water vapor measurements at similar heights as the EC towers (Schmid, 1994). This difference must be considered when working with EC and isotope data in agroecosystems.

The limitations of the MOD16 algorithm (Mu et al., 2011) could potentially be addressed and reduced, including the ET partitioning component based on soil moisture rather than relative humidity as it is in the current algorithm. In addition, modifying the current approach used in MOD16 to calculate stomatal conductance could improve some of the limitations that the data product faces in places with vegetation under water stress. Furthermore, after acknowledging the link between GPP and T with the stomatal conductance and common meteorological drivers, some recent studies proposed to use SIF for T monitoring (Li et al., 2018; Lu et al., 2018; Ma et al., 2018).

The vegetation transpiration model (VTM) (Alfieri et al., 2009) provides a simpler and accurate approach to estimating T in forest and grasslands agroecosystems (Alfieri et al., 2009), utilizing the biophysical relationship of GPP with T at the leaf level. The model estimates daily T as a function of GPP and water use efficiency (WUE = GPP/T, mol CO_2/mol H_2O). Literature review reported that the WUE for C_3 plants is 500 umol CO_2/umol H_2O) and WUE for C_4 plants is 250 umol CO_2/umol H_2O) (Taiz et al., 2015).

$$\mathrm{T} = 1/WUE^{*}\mathrm{GPP} \tag{4}$$

$$\mathrm{T(mm\,H_2O/day)} = 0.33\left(\mathrm{mm\,H_2O/g\,C/m^2/day}\right)^{*} \mathrm{GPP}\left(\mathrm{g\,C/m^2/day}\right) \text{ for } \mathrm{C_3} \text{ plants} \tag{5}$$

$$\mathrm{T(mm\,H_2O/day)} = 0.165\left(\mathrm{mm\,H_2O/g\,C/m^2/day}\right)^{*} \mathrm{GPP}\left(\mathrm{g\,C/m^2/day}\right) \text{ for } \mathrm{C_4} \text{ plants} \tag{6}$$

5 Unit Conversion of Energy, Carbon, and Water Fluxes

Scientific communication in agriculture, forestry, and meteorology requires a correct conversion of units and the appropriate use of equations (Allen et al., 1998). The interactions between atmospheric and ecosystem variables play an important role in agriculture; and many applications require variables that come from different sources in a variety of formats (Atzberger, 2013). The successful use of this information depends on accurate use of all equations and factors to work with the data in the same units (Nobel, 1999). For these reasons, this section includes the main conversion units for some of the main variables needed for this agricultural application, such as solar radiation, trace gases and water, carbon balance, evapotranspiration, and resource use efficiency.

5.1 Solar Radiation and PAR

Most weather stations provide information about the total incoming shortwave solar radiation (SW) typically W m^{-2} units. Photosynthetically active radiation (PAR) refers to a specific range of energy within the electromagnetic spectrum (4.0 to 7.0 μm). The conversion factor of SW units to PAR units is 1 W m^{-2} ≈ 4.57 μmol m^{-2} s^{-1} (Thimijan & Heins, 1983). However, since only ~ 45% of the energy of solar radiation is in the 4.0 to 7.0 μm range, the conversion factor of SW to PAR needs to be adjusted to account for only that amount of energy (Foken, 2008; Mavi & Tupper, 2004). Equation 7 shows the conversion factor of shortwave radiation (SR W m^{-2}) to PAR (μmol m^{-2} s^{-1}).

$$1 * \frac{W}{\mathrm{m}^2} = 2.02 \frac{\mu\,\mathrm{mol}}{\mathrm{m}^2\ \mathrm{s}} \tag{7}$$

Equation 8 displays the conversion factor of PAR (μmol m^{-2} s^{-1}) to PAR (mol m^{-2}time^{-1}). Equation 9 shows an example for a 30-min period data value.

$$1 * \frac{\mu\,\mathrm{mol}}{\mathrm{m}^2\ \mathrm{s}} = 10^{-6} \frac{\mathrm{mol}}{\mathrm{m}^2\ \mathrm{time}} \tag{8}$$

$$1 * \frac{\mu\,\mathrm{mol}}{\mathrm{m}^2\ \mathrm{s}} = 10^{-6} \frac{\mathrm{mol}}{\mathrm{m}^2\ \frac{1}{1800}} \Rightarrow 1800 * 10^{-6} \frac{\mathrm{mol}}{\mathrm{m}^2\ 30\,\mathrm{min}} \Rightarrow 0.0018 \frac{\mathrm{mol}}{\mathrm{m}^2\ 30\,\mathrm{min}} \tag{9}$$

$$1250 * \frac{\mu\,\mathrm{mol}}{\mathrm{m}^2\ \mathrm{s}} = 1250 * 0.0018 \frac{\mathrm{mol}}{\mathrm{m}^2\ 30\,\mathrm{min}} = 2.25 \frac{\mathrm{mol}}{\mathrm{m}^2\ 30\,\mathrm{min}} \tag{10}$$

5.2 Carbon Fluxes

The following conversion factors are important for operations related to the carbon balance and carbon fluxes like GPP. Equation 11 displays the conversion factor needed to convert carbon flux (μmol CO_2 m^{-2} s^{-1}) to (g C m^{-2}). Equation 12 displays the CO_2 assimilation rate (A) (μmol CO_2 m^{-2} s^{-1}) to (mg CO_2 m^{-2} s^{-1}).

$$\frac{\mu\,\text{mol CO}_2}{\text{m}^2\ \text{s}} \Rightarrow 10^{-6}\frac{\text{mol CO}_2}{\text{m}^2\ \text{time}} \Rightarrow \frac{10^{-6}*12\,\text{gC}}{\text{m}^2\ \text{time}} = 0.12*10^{-4}\frac{\text{gC}}{\text{m}^2} \tag{11}$$

$$A = \frac{\mu\,\text{mol CO}_2}{\text{m}^2\ \text{s}} \Rightarrow 10^{-6}\frac{\text{mol CO}_2}{\text{m}^2\ \text{s}} \Rightarrow \frac{10^{-6}*44\,\text{gC}}{\text{m}^2} * \frac{10^3\text{mg}}{1\text{g}} = 0.044\frac{\text{mg CO}_2}{\text{m}^2\ \text{s}} \tag{12}$$

Since most data are provided at 30-min intervals and need to be aggregated into daily intervals, Eq. 13 shows carbon flux conversion (μmol CO_2 m^{-2} s^{-1}) to (g C m^{-2}) utilizing a 30-min data interval and Eq. 14 exemplifies a 1-day interval. In addition, Eq. 15 can be used to determine the annual carbon totals per unit area from (g C m^{-2}) to (tC $area^{-1}$ $year^{-1}$). We used hectare (ha) given the common use of this unit area for agriculture.

$$1*\frac{\mu\,\text{mol CO}_2}{\text{m}^2\ \text{s}} \Rightarrow 10^{-6}\frac{\text{mol CO}_2}{\text{m}^2\frac{1}{1800}} \Rightarrow \frac{10^{-6}*12\,\text{gC}}{\text{m}^2\frac{1}{1800}} = 0.0216\frac{\text{gC}}{\text{m}^2} \tag{13}$$

$$1*\frac{\mu\,\text{mol CO}_2}{\text{m}^2\ \text{s}} \Rightarrow 10^{-6}\frac{\text{mol CO}_2}{\text{m}^2\frac{1}{86400}} \Rightarrow \frac{10^{-6}*12\,\text{gC}}{\text{m}^2\frac{1}{86400}} = 1.0368\frac{\text{gC}}{\text{m}^2} \tag{14}$$

$$1*\frac{\text{gC}}{\text{m}^2\ \text{year}} \Rightarrow 10^{-6}\frac{\text{tC}}{\frac{1}{10000}*\text{year}} = 10^{-2}\frac{\text{tC}}{\text{ha year}} \tag{15}$$

5.3 Water Fluxes (Transpiration and Evapotranspiration): Conversion of Energy Units to mm of H_2O

Transpiration rate can be provided in energy units or might need to be expressed in a different format; thus, it is important to highlight that to evaporate 1 g H_2O ~ 2450 Jules is necessary. Equation 16 shows the conversion factor of (mmol H_2O m^{-2} s^{-1}) to (W m^{-2}).

$$\begin{aligned}\frac{\text{mmol H}_2\text{O}}{\text{m}^2\text{s}} &= 10^{-3}\frac{\text{mol H}_2\text{O}}{\text{m}^2\ \text{s}} \Rightarrow 10^{-3}*18\frac{\text{g H}_2\text{O}}{\text{m}^2\ \text{s}}\\ &\Rightarrow 10^{-3}*18*2450\frac{\text{J}}{\text{m}^2\ \text{s}} \Rightarrow 44.1\frac{\text{W}}{\text{m}^2}\end{aligned} \tag{16}$$

The FAO Penman–Monteith method provides the ET of the reference surface. This method is crucial when working with the surface energy balance and using EC data. For this method, (Allen et al., 1998) defines a reference crop with a height of 12 cm, albedo of 0.23, and resistance of 70 s/m. The LE is assumed by some scholars to be actual ET and is used to directly estimate ET through the latent heat of vaporization (λ). According to Allen et al. (1998), λ (a single value of 2.45 MJ kg^{-1}) taken from the simplification of the P–M equation varies only slightly over normal temperature ranges as it is the LE flux for an air temperature of about 20 °C. Equation 17 shows the λ formula where T is the mean air temperature (°C), Eq. 18 displays the formula linking ET and LE, and Eq. 19 shows the use of the equation for a 30-min data interval, as is typically available at EC sites.

$$\lambda \frac{J}{Kg} = \frac{10^{-3}(2500 - 2.37T)}{10^{6}} \Rightarrow 10^{3}(2500 - 2.37T) \tag{17}$$

$$ET \frac{mm}{time} = \frac{LE}{\lambda} = \frac{\left(\frac{W}{m^2}\right)}{\left(\frac{J}{Kg}\right)} \Rightarrow \frac{\left(\frac{J}{m^2 s}\right)}{\left(\frac{J}{Kg}\right)} \Rightarrow \frac{J}{m^2 s} * \frac{Kg}{J} \Rightarrow \frac{Kg}{m^2 s} \Rightarrow \frac{mm}{s} \tag{18}$$

$$ET \frac{mm}{30\,min} = 1800 \frac{LE}{10^{3}(2500 - 2.37T)} \Rightarrow \frac{1.8*LE}{2500 - 2.37T} \tag{19}$$

6 Summary and Conclusion

Stomatal conductance links photosynthesis and transpiration at the leaf level (Farquhar & Sharkey, 1982; Irmak et al., 2008), enabling us to calculate T based on GPP and WUE, and allowing us to use the absorbed PAR from the chlorophyll to estimate the minimal stomatal conductance based on the Schulze energy-limited relation (Dorman & Sellers, 1989; Schulze et al., 1994). The P–M approach (Monteith, 1965; Penman, 1948)) combines the energy and mass transfer techniques as a tool for estimating actual evaporation or latent heat flux (Ivanov et al., 2004), and thus enabling us to calculate transpiration. However, the P–M technique requires extensive parametrization that increases the biases in the transpiration and ET estimates. New techniques following MOD16 (Mu et al., 2011) algorithm to calculate ET components that are controlled by soil moisture should consider including Deardorff's approach (Deardorff, 1978) to calculate plant transpiration and evaporation that accounts for different transpiration rates between older and newer leaves (Eltahir & Bras, 1996). However, despite these approaches being widely tested and proven to be successful to accurately estimating ET components (Ivanov et al., 2004; Ivanov et al., 2004; Méndez-Barroso et al., 2014; Vivoni et al., 2007b), the high level of complexity makes it difficult to apply at a global scale, regional, or field level in regions with strong soil heterogeneity and variable environmental conditions. For this reason, simple and accurate models like VTM can provide accurate and consistent transpiration data without the need for extensive parametrization. Furthermore, the combined use of Landsat 7, Landsat 8, Landsat 9, and Sentinel 2(A/B) to estimate high-resolution vegetation inputs and land cover type could enhance the precision of GPP and T estimates. Finally, approaches using SIF data and the isotopes technique to enhance models' predictions should be considered to support the development of new modeling tools.

Acknowledgements This literature review work was supported by research grants from NASA Geostationary Carbon Cycle Observatory mission (GeoCarb Contract #80LARC17C0001), NSF EPSCoR (IIA-1946093), and USDA NIFA (2020-67104-30935).

References

Abtew, W., & Melesse, A. (2012). *Evaporation and evapotranspiration: Measurements and estimations*. Springer Science & Business Media.

Alfieri, J. G., Xiao, X., Niyogi, D., Pielke, R. A., Chen, F., & LeMone, M. A. (2009). Satellite-based modeling of transpiration from the grasslands in the Southern Great Plains, USA. *Global and Planetary Change, 67*(1), 78–86. https://doi.org/10.1016/j.gloplacha.2008.12.003

Allen, R. G., Pereira, L. S., Raes, D., & Smith, M. (1998). Crop evapotranspiration-guidelines for computing crop water requirements-FAO Irrigation and drainage paper 56. *Fao Rome, 300*(9), D05109.

Amthor, J. S. (1995). Terrestrial higher-plant response to increasing atmospheric [CO_2] in relation to the global carbon cycle. *Global Change Biology, 1*(4), 243–274.

Anav, A., et al. (2015). Spatiotemporal patterns of terrestrial gross primary production: A review. *Reviews of Geophysics, 53*(3), 785–818.

Anderson, M. C. (2021). Level-3 evapotranspiration (ET_ALEXI) algorithm theoretical basis document, (pp. 26).

Atzberger, C. (2013). Advances in remote sensing of agriculture: Context description, existing operational monitoring systems and major information needs. *Remote Sensing, 5*(2), Art. no. 2. https://doi.org/10.3390/rs5020949.

Aubinet, M., Vesala, T., & Papale, D. (2012). *Eddy covariance: A practical guide to measurement and data analysis*. Springer Science & Business Media.

Bakker, D. C., et al. (2016). A multi-decade record of high-quality fCO_2 data in version 3 of the Surface Ocean CO_2 Atlas (SOCAT). *Earth System Science Data, 8*(2), 383–413.

Baldocchi, D. D. (2003). Assessing the eddy covariance technique for evaluating carbon dioxide exchange rates of ecosystems: Past, present and future. *Global Change Biology, 9*(4), 479–492.

Beer, C., et al. (2010). Terrestrial gross carbon dioxide uptake: Global distribution and covariation with climate. *Science, 329*(5993), 834–838.

Blackard, J. A., et al. (2008). Mapping US forest biomass using nationwide forest inventory data and moderate resolution information. *Remote Sensing of Environment, 112*(4), 1658–1677.

Blum, A. (2009). Effective use of water (EUW) and not water-use efficiency (WUE) is the target of crop yield improvement under drought stress. *Field Crops Research, 112*(2–3), 119–123.

Blyth, E., & Harding, R. J. (2011). Methods to separate observed global evapotranspiration into the interception, transpiration and soil surface evaporation components. *Hydrological Processes, 25*(26), 4063–4068.

Bondeau, A., et al. (2007). Modelling the role of agriculture for the 20th century global terrestrial carbon balance. *Global Change Biology, 13*(3), 679–706.

Boulet, G., Chehbouni, A., Braud, I., Vauclin, M., Haverkamp, R., & Zammit, C. (2000). A simple water and energy balance model designed for regionalization and remote sensing data utilization. *Agricultural and Forest Meteorology, 105*(1–3), 117–132.

Calvet, J.-C., et al. (1998). An interactive vegetation SVAT model tested against data from six contrasting sites. *Agricultural and Forest Meteorology, 92*(2), 73–95.

Campbell, G. S., & Norman, J. M. (1998). *Introduction to environmental biophysics* (2nd ed.). Springer.

Celis, J. A. et al. (2021). From standard weather stations to virtual micro-meteorological towers in ungauged sites: Modeling tool for surface energy fluxes, evapotranspiration, soil temperature, and soil moisture estimations. *Remote Sensing, 13*(7), Art. no. 7. https://doi.org/10.3390/rs13071271

Chang, Q., Xiao, X., Doughty, R., Wu, X., Jiao, W., & Qin, Y. (2021). Assessing variability of optimum air temperature for photosynthesis across site-years, sites and biomes and their effects on photosynthesis estimation. *Agricultural and Forest Meteorology, 298–299*, 108277. https://doi.org/10.1016/j.agrformet.2020.108277

Chen, J. M., Liu, J., Cihlar, J., & Goulden, M. L. (1999). Daily canopy photosynthesis model through temporal and spatial scaling for remote sensing applications. *Ecological Modelling, 124*(2–3), 99–119.

Chen, D.-X., Coughenour, M. B. (2004). Photosynthesis, transpiration, and primary productivity: Scaling up from leaves to canopies and regions using process models and remotely sensed data. *Global Biogeochemical Cycles, 18*(4). https://doi.org/10.1029/2002GB001979

Cleugh, H. A., Leuning, R., Mu, Q., & Running, S. W. (2007). Regional evaporation estimates from flux tower and MODIS satellite data. *Remote Sensing of Environment, 106*(3), 285–304.

Cox, P. M., Huntingford, C., & Harding, R. J. (1998). A canopy conductance and photosynthesis model for use in a GCM land surface scheme. *Journal of Hydrology, 212–213*, 79–94. https://doi.org/10.1016/S0022-1694(98)00203-0

Dash, J., & Curran, P. J. (2004). The MERIS terrestrial chlorophyll index.

Deardorff, J. W. (1978). Efficient prediction of ground surface temperature and moisture, with inclusion of a layer of vegetation. *Journal of Geophysical Research Oceans, 83*(C4), 1889–1903.

Dorman, J. L., & Sellers, P. J. (1989). A global climatology of Albedo, roughness length and stomatal resistance for atmospheric general circulation models as represented by the simple biosphere model (SiB). *Journal of Applied Meteorology, 28*, 833–855. https://doi.org/10.1175/1520-0450(1989)028%3c0833:AGCOAR%3e2.0.CO;2

Doughty, R., et al. (2018). Responses of gross primary production of grasslands and croplands under drought, pluvial, and irrigation conditions during 2010–2016, Oklahoma, USA. *Agricultural Water Management, 204*, 47–59. https://doi.org/10.1016/j.agwat.2018.04.001

Eltahir, E. A., & Bras, R. L. (1996). Precipitation recycling. *Reviews of Geophysics, 34*(3), 367–378.

Farquhar, G. D., von von Caemmerer, S., & Berry, J. A. (1980). A biochemical model of photosynthetic CO_2 assimilation in leaves of C3 species. *Planta, 149*(1), 78–90.

Farquhar, G. D., & Sharkey, T. D. (1982). Stomatal conductance and photosynthesis. *Annual Review of Plant Physiology, 33*(1), 317–345.

Fisher, J. B., Tu, K. P., & Baldocchi, D. D. (2008). Global estimates of the land–atmosphere water flux based on monthly AVHRR and ISLSCP-II data, validated at 16 FLUXNET sites. *Remote Sensing of Environment, 112*(3), 901–919.

Foken, T. (2008). The energy balance closure problem: An overview. *Ecological Applications, 18*(6), 1351–1367.

Foley, J. A., et al. (1996). An integrated biosphere model of land surface processes, terrestrial carbon balance, and vegetation dynamics. *Global Biogeochemical Cycles, 10*(4), 603–628.

Glenn, E. P., Huete, A. R., Nagler, P. L., Hirschboeck, K. K., & Brown, P. (2007). Integrating remote sensing and ground methods to estimate evapotranspiration. *Critical Reviews in Plant Sciences, 26*(3), 139–168.

Glenn, E. P., Huete, A. R., Nagler, P. L., & Nelson, S. G. (2008). Relationship between remotely-sensed vegetation indices, canopy attributes and plant physiological processes: What vegetation indices can and cannot tell us about the landscape. *Sensors, 8*(4), 2136–2160.

Gowda, P. H., Chavez, J. L., Colaizzi, P. D., Evett, S. R., Howell, T. A., & Tolk, J. A. (2008). ET mapping for agricultural water management: Present status and challenges. *Irrigation Science, 26*(3), 223–237.

Guanter, L., et al. (2014). Global and time-resolved monitoring of crop photosynthesis with chlorophyll fluorescence. *Proceedings of the National Academy of Sciences, 111*(14), E1327–E1333. https://doi.org/10.1073/pnas.1320008111

Harwood, K. G., Gillon, J. S., Roberts, A., & Griffiths, H. (1999). Determinants of isotopic coupling of CO_2 and water vapour within a Quercus petraea forest canopy. *Oecologia, 119*(1), 109–119.

Hazarika, M. K., Yasuoka, Y., Ito, A., & Dye, D. (2005). Estimation of net primary productivity by integrating remote sensing data with an ecosystem model. *Remote Sensing of Environment, 94*(3), 298–310.

He, M. et al. (2018). Regional crop gross primary productivity and yield estimation using fused landsat-MODIS data. *Remote Sensing, 10*(3), Art. no. 3. https://doi.org/10.3390/rs10030372

Huang, M., et al. (2019a). Air temperature optima of vegetation productivity across global biomes. *Nature Ecology and Evolution, 3*(5), 772–779.

Huang, X., Xiao, J., & Ma, M. (2019b). Evaluating the performance of satellite-derived vegetation indices for estimating gross primary productivity using FLUXNET observations across the globe. *Remote Sensing, 11*(15), 1823.

Hulley, G., Hook, S., Fisher, J.,& Lee, C. (2017). Ecostress, A NASA Earth-Ventures Instrument for studying links between the water cycle and plant health over the diurnal cycle. In *2017 IEEE International Geoscience and Remote Sensing Symposium (IGARSS)*, 2017, (pp. 5494–5496).

Huntzinger, D. N., et al. (2012). North American carbon program (NACP) regional interim synthesis: Terrestrial biospheric model intercomparison. *Ecological Modelling, 232*, 144–157.

Imhoff, M. L. (1995). Radar backscatter and biomass saturation: Ramifications for global biomass inventory. *IEEE Transactions on Geoscience and Remote Sensing, 33*(2), 511–518.

Irmak, S., et al. (2008). On the scaling up leaf stomatal resistance to canopy resistance using photosynthetic photon flux density. *Agricultural and Forest Meteorology, 148*(6–7), 1034–1044.

Ivanov, V. Y., Vivoni, E. R., Bras, R. L., & Entekhabi, D. (2004). Catchment hydrologic response with a fully distributed triangulated irregular network model. *Water Resources Research, 40*(11). https://doi.org/10.1029/2004WR003218

Jarvis, P. G., & McNaughton, K. G. (1986). Stomatal control of transpiration: Scaling up from leaf to region. *Advances in Ecological Research, 15*, 1–49.

Jocher, G., et al. (2017). Apparent winter CO_2 uptake by a boreal forest due to decoupling. *Agricultural and Forest Meteorology, 232*, 23–34.

Jung, M., Reichstein, M., & Bondeau, A. (2009). Towards global empirical upscaling of FLUXNET eddy covariance observations: Validation of a model tree ensemble approach using a biosphere model. *Biogeosciences, 6*(10), 2001–2013.

Kattge, J., Knorr, W., Raddatz, T., & Wirth, C. (2009). Quantifying photosynthetic capacity and its relationship to leaf nitrogen content for global-scale terrestrial biosphere models. *Global Change Biology, 15*, 976–991. https://doi.org/10.1111/j.1365-2486.2008.01744.x

Kondo, M., Ichii, K., Takagi, H., & Sasakawa, M. (2015). Comparison of the data-driven top-down and bottom-up global terrestrial CO_2 exchanges: GOSAT CO_2 inversion and empirical eddy flux upscaling. *Journal of Geophysical Research Biogeosciences, 120*(7), 1226–1245.

Kool, D., Agam, N., Lazarovitch, N., Heitman, J. L., Sauer, T. J., & Ben-Gal, A. (2014). A review of approaches for evapotranspiration partitioning. *Agricultural and Forest Meteorology, 184*, 56–70.

Kustas, W., & Anderson, M. (2009). Advances in thermal infrared remote sensing for land surface modeling. *Agricultural and Forest Meteorology, 149*(12), 2071–2081.

Lambers, H., Chapin III, F. S., & Pons, T. L. (2008). In *Plant physiological ecology*. Springer Science & Business Media.

Le Quéré, C., et al. (2018). Global carbon budget 2017. *Earth System Science Data, 10*(1), 405–448.

Le Toan, T., Beaudoin, A., Riom, J., & Guyon, D. (1992). Relating forest biomass to SAR data. *IEEE Transactions on Geoscience and Remote Sensing, 30*(2), 403–411.

Li, X., Xiao, J., & He, B. (2018). Chlorophyll fluorescence observed by OCO-2 is strongly related to gross primary productivity estimated from flux towers in temperate forests. *Remote Sensing of Environment, 204*, 659–671.

Liáng, L. L., Kirschbaum, M. U., Giltrap, D. L., Wall, A. M., & Campbell, D. I. (2020). Modelling the effects of pasture renewal on the carbon balance of grazed pastures. *Science Total Environment, 715*, 136917.

Liu, Y., et al. (2018). Satellite-derived LAI products exhibit large discrepancies and can lead to substantial uncertainty in simulated carbon and water fluxes. *Remote Sensing of Environment, 206*, 174–188.

Loescher, H. W., Oberbauer, S. F., Gholz, H. L., & Clark, D. B. (2003). Environmental controls on net ecosystem-level carbon exchange and productivity in a Central American tropical wet forest. *Global Change Biology, 9*(3), 396–412. https://doi.org/10.1046/j.1365-2486.2003.00599.x

Loescher, H. W., Law, B. E., Mahrt, L., Hollinger, D. Y., Campbell, J., & Wofsy, S. C. (2006). Uncertainties in, and interpretation of, carbon flux estimates using the eddy covariance technique. *Journal of Geophysical Research Atmospheres, 111*(D21). https://doi.org/10.1029/2005JD006932

Longdoz, B., & Granier, A. (2012). Eddy covariance measurements over forests. In *Eddy Covariance* (pp. 309–318). Springer.

Lu, X., et al. (2018). Potential of solar-induced chlorophyll fluorescence to estimate transpiration in a temperate forest. *Agricultural and Forest Meteorology, 252*, 75–87.

Lutz, F., et al. (2019). Simulating the effect of tillage practices with the global ecosystem model LPJmL (version 5.0-tillage). *Geoscience Model Development, 12*(6), 2419–2440.

Ma, J., Yan, X., Dong, W., & Chou, J. (2015). Gross primary production of global forest ecosystems has been overestimated. *Science and Reports, 5*(1), 1–9.

Ma, J., Xiao, X., Zhang, Y., Doughty, R., Chen, B., & Zhao, B. (2018). Spatial-temporal consistency between gross primary productivity and solar-induced chlorophyll fluorescence of vegetation in China during 2007–2014. *Science of the Total Environment, 639*, 1241–1253. https://doi.org/10.1016/j.scitotenv.2018.05.245

Mahfouf, J.-F., et al. (1996). Analysis of transpiration results from the RICE and PILPS workshop. *Global and Planetary Change, 13*(1–4), 73–88.

Maksyutov, S., et al. (2021). Technical note: A high-resolution inverse modelling technique for estimating surface CO_2 fluxes based on the NIES-TM–FLEXPART coupled transport model and its adjoint. *Atmospheric Chemistry and Physics, 21*(2), 1245–1266. https://doi.org/10.5194/acp-21-1245-2021

Matsui, T., Lakshmi, V., & Small, E. E. (2005). The effects of satellite-derived vegetation cover variability on simulated land–atmosphere interactions in the NAMS. *Journal of Climate, 18*(1), 21–40.

Mavi, H. S., & Tupper, G. J. (2004). In *Agrometeorology: Principles and applications of climate studies in agriculture.* CRC Press.

Medlyn, B. E. (1998). Physiological basis of the light use efficiency model. *Tree Physiology, 18*(3), 167–176. https://doi.org/10.1093/treephys/18.3.167

Melton, J. R., & Arora, V. K. (2016). Competition between plant functional types in the Canadian terrestrial ecosystem model (CTEM) v. 2.0. *Geoscientific Model Development, 9*(1), 323.

Mencuccini, M., Manzoni, S., & Christoffersen, B. (2019). Modelling water fluxes in plants: From tissues to biosphere. *New Phytologist, 222*(3), 1207–1222. https://doi.org/10.1111/nph.15681

Méndez-Barroso, L. A., et al. (2014). A modeling approach reveals differences in evapotranspiration and its partitioning in two semiarid ecosystems in Northwest Mexico. *Water Resources Research, 50*(4), 3229–3252.

Monteith, J. L. (1965). Evaporation and environment. *Symposia of the Society for Experimental Biology, 19*, 205–234.

Monteith, J. L., & Szeicz, G. (1960). The carbon-dioxide flux over a field of sugar beet. *Quarterly Journal Royal Meteorological Society, 86*(368), 205–214. https://doi.org/10.1002/qj.49708636810

Montgomery, R. B. (1948). Vertical eddy flux of heat in the atmosphere. *Journal of Atmospheric Science, 5*(6), 265–274.

Mu, Q., Heinsch, F. A., Zhao, M., & Running, S. W. (2007b). Development of a global evapotranspiration algorithm based on MODIS and global meteorology data. *Remote Sensing of Environment, 111*(4), 519–536.

Mu, Q., Zhao, M., & Running, S. W. (2011). Improvements to a MODIS global terrestrial evapotranspiration algorithm. *Remote Sensing of Environment, 115*(8), 1781–1800.

Mu, Q., Zhao, M., Heinsch, F. A., Liu, M., Tian, H., & Running, S. W. (2007a) Evaluating water stress controls on primary production in biogeochemical and remote sensing based models. *Jounal of Geophysical Reseach Biogeosciences, 112*(G1).

Myneni et al. (1999) Algorithm Theoretical Basis Document (pp. 130).

Nobel, P. S. (1999). In *Physicochemical and environmental plant physiology*. Academic Press.

Obukhov, A. M. (1951). Charakteristiki mikrostruktury vetra v prizemnom sloje atmosfery (Characteristics of the micro-structure of the wind in the surface layer of the atmosphere). *Izv SSSR Ser Geofiz, 3*, 49–68.

Pacheco-Labrador, J. et al. (2019). Multiple-constraint inversion of SCOPE. Evaluating the potential of GPP and SIF for the retrieval of plant functional traits. *Remote Sensing Environment, 234*, 111362. https://doi.org/10.1016/j.rse.2019.111362

Pastorello, G., et al. (2020). The FLUXNET2015 dataset and the ONEFlux processing pipeline for eddy covariance data. *Science Data, 7*(1), 1–27.

Penman, H. L. (1948). Natural evaporation from open water, bare soil and grass. *Proceedings of the Royal Society of London. Series A, Mathematical and Physical Sciences, 193*(1032), 120–145.

Ponce-Campos, G. E., et al. (2013). Ecosystem resilience despite large-scale altered hydroclimatic conditions. *Nature, 494*(7437), 349–352.

Poyatos, R., Granda, V., Molowny-Horas, R., Mencuccini, M., Steppe, K., & Martínez-Vilalta, J. (2016). SAPFLUXNET: Towards a global database of sap flow measurements. *Tree Physiology, 36*(12), 1449–1455.

Prentice, I. C., Dong, N., Gleason, S. M., Maire, V., & Wright, I. J. (2014). Balancing the costs of carbon gain and water transport: Testing a new theoretical framework for plant functional ecology. *Ecology Letters, 17*(1), 82–91.

Rahman, A. F., Sims, D. A., Cordova, V. D., & El-Masri, B. Z. (2005). Potential of MODIS EVI and surface temperature for directly estimating per-pixel ecosystem C fluxes. *Geophysical Research Letters, 32*(19).

Rayner, P. J., & O'Brien, D. M. (2001). The utility of remotely sensed CO_2 concentration data in surface source inversions. *Geophysical Research Letters, 28*(1), 175–178.

Reichstein, M., et al. (2002). Ecosystem respiration in two Mediterranean evergreen Holm Oak forests: Drought effects and decomposition dynamics. *Functional Ecology, 16*(1), 27–39.

Reichstein, M., et al. (2005). On the separation of net ecosystem exchange into assimilation and ecosystem respiration: Review and improved algorithm. *Global Change Biology, 11*(9), 1424–1439.

Restrepo-Coupe, N., et al. (2016). MODIS vegetation products as proxies of photosynthetic potential along a gradient of meteorologically and biologically driven ecosystem productivity. *Biogeosciences, 13*(19), 5587–5608.

Running, S. W., Nemani, R. R., Heinsch, F. A., Zhao, M., Reeves, M., & Hashimoto, H. (2004). A continuous satellite-derived measure of global terrestrial primary production. *BioScience, 54*(6), 547–560.

Saleska, S. R., et al. (2016). Dry-season greening of Amazon forests. *Nature, 531*(7594), E4–E5.

Scanlon, T. M., & Kustas, W. P. (2012). Partitioning evapotranspiration using an eddy covariance-based technique: Improved assessment of soil moisture and land–atmosphere exchange dynamics. *Vadose Zone Journal, 11*(3).

Scanlon, T. M., & Kustas, W. P. (2010). Partitioning carbon dioxide and water vapor fluxes using correlation analysis. *Agricultural and Forest Meteorology, 150*(1), 89–99.

Schmid, H. P. (1994). Source areas for scalars and scalar fluxes. *Bound.-Layer Meteorology, 67*(3), 293–318

Schulze, E.-D., Kelliher, F. M., Körner, C., Lloyd, J., & Leuning, R. (1994). Relationships among maximum stomatal conductance, ecosystem surface conductance, carbon assimilation rate, and plant nitrogen nutrition: A global ecology scaling exercise. *Annual Review of Ecology and Systematics, 25*(1), 629–662.

Schulze, E.-D., Leuning, R., & Kelliher, F. M. (1995). Environmental regulation of surface conductance for evaporation from vegetation. In *Global change and terrestrial ecosystems in Monsoon Asia* (pp. 79–87). Springer.

Sellers, P. J., Berry, J. A., Collatz, G. J., Field, C. B., & Hall, F. G. (1992). Canopy reflectance, photosynthesis, and transpiration. III. A reanalysis using improved leaf models and a new canopy integration scheme. *Remote Sensing of Environment, 42*(3), 187–216. https://doi.org/10.1016/0034-4257(92)90102-P

Sitch, S., et al. (2003). Evaluation of ecosystem dynamics, plant geography and terrestrial carbon cycling in the LPJ dynamic global vegetation model. *Global Change Biology, 9*(2), 161–185.

Song, Q., Zhang, G., & Zhu, X.-G. (2013). Optimal crop canopy architecture to maximise canopy photosynthetic CO_2 uptake under elevated CO_2–A theoretical study using a mechanistic model of canopy photosynthesis. *Functional Plant Biology, 40*(2), 108–124.

Song, Q., Wang, Y., Qu, M., Ort, D. R., & Zhu, X.-G. (2017). The impact of modifying photosystem antenna size on canopy photosynthetic efficiency—Development of a new canopy photosynthesis model scaling from metabolism to canopy level processes. *Plant, Cell and Environment, 40*(12), 2946–2957. https://doi.org/10.1111/pce.13041

Sperry, J. S., et al. (2017). Predicting stomatal responses to the environment from the optimization of photosynthetic gain and hydraulic cost. *Plant, Cell and Environment, 40*(6), 816–830.

Stanhill, G. (1986). Water use efficiency. In N. C. Brady (Ed.), *Advances in agronomy* (Vol. 39, pp. 53–85). Academic Press. https://doi.org/10.1016/S0065-2113(08)60465-4

Stewart, J. B., & Verma, S. B. (1992). Comparison of surface fluxes and conductances at two contrasting sites within the FIFE area. *Journal of Geophysical Research Atmospheres, 97*(D17), 18623–18628.

Stoy, P. C., et al. (2013). A data-driven analysis of energy balance closure across FLUXNET research sites: The role of landscape scale heterogeneity. *Agricultural and Forest Meteorology, 171*, 137–152.

Stoy, P. C., et al. (2019). Reviews and syntheses: Turning the challenges of partitioning ecosystem evaporation and transpiration into opportunities. *Biogeosciences, 16*(19), 3747–3775. https://doi.org/10.5194/bg-16-3747-2019

Swinbank, W. C. (1951). The measurement of vertical transfer of heat and water vapor by eddies in the lower atmosphere. *Journal of Atmospheric Science, 8*(3), 135–145.

Taiz, L., Zeiger, E., Møller, I. M., & Murphy, A. (2015). Plant physiology and development. In *Plant Physiology Development*, edn 6. Accessed 04 Sep 2022. [Online]. Available: https://www.cabdirect.org/cabdirect/abstract/20173165866

Taiz, L., & Zeiger, E. (2002). *Plant physiology* (3rd ed.). Sinauer Associates.

Thimijan, R. W., & Heins, R. D. (1983). Photometric, radiometric, and quantum light units of measure: A review of procedures for interconversion. *HortScience, 18*(6), 818–822.

Tucker, C. J., Slayback, D. A., Pinzon, J. E., Los, S. O., Myneni, R. B., & Taylor, M. G. (2001). Higher northern latitude normalized difference vegetation index and growing season trends from 1982 to 1999. *International Journal of Biometeorology, 45*(4), 184–190.

van der Ent, R. J., Savenije, H. H. G., Schaefli, B., & Steele-Dunne, S. C. (2010). Origin and fate of atmospheric moisture over continents. *Water Resources Research, 46*(9). https://doi.org/10.1029/2010WR009127

Vivoni, E. R., et al. (2007a). Variation of hydrometeorological conditions along a topographic transect in Northwestern Mexico during the North American monsoon. *Journal of Climate, 20*(9), 1792–1809.

Vivoni, E. R., Entekhabi, D., Bras, R. L., & Ivanov, V. Y. (2007b). Controls on runoff generation and scale-dependence in a distributed hydrologic model.

Vivoni, E. R. et al. (2008). Observed relation between evapotranspiration and soil moisture in the North American monsoon region. *Geophysical Research Letters, 35*(22). https://doi.org/10.1029/2008GL036001

Wagle, P., Raghav, P., Kumar, M., & Gunter, S. A. (2023). Influence of water use efficiency parameterizations on flux variance similarity-based partitioning of evapotranspiration. *Agricultural and Forest Meteorology, 328*, 109254. https://doi.org/10.1016/j.agrformet.2022.109254

Wei, S., Yi, C., Fang, W., & Hendrey, G. (2017a). A global study of GPP focusing on light-use efficiency in a random forest regression model. *Ecosphere, 8*(5), e01724.

Wei, Z., Yoshimura, K., Wang, L., Miralles, D. G., Jasechko, S., & Lee, X. (2017b). Revisiting the contribution of transpiration to global terrestrial evapotranspiration. *Geophysical Research Letters, 44*(6), 2792–2801.

White, M. A., Thornton, P. E., Running, S. W., & Nemani, R. R. (2000). Parameterization and sensitivity analysis of the BIOME–BGC terrestrial ecosystem model: Net primary production controls. *Earth Interactions, 4*(3), 1–85.

Williams, D. G., et al. (2004). Evapotranspiration components determined by stable isotope, sap flow and eddy covariance techniques. *Agricultural and Forest Meteorology, 125*(3), 241–258. https://doi.org/10.1016/j.agrformet.2004.04.008

Wu, J., et al. (2016). Leaf development and demography explain photosynthetic seasonality in Amazon evergreen forests. *Science, 351*(6276), 972–976. https://doi.org/10.1126/science.aad5068

Wu, X., et al. (2018). Spatiotemporal consistency of four gross primary production products and solar-induced chlorophyll fluorescence in response to climate extremes across CONUS in 2012. *Journal of Geophysical Research Biogeosciences, 123*(10), 3140–3161.

Wu, X., Xiao, X., Yang, Z., Wang, J., Steiner, J., & Bajgain, R. (2020). Spatial-temporal dynamics of maize and soybean planted area, harvested area, gross primary production, and grain production in the Contiguous United States during 2008–2018. *Agricultural Forest Meteorology*, 108240.

Xiao, X. (2006). Light absorption by leaf chlorophyll and maximum light use efficiency. *IEEE Transactions on Geoscience and Remote Sensing, 44*(7), 1933–1935.

Xiao, X., et al. (2004a). Satellite-based modeling of gross primary production in an evergreen needleleaf forest. *Remote Sensing of Environment, 89*(4), 519–534.

Xiao, X., et al. (2004b). Modeling gross primary production of temperate deciduous broadleaf forest using satellite images and climate data. *Remote Sensing of Environment, 91*(2), 256–270.

Xiao, X., et al. (2005). Mapping paddy rice agriculture in southern China using multi-temporal MODIS images. *Remote Sensing of Environment, 95*(4), 480–492.

Xiao, J. et al. (2019). Remote sensing of the terrestrial carbon cycle: A review of advances over 50 years. *Remote Sensing Environment, 233*, 111383. https://doi.org/10.1016/j.rse.2019.111383

Yan, H., et al. (2015). Improved global simulations of gross primary product based on a new definition of water stress factor and a separate treatment of C3 and C4 plants. *Ecological Modelling, 297*, 42–59.

Yepez, E. A., Williams, D. G., Scott, R. L., & Lin, G. (2003). Partitioning overstory and understory evapotranspiration in a semiarid savanna woodland from the isotopic composition of water vapor. *Agricultural and Forest Meteorology, 119*(1–2), 53–68.

Zhang, Y., et al. (2014). Estimation of vegetation photosynthetic capacity from space-based measurements of chlorophyll fluorescence for terrestrial biosphere models. *Global Change Biology, 20*(12), 3727–3742.

Zhang, Y. et al. (2017). A global moderate resolution dataset of gross primary production of vegetation for 2000–2016. *Science Data, 4*, 170165.

Zhao, M., & Running, S. W. (2010). Drought-induced reduction in global terrestrial net primary production from 2000 through 2009. *Science, 329*(5994), 940–943. https://doi.org/10.1126/science.1192666

Zhao, M., Heinsch, F. A., Nemani, R. R., & Running, S. W. (2005). Improvements of the MODIS terrestrial gross and net primary production global data set. *Remote Sensing of Environment, 95*(2), 164–176. https://doi.org/10.1016/j.rse.2004.12.011

Zhou, Y. et al. (2019). Climate effects on tallgrass prairie responses to continuous and rotational grazing. *Agronomy*, *9*(5), Art. no. 5. https://doi.org/10.3390/agronomy9050219

Role of Virtual Plants in Digital Agriculture

Suchitra M. Patil, Michael Henke, Magesh Chandramouli, and Adinarayana Jagarlapudi

Abstract Virtual plants (VP) are computer-simulated three-dimensional (3D) models of plants or trees. One approach to create VPs is so-called functional-structural plant models (FSPM), which are used to model an accurate plant shape and architecture and combines it with physiological processes. Static or dynamic FSPMs are a well-established approach to serve as a versatile tool for predicting crop growth patterns in response to variations in environmental conditions. Using VP, crop growth can be simulated (in silico) rapidly compared to its natural growing season, which can take up to a few months or longer. Crop researchers and breeders employ in silico as an alternative to time-consuming, labor-intensive actual field trials to explore the focused growth of crops in specific environments and swiftly quantify essential attributes. This facilitates accelerate selective crop breeding, a promising strategy to address the impending challenge of food security. In VP modeling, first step is to select the study crop and decide on what traits to be examined, then set up a field experiment, gathering temporal data on crop growth, perform statistical analysis of this data to explore the relationship among various parameters, mathematical modeling to establish the growth rules for plant parameters. Finally, the computer simulation of FSPM is used to visualized the study crop at the level of individual stand or canopy. After getting the expected output, researchers can parameterize this model to generate different test scenarios. In-depth analysis of these test cases offers an ideal method to be used in evaluating the performance of the resultant test breed in terms of growth time, yield, resistance to biotic and abiotic stress, etc., as per application. This technique can potentially augment and expedite the existing high-performance plant phenotyping research by generating synthetic datasets to sort out the imbalance in the existing actual plant phenotyping datasets; however, it has a few

S. M. Patil (✉) · A. Jagarlapudi
Centre of Studies in Resources Engineering (CSRE), Indian Institute of Technology Bombay, Mumbai 400076, India
e-mail: 204314001@iitb.ac.in

M. Henke
School of Agronomy, Hunan Agricultural University, Changsha 410128, People's Republic of China

M. Chandramouli
Computer Graphics Technology, Purdue University NW, Hammond, IN 46323, USA

© The Author(s), under exclusive license to Springer Nature Singapore Pte Ltd. 2023
S. Chaudhary et al. (eds.), *Digital Ecosystem for Innovation in Agriculture*,
Studies in Big Data 121, https://doi.org/10.1007/978-981-99-0577-5_8

challenges in its application. This chapter will cover the concept of VP modeling, its applications, and some challenges in its application.

Keywords Functional structural plant modeling (FSPM) · Virtual plant · Crop modeling · Plant phenotyping

1 Introduction

The globe now faces a food security crisis because of two significant factors: a growing population and unfavorable climate conditions (Ehrlich et al., 1993). According to the United Nation's world population outlook, by 2100, the world's population will be around 11 billion. It is a big challenge for crop scientists and breeders to work on this serious issue and find a solution to provide a nutritious and sufficient quantity of food and fodder to the growing population. Breeders and agricultural research groups are working to improve crop cultivars and farming practices to provide enough wholesome food to meet the increasing demand (Danzi et al., 2019). The primary objective of breeders is to breed new crop cultivars that produce high-yield, good-quality food, and withstand climate changes (such as saline soil, and high temperatures.). Humans have been experimenting with crops and plants for a long time to monitor their growth and development. Unlike traditional agronomic methods, where biomass production is one of the primary goals, this aims to predict vegetation condition and yield and understand plant growth and development while considering societal food security challenges constantly emerging along with changing environmental conditions. One traditional way to determine the precise impact of various plant trait combinations along with changing needs is to set up an actual experiment with these specifications and then monitor the plant's growth; nevertheless, this is a cumbersome and time-consuming process. Due to the complexity of these challenges, there is great demand for an advanced multidisciplinary scientific technique that controls and aids in understanding the development of a plant throughout its lifetime and tests how it adapts to changes in a growth environment. In the agricultural industry, this notion of modeling and simulation is often known as in silico (Marshall-Colon et al., 2017). The current state of the art is where scientists create a computer simulation of crop growth that resembles the near real one. Researchers use these simulations for plant phenotyping: to determine structural parameters like leaf count, leaf width, length, leaf angle, and plant height, and functional characteristics like light interception, photosynthesis, carbon (C) accumulation and distribution, etc. that are most suited to meet the targeted growth of study crop. Additionally, the VP model built with a high level of precision and accuracy with the ability to produce a near-real plant can simulate plant growth in lesser time compared to the entire growth period of the plant during its conventional growth season. Hence, this approach with the near-real VP model speeds up the selective breeding process. In breeding experimentation, multiple traits of many cultivars are grown with different management practices in a varying growth environment. Genotype × Environment × Management (***G* x *E* x *M***) is performed to find out desired features of cultivars and use them to form a new cultivar, the ideotype (Donald, 1968), a conceptual model that grows in a predefined manner in a controlled environment and yields a higher quality and quantity of grain, oil, or any other product expected from it (Martre et al., 2015).

In the process of VP modeling, the number of parameters or degree of details to be used depends upon the objectives, i.e., whether it is used for either decision-making or as a research tool (Seleznyova et al., 2018). When used as a research tool, it incorporates high-level degree of details to obtain a clear insight into plant architecture (thickness, length of leaves and stems, etc.) and functional mechanism (e.g., exact accumulation and distribution of the carbon assimilate to different organs as per demand) of the plant. Furthermore, if it is used in decision-making on planting and management strategies, the level of detail required is low. For example, to enhance the red:far-red ratio for the production of taller and stronger stem plants, a row distance management study like the one discussed by (Hitz et al., 2019) requires basic architectural details. The next step is the statistical analysis of collected data to determine the mathematical relationships between the growth parameters and build the plant growth rules. For example, sampling the data of growth of stem in the case of C4 crops like maize, the first three to four internodes will be of small length, middle internodes with numbers five to nine grow longer in size, and further internodes on the top portion of the plant, internode number ten to fourteen grows steadily showing the same sigmoidal growth pattern of internodes lengths in a plant like its leaves. Further, the increase in internode length is a function of thermal time measured in growing degree days (GDD), so the growth rule here decides the size of the internode as a function of growing degree days. For example, the growth of wheat internode number eight over time, as shown in Fig. 1, is a sigmoidal growth pattern. The equation of this curve is used as the growth function or mathematical rule for the internode growth simulation. Growth equations for all the plant organs for study-plant are formulated in this manner by performing statistical analysis of the collected data to estimate the growth pattern of these organs, and hence the entire plant. Finally, researchers build the virtual crop model using these mathematical rules and simulate the crop growth at the single plant level or the canopy level. Simulation of systematic arrangement of such single plant stands in rows and columns replicated from the virtual canopy. Parameterization of the VP model is used to generate different simulations (in silicos). The inputs (number of leaves, plant height, leaf angle, etc.) are provided to the VP model during parametrization. Based on these parameters, the different architectures of the plant under study are simulated. VP model-based simulations and in silico are used interchangeably. With these simulations, crop scientists and breeders can perform quantitative assessments of plant traits on screen. In addition, the data can be manipulated to test which combination results in a better-quality breed in terms of fastest growth, yield, resistance to biotic and abiotic stress, etc. Quantitative assessment of the traits (e.g., trials with different combinations of features like leaf angles and plant height) is used to calculate value

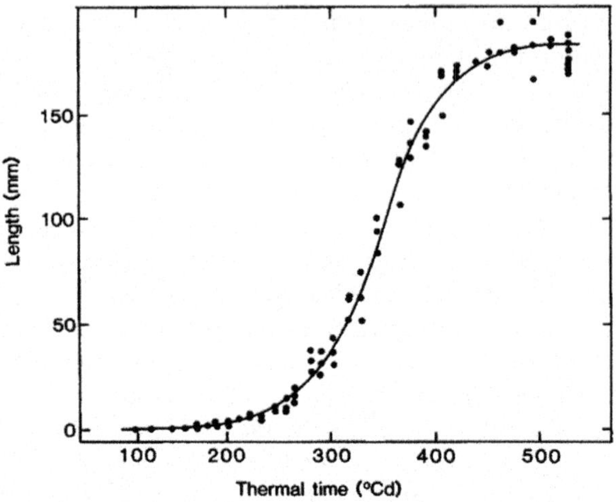

Fig. 1 Length of internode eight versus thermal time in wheat crop. *Source* Kirby (1988)

combinations for maximizing harvests with available sunlight so that the targeted photosynthesis and the yield (qualitative assessment) are met.

FSPM is a paradigm where the plant architecture (Godin et al., 1999; Vos et al., 2009) (topology and geometry of organs) comprising of number, shape, size, and arrangement of plant organs is integrated with plant physiology functions to simulate and visualize plant growth on a computer screen. FSPMs have gained more attention in the last two decades in phenotyping for breeding trials, and FSPM is one of the ways to create VP. In the subsequent sections, the concept of VP, its applications, and challenges in use are discussed. This chapter aims to explain the systematic way to develop the architectural model of the plant: the foundation of VP design, a brief introduction to the process-based model (PBM) (de Wit, 1982), and FSPM used for implementing VP using simulation wherein integration of architectural and PBM is sought. The chapter started with the first section comprising an introduction to VP. The second covered the concept of VP, and the third section provided an overview of the crop models VP modeling. The fourth section presented a few applications for VP. The fifth section covered the design of an architectural model using a case study of the arabidopsis plant. The sixth section highlighted the challenges in the agronomic applications of VP. Finally, the chapter concluded with the summary section.

2 Virtual Plant: The Concept

VP represents a computer-based simulation of a plant where the natural plant is visualized on screen in 3D. In computer-based crop modeling, the plants are considered as individual entity, and their growth is simulated with respect to spatial characteristics to obtain the plant 3D architecture as an output. 3D architectures of a few plant species are shown in Fig. 2. Though understanding visual output offers notable

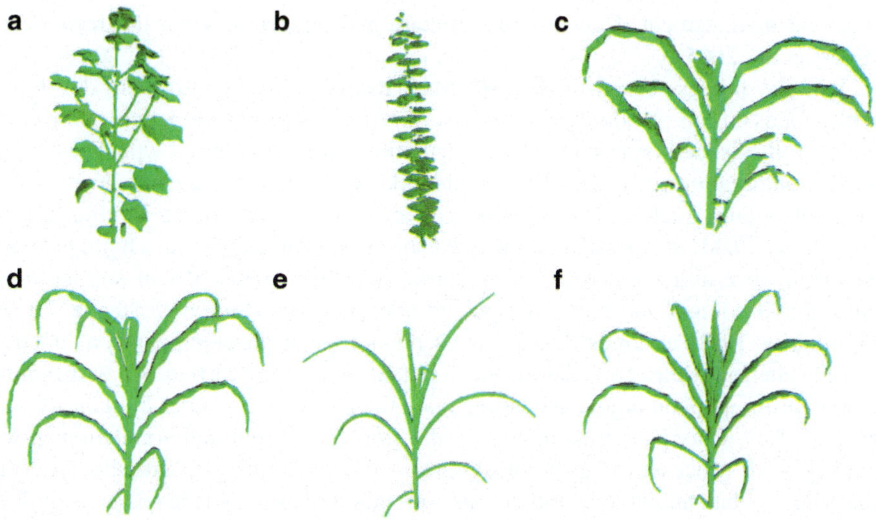

Fig. 2 3D architectures of different plant species: cotton (**a**), apple tree (**b**), sorghum (**c**), and maize (**d–f**). *Source* Artzet et al., (2019)

advantages than the classic output mode, which uses textual description or mathematical equations (Chandramouli et al., 2013), and no one can defeat the model built using mathematical equations formulated with higher precision. Visualization-based VP models serve as a powerful tool for modeling plant architecture and growth processes like light interception, photosynthesis, biomass production, etc. These growth processes are, in turn, driven by the 3D architecture plant (both shoot and root) responsible for collecting the required resources from the growth environment. For instance, light interception depends on the canopy extent and the way it is arranged. Similarly, water and nutrient absorption depend on root architecture. Such a complex design model usually needs numerous microscopic inputs for selected design parameters. Furthermore, the implementation necessitates efficient hardware (GPU for accelerating process) and software supporting object-oriented language with GPU compatibility.

3 Crop Models and Virtual Plant Modeling

The development of computer models of plants/crops was initiated in the mid1960s by (de Wit, 1982) and his colleagues to measure and understand the growth of plants quantitatively. The main objectives behind the crop modeling are to perform (1) integration of existing knowledge to expand the thinking capability of the human brain, (2) test the hypothesis quantitatively, (3) extrapolate beyond experimental range factors effects, (4) highlight the research gaps in phenomics, and (5) promising practical management decisions like climate and nutrient control in greenhouses (Evers et al., 2005).

Initially, process-based models (PBMs) (Buck-Sorlin, 2013b) were developed with a focus on eco-physiological processes like nutrient absorption, light harvesting, photosynthesis, carbon assimilation, distribution, etc., for plant growth. These were used in simulating plant growth variables like yield considering the development of plant organs, such as leaf weight, leaf area index, etc., under varying growth conditions. PBM, independent of the plant structure, works fine with a homogeneous canopy that assumes constant architectural growth in terms of leaf surface area, chlorophyll content and biomass, specific leaf area, etc. (Simon & Steppe, 2019). Summarily, PBM assumes that a plant's phenological development is a function of only thermal time, considering the constant amount of carbon (C) partitioning among different plant organ, depending upon the phenological stage of plants (Evers et al., 2005). Classical PBM addresses the implementation and simulation of the plant growth processes (e.g., Photosynthesis, C-N assimilation, and allocation) at the whole plant canopy level rather than individual plant organs level. Since it does not focus on the plant's architecture, it does not consider the interactions between plant growth processes and architecture development as per changing environment; hence alone, it is not so effective (de Reffye et al., 1997). But as its focus is mainly on dry matter production, such a model, along with expert systems support, proved to be helpful in the case of farm production management (Guo & Li, 2001) wherein they are used for predicting net farm production, assess land productivity at crop level from available data. TOMSIM crop simulator (Heuvelink, 1996, 1999) is an example of PBM, where the leaf area index (LAI) in tomato plants is updated in every growth cycle based on feedback from the leaf biomass module.

On the other hand, in the case of the pure architectural model, the focus is on the plant's architecture and not on plant processes. Knowledge of plant architecture helps to understand the plant's characteristics individually and on a broader scale as a community. For instance, knowledge of plant architecture can facilitate better fruit pruning practices and canopy improvement for enhanced interception of solar radiation for good-quality fruits production. Use of VP for citrus fruit pruning in the early stage when the fruit is of smaller size is applied, and the rendering of architecture changes immediately after pruning is used to demonstrate fruit pruning effects to the farmers (Lv et al., 2015). Pesticide sprinkling is another example, where the knowledge of the exact plant architecture and canopy and the site of insect attack and extent can facilitate spraying of the pesticide droplets at the exact place to control the infection (Room et al., 1996). But pure architectural crop models could not integrate plant physiological or functional processes with architecture. Hence, there was a strong need for a model that integrates architecture with physiological functions. As a result, the FSPM model is introduced. In this paradigm, a computer-based model of the plant is created using an object-oriented approach with the integration of plant structure with a defined level of details and functional processes like photosynthesis, carbon assimilation, etc. (one cannot include all plant available processes in the same model as it will turn into more complex model and challenging to implement, handle, and operate) applied at certain hierarchical levels such as organ, plant, or canopy. These processes are modulated by the local environment.

The primary objective behind developing computer models of plant functioning and growth is to get deeper insights into the complex interconnection between plant architecture and the physical and biological processes and visualize how they affect the development of plants at various temporal and spatial scales. The workflow for VP design is shown in Fig. 3. Being an interdisciplinary area, VP modeling combines knowledge of biology, curve fitting, mathematical modeling, and computer graphics techniques. VP must show growth and appearance-related similarities with the natural plants. This enables realistic visualization for plant scientists and virtually carries out different measurements and experimentations, which are impossible in the real world due to time, cost, and feasibility constraints (Chelle et al., 2007). VP developed using FSPM can respond to varying climate conditions, which is the main requirement of many agronomy applications. PBMs capable of determining biomass production of plants with changing environmental conditions, the number and types of plant organs, and their growth over time can serve as inputs for FSPM. A VP can be created using this FSPM that generates the structural information, simulates the climatic conditions for the canopy, and computes the different resources required for the physiological process models to affect the plant growth on a computer. The OZCOT, a virtual cotton model integrated with insects' performance to predict the structured pest activities on cotton using a virtual laboratory (VLab) (Hanan & Room, 1997; Federl & Prusinkiewicz, 1999), is another example of VP. An example of an FSPM-created VP is using botany and modeling of plant and vegetation architecture (AMAP) software along with the GOSSYM (McKinion et al., 1989) to simulate cotton growth under different environmental stresses (de Reffye et al., 1997). The main drawback of this model is that though it can affect the carbon partitioning among other organs, it cannot specify the carbon allocation to other organs based on geometry and topology. Another well-known example of FSPM-based Virtual plant is L-PEACH (Allen et al., 2006). The L-PEACH dynamic model demonstrates the growth and development of peach trees. It uses powerful formal language and L-systems to build growth rules of FSPM and solve differential equations for carbon (C) flow and allocation. Source-sink interactions between tree components drive the carbon allocation model used here. The annual life cycle of a tree is considered for the storage and utilization of carbohydrates. The simulation of carbohydrate production in the leaves is based on water and light availability. Furthermore, local supply of carbohydrates decides the growth of fruits, internodes, and apices. The output of L-PEACH (Lopez et al., 2010) is an animated 3D visual representation of the growing tree and user-specified statistics that characterize selected stages of plant development. L-PEACH can assist in horticultural decision-making processes after being calibrated to specific fruit trees. In short, FSPM integrates an architectural model with PBM. The full-fledged FSPM is generally dynamic in nature. The next section provides a brief introduction to static and dynamic plant models.

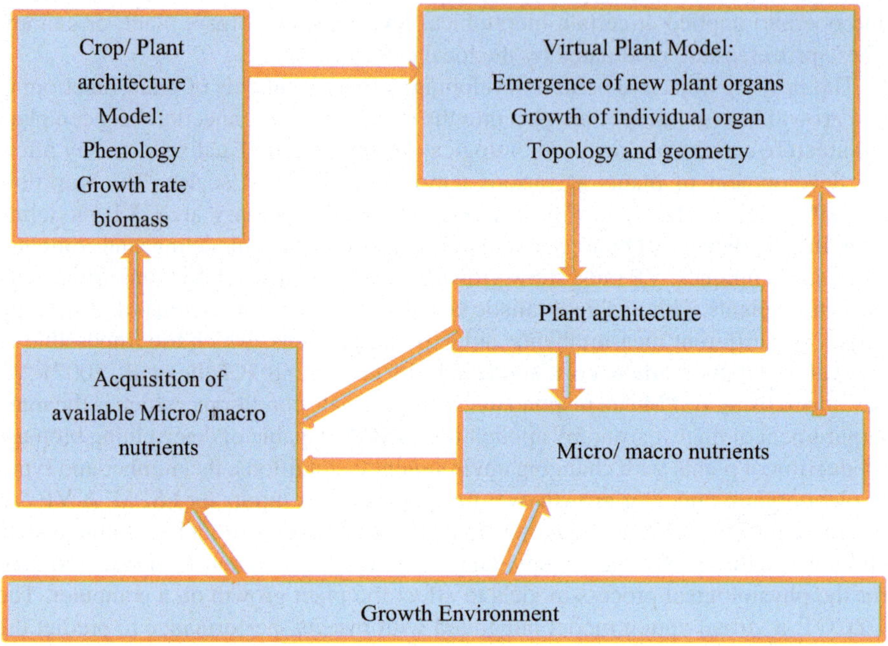

Fig. 3 Design of virtual plant model. *Source* https://www1.biologie.uni-hamburg.de/b-online/virtualplants/ipivp.html

3.1 *Static and Dynamic Crop Models*

The crop models are proposed and reviewed in terms of various system representations of crops and are broadly classified as static and dynamic crop models. In the case of the dormant plant model, a plant is represented as a fixed or steady system that does not change with time. The dynamic model of the plant captures variations in plant growth over time and ensures that the output changes with time. Static models are easy to implement relative to dynamic models. Dynamic models can also be represented as simulations using a sequence of static models at different growth stages (Simon & Steppe, 2019). In biological research, 3D static models are generally generated using digitizers. Dynamic models simulate plant morphogenesis using mathematical equations and metamorphic growth rules. These models are not linked to the plant physiological models and can be used for applications like computation of light interception and distribution (Chelle & Andrieu, 1998; Dauzat & Eroy, 1997). But the functional, structural model can perform the dynamic simulation of plant morphogenesis. A few examples are LIGNUM model (Perttunen et al., 1998), GroGra (Kurth & Sloboda, 1997), AmapHydro (de Reffye et al., 1999), and Cotons (Jallas et al., 2000) which integrates physiological model Gossym (McKinion et al., 1989). The following section discusses these two categories of crop models in detail.

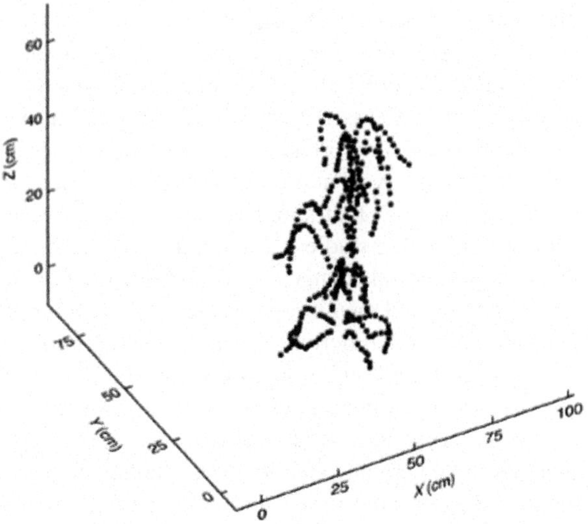

Fig. 4 3D point cloud obtained using a 3D digitizer. *Source* Evers et al., (2005)

3.2 Static Crop Models

The static model considers a fixed age or growth stage plant and resembles an actual plant at that stage. An example of kiwi fruit architecture (Smith et al., 1992) was reconstructed with the handheld digitizer gathered 3D coordinates of kiwi fruit vines parts using different intensities to represent soluble solid concentration. This was used to evaluate the fruit quality and plant architecture relationships. A 3D maize plant model constructed using 3D coordinate points data is shown below in Fig. 4. The images of the maize canopy may be viewed from any angle, and this model investigates the effect of the top leaves angle on light interception (Guo & Li, 2001). Static models can produce an exact replica of the natural vegetation structure at a fixed point in time and are further used to analyze plant architecture's qualitative and quantitative characteristics at that instance. The primary uses of static plant models include: (1) Assessing the effect of plant architectures on remote sensing application's accuracy (which architecture reflects better) and (2) Investigating the architecture-dependent biophysical processes like solar radiation distribution in space, transpiration, etc. The primary disadvantage of static models is that it requires too many data points measurements (Guo & Li, 2001). These models are independent of crop architecture-based growth-related functional characteristics (Guo & Li, 2001).

3.3 Dynamic Crop Models

Unlike static models, dynamic models represent the plant's growth over time. Experimental observations on geometrical and topological feature changes over time at different growth stages are performed to build a dynamic model. The growth rules are formulated from the statistical analysis of these observations. Further, using these rules, the crop models of plants are developed. The main objective of this type of crop model is to perform the simulation of variations in plant architecture as a function of time. Plant structure growth-related data collection is commenced either destructively or non-destructively at certain time intervals. The dynamic crop model of tomatoes that simulates the dry matter partitioning and growth has been addressed in earlier studies (Heuvelink, 1996, 1999). Dynamic representation of plant morphogenesis and changes in the architecture of a plant using minimum mathematical equations and metamorphic rules was explored in GreenLab (Yan, 2004) model with no claim for functional processes integration. The morphological descriptors are captured based on the measurements taken with the sample plant over its entire growth and development period. Some biological growth-related assumptions from crop physiology were used to construct the GreenLab model. It was compatible with identical or divergent phenotypes for either the same or different growth environment using the same rules and parameters for modeling, especially the changes in the architecture in response to fluctuations in the accumulated biomass.

The static or dynamic crop models cannot be generalized to model all species as each plant has its own architecture and growth mechanism. L-Systems (Lindenmayer, 1968a, b) are parallel graph rewriting systems based on grammar with a solid formal mathematical base to describe growth and branching pattern. The growth rules in L-System are generally a repetition of the basic design units and are rewritten with parallel replacement during execution. VLab (Virtual Lab in Botany), developed by the University of Calgary (Federal & Prusinkiewicz, 1999; Prusinkiewicz, 2004) in Canada, has L-System embedded in it and was used for creating a few models for some species. Still, it could not develop any significant plant architecture as such because every plant has a complex structure that increases the complexities of growth rules. Further, de Reffye and a group from CIRAD developed AMAP and created around 30 general plant architecture models. These AMAP models were meant for helping simulation of trees or large plants (de Reffye et al., 1997; Guo & Li, 2001).

Alternatively, an open L-system can be used for crop modeling. With available L-system provision for modeling, bidirectional information exchange between plants and their environment is possible; it is used as a key to bring together environmental interaction mechanisms like light interception, transpiration, etc., in VP models. Simulation of the effect of climate on plant growth using this model, for example, one can take into consideration how the tree branches development is restricted to avoid the smashes among branches of neighboring trees and the competition among these to trees for sunlight harvesting (Mech & Prusinkiewicz, 1996), but cannot integrate back the fundamental mechanism of feedback of architecture growth on growth processes and vice versa required for measuring plant development, like in above example if any branch grows and harvests more light then it will contribute more to photosynthesis and will grow faster.

Of late, FSPM for VP modeling has integrated the plant architecture with physiology functions to simulate and visualize plant growth on a computer screen and this paradigm is gaining more attention. This is due to its ability to serve as a versatile tool for predicting the effect of change in plant characteristics as per variation in the growth environment conditions. FSPM explicitly captures the communication and information exchange between architecture and function and addresses the feedback between them at the individual organ level (local) as well as at the whole plant (global) level structure, which is further determined by environment, and plant's adaption to the domain is as per structure. FSPM serves as a toolbox for plant scientists, where though plant organs are the main objects for simulation, one can include different and interacting plants in one simulation. The final structure of each model is defined by study objectives (Vos et al., 2009).

From the above discussion on plant models, it is evident that there is a need to collect data about the architectural and functional development of the study plant or crop, measured over time across its entire growth cycle. In the case of VP, the output is the graphical representation of the plant structure. So, modeling architecture is the critical step in FSPM. The plant's architecture cannot be generalized for different species as it is qualitative for each plant. Still, the functional mechanism, e.g., photosynthesis or C-assimilation, which produces quantitative output, can be generalized with a variable relationship with each crop or plant type. Hence, it is essential to study the modeling of the plant architecture in detail so that the VP developed will resemble the natural plant. A case study on architectural modeling is presented in the sect 4 with a detailed explanation to facilitate an understanding of the plant's architectural design..

4 Case Study on Static Architectural Model

The growth modeling and simulation of arabidopsis with quantitative modeling (Mundermann et al., 2005) of the relationship between the plant parameters like leaf angle, leaf length, stem size, plastic horn, etc. is illustrated in this section.

Building an architectural or structural plant model is essential in developing a VP. Plant growth and development is a dynamic process where the geometry and topology of the plant change over time. To understand it quantitatively, a few key traits are measured at a decided frequency during the entire growth period to specify the plant's growth pattern. The inputs taken by model are used as guidelines to proceed with data collection, and the final choice of parameters is validated by the final outcome of the model appearance (Mundermann et al., 2005). Majorly the manual measurement method opts for data collection of the required parameters; sometimes, it is invasive too. Other approaches to non-invasive data collection include using hand-held 3D digitizers that capture the 3D structure of the plant and produce a dense 3D point cloud (e.g., shown in Fig. 4) which can be used to generate a mesh and create the plant structure. Also, using plants imaged in 2D (images of plants) and applying image processing for 2D measurements from these images to get the key traits is another non-invasive approach to dimension-related 2D measurements.

An empirical model of development and growth of the arabidopsis (*Landsberg erecta*) plant developed in (Mundermann et al., 2005) represents the growth from seedling (the initial stage of 1 mm approx.) to maturity (final growth) and evolution of individual organs considering its size, shape, and position as they develop. For data collection, non-destructive measurement on destructively sampled plants is used. Few measurements of organs during growth stages are taken and subsequently interpolation is used to model the growth of plant in continuous time. The main focus of this model is the choice of parameters, count, nature, and their growth scale to capture the real-time growth structures.

Design of Arabidopsis Plant Shoot The architectural design of the arabidopsis plant shoot is shown in Fig. 5. The basic design unit is a metamer (m) comprising the auxiliary meristem, subtending leaf (if any), and supporting internode. Repeated use of metamer with scale applied over time is used to simulate the arabidopsis shoot. m0 and m1 are two cotyledons. The leaf angle is the angle subtended by the leaf axis with the stem, and the branching angle is the angle subtended by the first internode of the branch with the main stem. The scale factor to model the continuous growth of the module in time is determined in terms of the length or width of organs in the module. Data measurements from five sample plants are taken, and an average of these is taken to construct this architectural model.

Stem: An internode length is used as a scale parameter for the branch. Internodes m0 to m8, comparatively grow lesser; hence, m9 to m15, where a significant increase

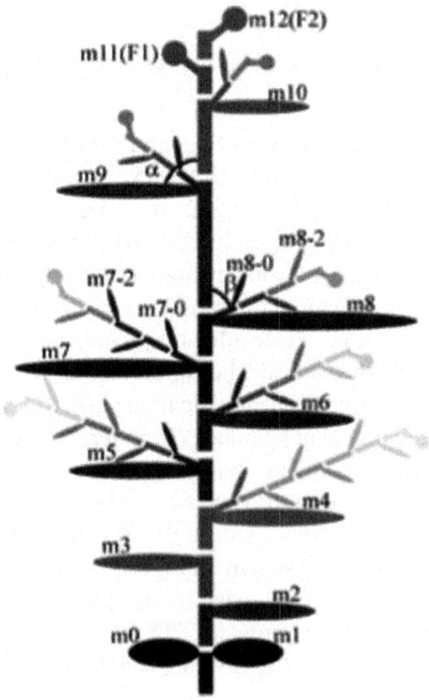

Fig. 5 Schematic diagram of arabidopsis shoot, α is leaf angle, and β is branching angle, m_i and $m_{i,j}$ are metamere numbers. *Source* Mundermann et al., (2005)

in length is seen, are measured regularly for the data collection on an hourly basis from sowing time. Then this data is fitted using logarithmic and exponential functions, but the best fit with $r^2 > 0.9$ was found using the sigmoidal Boltzmann function (computed using Origin software). This function has the form:

$$y(t) = \frac{-A}{\left(1 + e^{k(t-t_{m)}}\right)} + A$$

where A is the final (maximal) size, t_m is the time when the organ attains half of its maximal size (corresponding to the inflection point in linear plots), and k controls the relative elemental growth rate, also known as the specific growth rate (Richards & Kavanagh, 1943). The internode data from five plants was fitted with Boltzmann functions, allowing averages for each of the parameters (A, t_m, and k), with use of $k = 0.06$ +/- 0.007 h^{-1}, the doubling up time is 11.3 +/- 1.2 h (doubling time = ln $2/k$). Another relationship used was the length-to-width ratio of internodes. Internodes being cylindrical, the data points were fitted along the straight line with mean slope of the plots as 0.2 +/- 0.08, which is constant, indicating that all internodes from m9 onward grew in length about five times faster than its width (Mundermann et al., 2005).

Leaves: The total number of observed leaves (leaf number) is 11, with m0 and m1 being two cotyledons, m2 to m8 are seven leaves forming arabidopsis rosette, and m9 and m10 being two cauline leaves. Leaf width was the most appropriate scaling factor for leaf growth as it was easily measured from images of plants photographed via a digital camera. Leaf width measurements were taken at the point of maximum width, starting from the leaf size of 1 mm to its final size. Like internode length data, leaf width data could be fitted with the Boltzmann function. During the exponential phase, the growth rate in leaf width was similar for m2 to m10 ($k = 0.03$ +/- 0.007 h^{-1}, corresponding to a doubling time of about 23 h). The exact process and functions of the stem are followed for lateral branches, and the results were similar.

Plastochorn and phyllotactic angle: Time duration of scaling factors for leaf and flower growth was used to estimate the time interval (plastochron) between the development of two successive metamers. For leaves bearing metameres (m1–m10), the time at which leaf width attained a particular value ($\log_{10}$ of leaf width (in mm) = 0.1) was calculated according to the relevant growth function. The difference in this time for successive metamers gave an estimate of the plastochron. A phyllotactic angle refers to the angle formed by straight leaves or flowers. These angles were estimated from digital images and averaged across five plants for the first 14 metamers. The cotyledons and first pair of leaves (metamers m1–m5) appear in a roughly decussate arrangement, which gradually changes to spiral phyllotaxy as the divergence angle approaches 138.2°, close to the golden tip of 137.5°. Then angles are inversely correlated with plastochron; the short plastochorn shows a large divergence angle.

Flowers: The first five flowers of the main branch (m11–m15) from five plants were photographed and measured for the flowers on the main stem. During the early stages (before flower opening), bud width was chosen as a convenient scaling factor because it increased exponentially with time ($k = 0.004$ +/- 0.0002 h^{-1}). Bud width did not change significantly after flowering and thus ceased to be an excellent scaling factor. For later stages of flower development, pedicel length was used as a scaling factor. Changes in pedicel length could be measured just before flower opening when the pedicels are a few millimeters long. The Boltzmann function was found to fit pedicel length growth from 4 mm onwards. In the case of pedicel length, the switch from a low growth rate (k = approx. 0.005 h^{-1}) to a high growth rate (k = approximately 0.06 h^{-1}) occurred at about 20 h before flower opening (Mundermann et al., 2005).

Summarily, it can be stated that the model described represents several aspects of arabidopsis morphology, growth, and development in an integrated manner. Mundermann's model captures variation in leaf shapes in space (along the stem and lateral branches) and overtime at the organ level. This model built in (Mundermann et al., 2005) simulates and realistically visualizes the plant's development as shown in Fig. 6 (central axis and first-order branches), with individual organs described from early stages (approximately 1 mm in size) to maturity (Mundermann et al., 2005).

Fig. 6 Comparison of sample actual arabidopsis plants (**a**, **c**, **e**) with the model-generated plant (**b**, **d**, **f**) at different hfs. **a**, **b** at 264 hfs; **c**, **d** at 417 hfs; **e**, **f** at 491 hfs. *Source* Mundermann et al., (2005)

The model considered a large amount of experimental data, including measurements of the sizes and shapes of individual organs (internodes, leaves, and flower organs) at regular time intervals. The problem of interpolating experimental data arose due to constructing a model operating in continuous time. This interpolation was accomplished in the case of scalar measurements, such as lengths or widths, by fitting growth curves to the data. Parameters involved in module development may represent many features inherent in a module. These could be features such as the length of an internode, the size of a bud, the magnitude of a branching angle, or external factors such as time, temperature, temperature sum, day length, or light sum. Gradual changes in parameter values may be specified using differential or algebraic functions of time called growth functions (Huxley, 1932). In empirical models, growth functions are found by fitting mathematical functions to data using statistical methods.

After getting the plant organ's growth relationship, the rules were formulated using the L-System, and a final model was developed using C++ with L-System rules embedded in it. Parameterization of the model was possible by varying the leaf number, internode count, and leaf width if required. Figure 6a–e illustrates a comparison of the original and model-generated developmental stages of the arabidopsis plant, and it can be seen that the model fairly captures the architecture of a growing arabidopsis plant. Some discrepancies were seen, as this was an average plant model constructed from average measurements of five sample plants. Further, a few finer details on venation were not seen in the leaves; the stem nutation was also missing in the model.

The case study can be summarized as follows. It is a pure static structural crop model used to monitor the growth of the arabidopsis plant. The concept of phytomers ranking is used to differentiate between phytomere and leaves numbers. Destructive sampling-based measurements of field data started when organs were of approx. 1 mm in length and it continued till the organs reached maturity. For visualization, interpolation is used to generate sufficient number of data points. Parameters representing architectural traits of the growth module, such as internode length, leaf size, and branching angles, often increase according to sigmoidal functions, which means that they initially increase in value slowly, then accelerate, and eventually level off near or at the maximum.

5 Applications of Virtual Plants

VPs can serve as a tool for the visualization of 3D features of plants concerning the analysis of agronomic characteristics like biomass and energy changes in an ecosystem. In the past, it was challenging to understand the 3D distribution of sunlight in the canopy just by simulation or experimentation. With VP modeling, it is possible to construct a 3D architecture of a plant to simulate and visualize the light interaction with the canopy and thus allow systematic assessment of light interception by each leaf of the study plant. An added benefit of VP for the crop scientist is that they can see and analyze the complex morphological characteristics like leaf area, canopy development, and change in red to far-red ratio (R:FR) (Evers et al., 2006). An application of R:FR ratio change due to scattering of the sun radiation (Evers et al., 2005) due to one crop and re-interception of this scattered light along with interception of direct sunlight by the other crops with different heights is explored and analyzed using intercropping of short and taller plants like pea and wheat respectively (Barillot et al., 2010).

It is possible to parameterize the VP to get the sequence of different growth stages. With a different number of leaves given as an input, it is possible to get the VP representation of a single strand of plant with a different number of leaves. Also, by applying rotation of the stem (for plants like maize having alternate phyllotaxy), the front, side, or specific angle view of the plant can be obtained. These views of VP are saved as images and utilized as a synthetic dataset in image-based plant phenotyping applications. Leaf count in the early growth stage indicates the vigor of the crop. An attempt to implement the automated leaf counting for maize and sorghum using a natural plant image dataset augmented with VP generated synthetic image dataset for border cases using deep learning is attempted in (Miao et al., 2021).

Precision farming, remote sensing of plant growth, landscape design, and educational training or demonstration to farmers on planting and planning strategies are just some of the applications for these models (Guo & Li, 2001).

VP models have demonstrated potential in simulating the growth process of the above-ground portion of a plant. They are being used in planning designs of the parks, gardens, and open space plantations in residential areas with analysis of various simulations to test different planting patterns.

VP as a tool predicts the future growth status of modeled plants for upcoming years and hence helps in getting an idea about the probable density and hence, allows to work on evaluating and optimizing the density to decide on various planting strategies in advance. In agronomy, growers adjust the surrounding environmental conditions to get the highest possible yield. Such VP model can be utilized as a handy tool, and simulations of crop growth under different cropping scenarios, can be performed to evaluate the best possible outcome. This is a well-proven and widely used precision farming strategy.

The growth and development of the crop in 3D concerning time are simulated and visualized. Precision farming tasks like efficient irrigation, fertigation, and pest control can be performed and tested on the computer screen before the actual field trial. Virtual farm modeling and 360° rotation for monitoring of fields, controlled nutrients, and pest control etc., can be collectively used to predict the crop performance considering various traits. Such VP model helps in 'blended learning' with this immersive technology and as an innovative new media tool educating and training the crop researchers and farmers.

In the remote sensing, plant architecture research is essential for improving image interpretation by exploring the effects of architecture and leaf orientation on reflectance (Guo & Li, 2001). With use of geospatial technologies like geographical information systems (GIS) and global positioning systems (GPS), the tested approach of VP can be executed in the field smoothly.

Two problems in virtual crop applications need to be further studied and improved: the interaction between virtual crops and the environment and the virtual root system of yields.

6 Challenges in Agronomic Applications of Virtual Plants

1. The Virtual plant model assumes Beer-Lambert law-based (Swinehart, 1962) light interception. Since this law takes a random spatial distribution of leaves that is not enough and introduces significant errors for super high-yield breed variety, improving the crop architecture to enhance light interception is essential. A proper understanding of light interception is necessary for plant breeders as they simulate the spatial radiation interception using the VP model and then search for optimal crop architecture. If competition between the plants for space and resources is to be analyzed, then VPs are the best alternative to actual field trials of the intercropping system. The best planting strategy can be proposed with these virtual experimentations using VPs. This knowledge can improve the planting and intercrop according to the selected plant's structural characteristics considering individual competitive ability.
2. VP experimentation is used to find the optimal controlled climate in greenhouses for profitable production.
3. Each plant has different morphology, appearance, and growth functions with varying intensities for each species. Generalization of the VP model for other plant species is not possible. There are different development platforms like L-studio (Karwowski and Lane, 2004; Radoslaw & Przemyslaw, 2004) which uses L + C (C++) (Karwowski & Prusinkiewicz, 2003) language, GroIMP (Kniemeyer et al., 2007), which issues a Java-based language called XL, and OpenAlea (Pradal et al., 2008) which is Python based. Each one of the aforementioned examples has developed ample models using a module-based approach. There is an imminent need to promote the incorporation and interoperability of heterogeneous models and data structures from various scientific disciplines. This will make it easier and faster for new computational methods to spread as they become accessible from multiple research groups. This type of software environment is aimed not only at programmers and computer scientists but also biologists and other scientists, as they can assemble models while minimizing the programming effort.
4. Simulation of root architecture: Research done till now in virtual plant modeling has focused more on building virtual plant models for above ground part of the plant. But in reality, for survival, growth, plants uptake water and other nutrients through their root. For efficiently collecting these resources, the root must have enough surface area and total length also, the spatial distribution of root segments

must be large enough to compete with the neighboring root systems. Since roots are hidden in the soil, no other method than excavating roots from the ground is used for field observation. It has some challenges such as, primarily, it is labor-consuming; secondly, there is root loss during excavation; and the last one is the plasticity of roots which may change the original architecture during or after some time of excavation. Few root models are also seen in the literature. Few 3D models for origins are developed. A spatially explicit model that simulates the dynamic interaction between root growth and soil water status is developed by (Clausnitzer & Hopmans, 1994). This model combines the 3D root growth model with a transient soil water flow model. None of these models simulated the morphological features of root segments. The radial features and surface areas significantly affect water and nutrient uptake processes. For a more accurate relationship between the spatial deployment of the root system with the spatial deployment of root functions, (Lynch et al., 1997) developed the SimRoot model. But soil heterogeneity and competition between neighboring root systems are not included in this model.

Currently, there is a critical need to develop a fully functional virtual plant model with the integration of root and shoot and the focused research is going on. It is essential as the activity of the individual plant is a function of the whole plant (root and shoot). It constantly adjusts the shoot–root relation to adapt to the constantly varying environment and incurring stress.

7 Preliminary Results

One of the significant applications of VP modeling is experimenting with the different traits of the plant to come up with an ideal combination (ideotype). The objective behind designing such ideotypes is to find an optimal architecture with ideal features like plant height and leaf angles which contribute to the formulation of a plant canopy that can harvest the maximum from available sun radiation and hence can enhance the photosynthesis production (carbon accumulation) and finally lead to a higher yield.

Plant architecture with wider leaf angles at the bottom or in the middle of the plant canopy and shallow leaf angles in the top plant canopy helps penetration of sunlight deep into the dense canopy and reach the mature leaves, which generally contribute majorly to photosynthesis by utilizing the harvested sun radiation and maximizing the photosynthesis. Manually altering the leaf angles of the plant is challenging and requires many chemicals to be sprayed to adjust and maintain the desired leaf inclination angle. If this angle is not performing as per expectation, then there is a need to repeat the entire laborious process to set up the new angle. This limits the number of combinations of leaf angles that can be tried in real-world experiments. Using in silico, the crop researchers can set up a virtual light source of desired intensity to simulate the available solar radiation at a given point of time and use the VP model of the selected plant, parameterize leaf angles, and based on it can compute the total light harvest by the virtual plant. With the VP model, thousands of such combinations can be easily validated in short time using computer simulations. Finally, one combination of height and leaf angle can be selected. For example, a tomato architecture model was discussed by (Zhang et al., 2021). To breed for a new cultivar, for example, the one with a medium desired leaf angle in the top canopy, crop breeders can perform crosses with varieties with broader and shallower leaf angles at the top canopy. The new cultivar will harvest enough light and allow penetrating more sunlight to mature leaves in the middle and bottom top and helps to enhance photosynthesis.

GroIMP, a Java-based interactive modeling platform for VP modeling, has two main implementations for light calculation; the first one is CPU-based, and the second is GPU-based (GPUFlux). These implementations can simulate the full spectrum of visible light using reversed path tracer algorithms with Monte Carlo integration (Huwe and Hemmerling, 2008; Henke & Buck-Sorlin, 2017). To do light simulations, the following three aspects have to be defined: (1) The light source, (2) The optical properties of the objects, and (3) The light model. To determine the light sources for a typical field simulation, a diffuse sky, and a direct light source to simulate the sun are used. In contrast, in greenhouse simulations, all kinds of additional light sources, like LEDs, are common. The intensity, duration, and light quality (spectrum/color) for each light source must be defined. The optical properties of the scene objects and plants are taken using real-world measurements and used to parameterize the shader for each geometric object. The light model that performs the simulations needs to "know" how many light rays will be simulated. During the light simulation, the emission of millions of individual rays from a light source is simulated. When rays hit an object in the virtual scene, they are either absorbed, reflected, or transmitted according to the object's optical properties. The actual number of light rays that needed to be simulated mainly depends on the desired accuracy and the complexity of the scene (Henke & Buck-Sorlin, 2017), where about 200 million is a good start value. Each plant organ is represented as a separate geometric primitive object in GroIMP, where the amount of light absorbed by them can be obtained using the provided light model's methods. Hence if the architecture of the plant is altered, then as per the arrangement of the leaves in the plant, the amount of the light intercepted by the new architecture of the crop canopy changes too and can be computed using these methods of light model. Besides other implementations, GroIMP, mainly uses ray tracers to calculate the amount of light reflected, absorbed, and transmitted by the objects in a 3D scene. Light available for the plant canopy is an essential parameter that the photosynthesis model uses to decide the development and growth of the plant. But for modeling photosynthesis with a high level of accuracy, not only light but also information on other parameters like temperature, CO_2 concentration, microclimate, humidity, and leaf age are equally important.

In this experiment, data from the maize plants was collected from maize testbed at Rajendranagar, Telangana, India (17.3220° N, 78.4023° E) in March 2021. The reading of plant height, leaf lengths, and leaf angles were measured with measure tape and a mobile app clinometer.

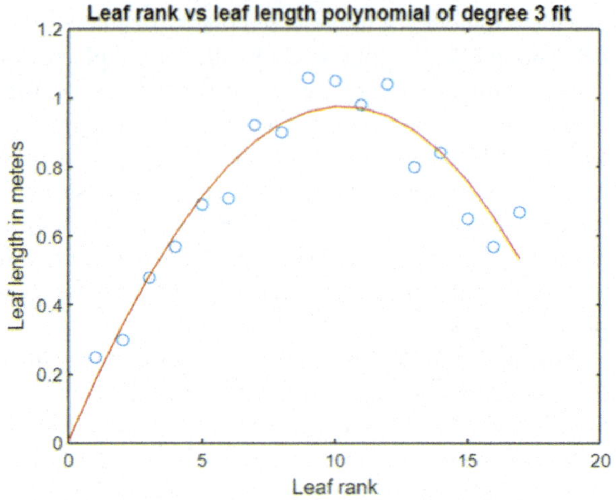

Fig. 7 Polynomial fit for data of leaf rank versus leaf length

Further, the leaf length data was analyzed using MATLAB 2020a (MathWorks, Inc.), and it was found that a polynomial of degree 3 was the best fit for this data, as shown in Fig. 7. The coefficients of this polynomial a3, a2, a1, and a0 were used to get the length of each leaf as a function of leaf rank when the maize model was being constructed. The maize plant was 2.2 m tall and with seventeen fully grown visible leaves. In this VP model, a static architectural model of a maize plant with seventeen leaves was constructed. A spherical stationary point light source with emitting 100,000 rays, placed perpendicular to the virtual maize plant, was simulated with GroIMP, as shown in Fig. 8.

Using the *getAbsorbedPower3d( )* method of the light model in GroIMP, the amount of light absorbed by each leaf (3D object in the scene) was computed, and the sum of light absorbed by all leaves was recorded.

In the test simulation scenario, the height of the plant was varied with a step size of 10 cm from 2.5 m down to 1.5 m for this static structural virtual maize model for two cases, first shown in Fig. 9a the architecture, using the actual values of top = 48, bottom = 69, and middle = 42 leaf angles, the average of the six bottom leaf angles, and six middle leaf angles. In Fig. 9b where except for the leaf angle values, all other readings mentioned above were taken from real readings, a leaf angle values top = 10, middle = 24, and bottom = 45 were used so that an architecture with a much narrower top leaf angle allowing more light penetration in the deep canopy can be formed. Both architectures recorded the amount of light intercepted by the entire plant in trials with different heights. And from both graphs, it was clear that the light intercepted by the plant with variation in its height forms a sigmoid pattern. But after attaining the height of 1.8–2.4 m, this variation turned steep, so in as per this experiment it is suggested that between this range, the plant height can be selected to get the the cultivar that intercepts maximum sunlight and enhances the yield.

This preliminary experimental result has proposed the potential of using the VP model or FSPM for getting new ideotype breeding under such experimentation.

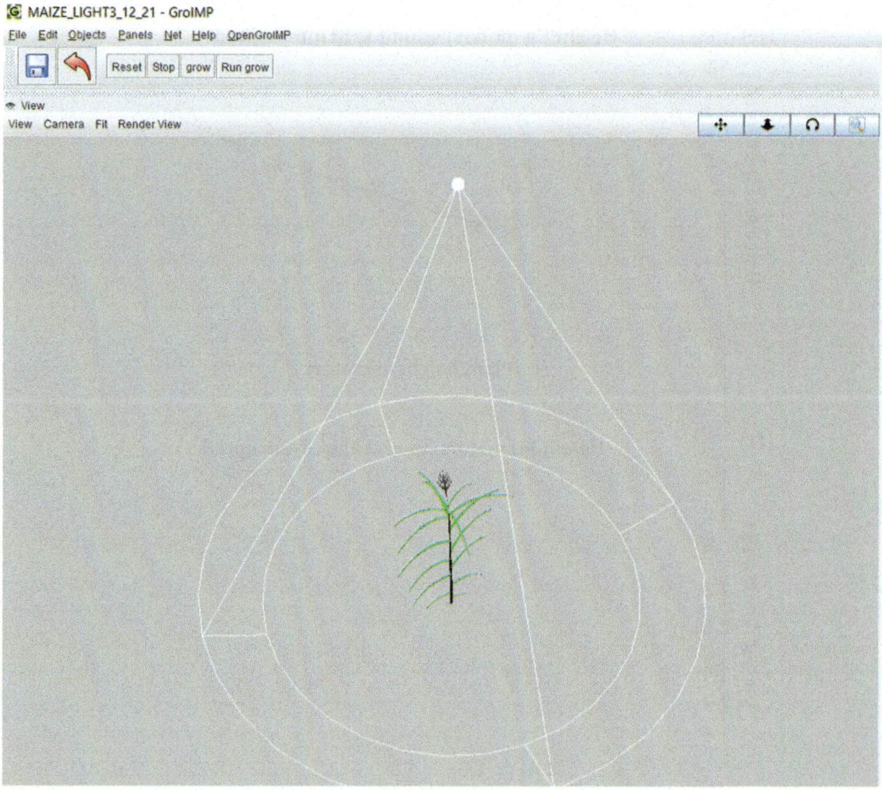

Fig. 8 Screenshot of a 3D visualization of a static architectural maize model with a single point light source within the GroIMP modeling platform

8 Summary

VPs are a versatile tool for understanding the growth of plants. VPs are computer simulations of the 3D architecture of the plant. The simulation generates the 3D architecture of the plant along with valuable information on the plant parameters that can attract the breeders.

VP modeling started with PBMs, where the main focus was on simulating the plant's physiological processes like light interception, photosynthesis, dry matter accumulation, distribution, etc. Such models target only plant processes and not plant structures. Plant structures were modeled separately without considering feedback from plant processes. Development in both these models was either deterministic or stochastic based on repetitive growth development rules written using L-Systems, a well establish formalism.

Integration of plant architecture with plant processes which is seen in functional structural plant modeling and is used for VP development plays a vital role, as

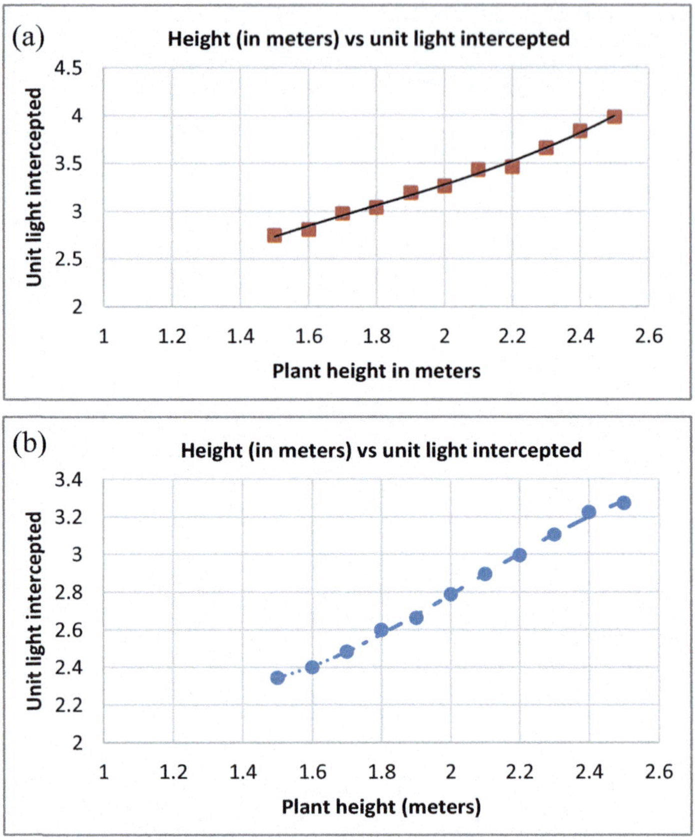

Fig. 9 Plant height vs unit light intercepted for 17 leaves, 2.2 m tall fully grown maize plant. **a** Maize model with natural leaf angle data and natural leaf length, top leaf angle = 48, middle leaf angle = 42 and bottom leaf angle = 69, and **b** maize model with dummy angle data, top leaf angle = 10, middle leaf angle = 24, and bottom leaf angle = 45 (ideal for harvesting more light)

the architecture of the plant is considered a plant structure in 3D space. It gathers resources from the environment, such as canopy light interception and competition for resources like water and nutrients among nearby plants. ADEL-maize is an L-system-based model that integrates architecture with the growth process considering individual organ to canopy (Fournier & Andrieu, 1999). On the other hand, the production and partition of biomass through photosynthesis determines the growth rates of different organs and hence the architecture.

VP modeling can be broadly classified as static and dynamic. In the static plant model, the focus is to digitize a predefined stage of the plant and construct a model which is a replica of that plant. In the dynamic plant model, time series data is collected and stored in the database, and growth rules are developed, so in this case, plant architecture changes with time. A series of static plant architectures can also be used to show the dynamic development of plant structure.

In VP model building, first, the characteristics of the plant to study are decided at first. Then, from field experimentation, collection of the timely data about the variation of these parameters (e.g., leaf angle, leaf width, internode length, etc.). Proper statistical analysis of this data and fitting it into the appropriate function to derive parametric relationships among these is required. Using this relationship, further growth rules using L-System, Relational growth grammars are formulated. Then using available simulation platforms like L-studio (Karwowski and Lane, 2004; Radoslaw & Przemyslaw, 2004), GroIMP (Kniemeyer et al., 2007), GreenLab (Kang et al., 2007), VLab (Hanan & Room, 1997), etc., the virtual plant growth is simulated.

The primary area of application of VP can be in crop breeding. Use of VP-based computer simulations offers an excellent alternative for laborious and time-consuming real-field trials for G × E × M. For instance, in case of changing leaf angles in real field trails needs spraying chemicals to maintain the desired ones. Just parameterizing the VP model with desired leaf angle as an input can be accomplished quickly. In these simulations, a breeder can try to produce a super yield and stress-resistant variety of plants manipulates the plant architecture and tests its effect on different physiological functions. For example, a test can be performed by changing the architecture of the plants in the canopy to check if it intercepts maximum solar radiation, improves the R:FR ratio to improve photosynthesis, and increase the yield. VP is the best alternative for this as using VP 3D architecture of the plant can be altered and explored to harvest maximum radiation. The optimal architecture is selected, and parameters for such variants are noted.

In conclusion, it can be stated that virtual plants are versatile tools for evaluating the impact of traits variation along with a change in environmental conditions on plant growth and development.

References

Allen, M., DeJong, T., & Prusinkiewicz, P. (2006). L-PEACH, an L-systems-based model for simulating the architecture and carbon partitioning of growing fruit trees. *Acta Horticulturae, 707*, 71–76. https://doi.org/10.17660/actahortic.2006.707.8

Artzet, S., Chen, T.-W., Chopard, J., Brichet, N., Mielewczik, M., Cohen-Boulakia, S., Cabrera-Bosquet, L., Tardieu, F., Fournier, C. and Pradal, C. (2019). Phenomenal: An automatic open source li-brary for 3D shoot architecture reconstruction and analysis for im-age-based plant phenotyping. https://doi.org/10.1101/805739

Barillot, R., Combes, D., Huynh, P. and Escobar-Gutiérrez, A.J. (2010). Analysing light sharing in cereal/legume intercropping sys-tems through Functional Structural Plant Models. In: 6th Interna-tional Workshop on Functional-Structural Plant Models,University of California, Davis.

Buck-Sorlin, G. (2013a). Functional-structural plant modeling. *Encyclopedia of Systems Biology*, 778–781. https://doi.org/10.1007/978-1-4419-9863-7_1479

Buck-Sorlin, G. (2013b). Process-based model. In *Encyclopedia of systems biology* (pp. 1755–1755). https://doi.org/10.1007/978-1-4419-9863-7_1545

Chandramouli, M., Narayanan, B. & Bertoline, G. R. (2013). A graphics design framework to visualize multi-dimensional eco-nomic datasets. *The Engineering Design Graphics Journal, 77*(3).

Chelle, M., & Andrieu, B. (1998). The nested radiosity model for the distribution of light within plant canopies. *Ecological Modelling, 111*(1), 75–91. https://doi.org/10.1016/s0304-3800(98)00100-8

Chelle, M., Evers, J. B., Combes, D., Varlet-Grancher, C., Vos, J., & Andrieu, B. (2007). Simulation of the three-dimensional distribution of the red:far-red ratio within crop canopies. *New Phytologist, 176*(1), 223–234. https://doi.org/10.1111/j.1469-8137.2007.02161.x

Clausnitzer, V., & Hopmans, J. W. (1994). Simultaneous modeling of transient three-dimensional root growth and soil water flow. *Plant and Soil, 164*(2), 299–314. https://doi.org/10.1007/bf00010082

Dauzat, J., & Eroy, M. N. (1997). I am simulating light regime and intercrop yields in coconut-based farming systems. *European Journal of Agronomy, 7*(1–3), 63–74. https://doi.org/10.1016/s1161-0301(97)00029-4

Danzi, D., Briglia, N., Petrozza, A., Summerer, S., Povero, G., Stivaletta, A., Cellini, F., Pignone, D., De Paola, D. and Janni, M. (2019). Can High Throughput Phenotyping Help Food Security in the Mediterranean Area? *Frontiers in Plant Science, 10.* https://doi.org/10.3389/fpls.2019.00015

de Reffye, P., Barthélémy, D., Blaise, F., Fourcaud, T. and Houllier, F. (1997). A functional model of tree growth and tree architecture. *Silva Fennica, 31*(3). https://doi.org/10.14214/sf.a8529

de Reffye, P., Blaise, F., Chemouny, S., Jaffuel, S., Fourcaud, T., & Houllier, F. (1999). Calibration of a hydraulic architecture-based growth model of cotton plants. *Agronomie, 19*(3–4), 265–280. https://doi.org/10.1051/agro:19990307

de Reffye, P., Blaise, F., & Houllier, F. (1998). Modeling plant growth and architecture: recent advances and applications to agronomy and forestry. *Acta Horticulturae, 456*, 105–116. https://doi.org/10.17660/actahortic.1998.456.12

de Wit, C. T. (1982). Simulation of living systems. *In Simulation of plant growth and crop production Pudoc* (pp. 3–8).

Donald, C. M. (1968). The breeding of crop ideotypes. *Euphytica, 17*(3), 385–403. https://doi.org/10.1007/bf00056241

Ehrlich, P. R., Ehrlich, A. H., & Daily, G. C. (1993). Food security, population and environment. *Population and Development Review, 19*(1). https://doi.org/10.2307/2938383

Evers, J.B., Vos, J., Fournier, C., Andrieu, B., Chelle, M. and Stru-ik, P.C. (2005). Towards a generic architectural model of tillering in Gramineae, as exemplified by spring wheat (Triticum aestivum). *New Phytologist, 166*(3), pp.801–812. https://doi.org/10.1111/j.1469-8137.2005.01337.x

Evers, J. B., Vos, J., Andrieu, B., & Struik, P. C. (2006). Cessation of tillering in spring wheat about light interception and red: far-red ratio. *Annals of Botany, 97*(4), 649–658. https://doi.org/10.1093/aob/mcl020

Federl, P., & Prusinkiewicz, P. (1999). Virtual laboratory: an interactive software environment for computer graphics. In *Proceedings—Computer Graphics International, CGI.*

Fournier, C., & Andrieu, B. (1999). ADEL-maize: an L-system-based model for integrating growth processes from the organ to the canopy. Application to the regulation of morphogenesis by light availability. *Agronomie, 19*(3–4), 313–327. https://doi.org/10.1051/agro:19990311

Godin, C., Costes, E., & Sinoquet, H. (1999). A method for describing plant architecture which integrates topology and geometry. *Annals of Botany, 84*(3), 343–357. https://doi.org/10.1006/anbo.1999.0923

Godin, C., & Sinoquet, H. (2005). Functional-structural plant modelling. *New Phytologist, 166*(3), 705–708. https://doi.org/10.1111/j.1469-8137.2005.01445.x

Guo, Y., & Li, B. (2001). New advances in virtual plant research. *Chinese Science Bulletin, 46*(11), 888–894. https://doi.org/10.1007/bf02900459

Hanan, J.S. and Room, P.M. (1997). Practical aspects of virtual plant research. In: In: Plants to Ecosystems - Advances in Computa-tional Life Sciences. [online] CSIRO Publishing, p.Chapter 2:28-44; 25 refs; illus. Available at: http://hdl.handle.net/102.100.100/220457?index=1

Henke, M., & Buck-Sorlin, G. H. (2017). Using a full spectral raytracer for calculating light microclimate in functional-structural plant modelling. *Computing and Informatics, 36*(6), 1492–1522. Available via DIALOG. https://www.cai.sk/ojs/index.php/cai/article/view/2017_6_1492

Heuvelink, E. (1996). Dry matter partitioning in tomato: Validation of a dynamic simulation model. *Annals of Botany, 77*(1), 71–80. https://doi.org/10.1006/anbo.1996.0009

Heuvelink, E. (1999). Evaluation of a dynamic simulation model for tomato crop growth and development. *Annals of Botany, 83*(4), 413–422. https://doi.org/10.1006/anbo.1998.0832

Hitz, T., Henke, M., Graeff-Hönninger, S., & Munz, S. (2019). Three-dimensional simulation of light spectrum and intensity within an LED growth chamber. *Computers and Electronics in Agriculture, 156*, 540–548. https://doi.org/10.1016/j.compag.2018.11.043

Huwe, T. and Hemmerling, R. (2008). Stochastic path tracing on consumer graphics cards. Proceedings of the 24th Spring Confer-ence on Computer Graphics. https://doi.org/10.1145/1921264.1921287

Huxley, J. S. (1932). *Problems of relative growth* (p. 273). Johns Hopkins University Press.

Jallas, E., Martin, P., Sequeira, R., Turner, S., Cretenet, M., & Gérardeaux, E. (2000). Virtual COTONS®, the firstborn of the next generation of simulation model. In *Virtual worlds* (pp. 235–244). https://doi.org/10.1007/3-540-45016-5_22

Kang, M., Evers, J. B., Vos, J., & de Reffye, P. (2007). The derivation of sink functions of wheat organs using the GreenLab model. *Annals of Botany, 101*(8), 1099–1108. https://doi.org/10.1093/aob/mcm212

Karwowski, R. and Lane, B. (2004). L-studio 4.0 user's guide. [online] Available at: http://www.cpsc.ucalgary.ca/Research/bmv/lstudio

Karwowski, R., & Prusinkiewicz, P. (2003). Design and implementation of the L+C modeling language. *Electronic Notes in Theoretical Computer Science, 86*(2), 134–152. https://doi.org/10.1016/s1571-0661(04)80680-7

Kirby, E.J.M. (1988). Analysis of leaf, stem and ear growth in wheat from terminal spikelet stage to anthesis. *Field Crops Re-search, 18*(2–3), pp.127–140. https://doi.org/10.1016/0378-4290(88)90004-4

Kniemeyer, O., Buck-Sorlin, G. and Kurth, W. (2007). The GroIMP is a platform for the functional-structural modelling of plants. In *Functional-structural plant modelling in crop production* (pp. 43–52). https://doi.org/10.1007/1-4020-6034-3_4

Kurth, W., & Sloboda, B. (1997). Growth grammars simulate trees—An extension of L-systems incorporating local variables and sensitivity. *Silva Fennica, 31*(3). https://doi.org/10.14214/sf.a8527

Lindenmayer, A. (1968a). Mathematical models for cellular interactions in development I. Filaments with one-sided inputs. *Journal of Theoretical Biology, 18*(3), 280–299. https://doi.org/10.1016/0022-5193(68)90079-9

Lindenmayer, A. (1968b). Mathematical models for cellular interactions in development II. Simple and branching filaments with two-sided inputs. *Journal of Theoretical Biology, 18*(3), 300–315. https://doi.org/10.1016/0022-5193(68)90080-5

Lopez, G., Favreau, R.R., Smith, C. and DeJong, T.M. (2010). L-PEACH: A Computer-based Model to Understand How Peach Trees Grow. *HortTechnology, 20*(6), pp.983–990. https://doi.org/10.21273/hortsci.20.6.983

Lv, M. M., Lu, S. L., Guo, X. Y. (2015). Interactive virtual fruit tree pruning simulation. In: *Proceedings of the 2015 International Conference on Electrical, Automation and Mechanical Engineering* (pp. 78–681). Atlantis Press.

Lynch, J. P., Nielsen, K. L., Davis, R. D., & Jablokow, A. G. (1997). SimRoot: Modelling and visualization of root systems. *Plant and Soil, 188*(1), 139–151.

Marshall-Colon, A., Long, S. P., Allen, D. K., Allen, G., Beard, D. A., Benes, B., von Caemmerer, S., Christensen, A. J., Cox, D. J., Hart, J. C., Hirst, P. M., Kannan, K., Katz, D. S., Lynch, J. P., Millar, A. J., Panneerselvam, B., Price, N. D., Prusinkiewicz, P., Raila, D., & Shekar, R. G. (2017). Crops in silico: Generating virtual crops using an integrative and multi-scale modeling platform. *Frontiers in Plant Science, 8*, 786. https://doi.org/10.3389/fpls.2017.00786

Martre, P., Quilot-Turion, B., Luquet, D., Memmah, M.-M.O.-S., Chenu, K., & Debaeke, P. (2015). Model-assisted phenotyping and ideotype design. In *Crop physiology* (pp. 349–373). https://doi.org/10.1016/b978-0-12-417104-6.00014-5

McKinnon, J. M., Baker, D. N., Whisler, F. D., & Lambert, J. R. (1989). Application of the GOSSYM/COMAX system to cotton crop management. *Agricultural Systems, 31*(1), 55–65. https://doi.org/10.1016/0308-521x(89)90012-7

Mech, R., Prusinkiewicz, P. (1996). Visual models of plants interacting with their environment. In *Computer Graphics Proceedings, Annual Conference Series*. New York: ACM SIGGRAPH.

Miao, C., Guo, A., Thompson, A. M., Yang, J., Ge, Y., Schnable, J. C. (2021). Automation of leaf counting in maize and sorghum using deep learning. *The Plant Phenome Journal, 4*(1). https://doi.org/10.1002/ppj2.20022

Mündermann, L., Erasmus, Y., Lane, B., Coen, E., & Prusinkiewicz, P. (2005). Quantitative modeling of arabidopsis development. *Plant Physiology, 139*(2), 960–968. https://doi.org/10.1104/pp.105.060483

Perttunen, J., Sievänen, R., & Nikinmaa, E. (1998). LIGNUM: A model is combining the structure and the functioning of trees. *Ecological Modelling, 108*(1–3), 189–198. https://doi.org/10.1016/s0304-3800(98)00028-3

Pradal, C., Dufour-Kowalski, S., Boudon, F., Fournier, C., & Godin, C. (2008). OpenAlea: Visual programming and component-based software platform for plant modeling. *Functional Plant Biology, 35*(10), 751. https://doi.org/10.1071/fp08084

Prusinkiewicz, P. (2004). Art and science of life: designing and growing virtual plants with L-systems. *Acta Horticulturae, 630*, 15–28. https://doi.org/10.17660/actahortic.2004.630.1

Radoslaw, K., & Przemyslaw, P. (2004). The L-System-based plant-modeling environment L-Studio 4.0. In *Proceedings of the 4th International Workshop on Functional and Structural Plant Models*, Montpellier, France (pp. 403–405).

Richards, O. W., & Kavanagh, A. J. (1943). The analysis of the relative growth gradients and changing form of growing organisms: Illustrated by the tobacco leaf. *The American Naturalist, 77*(772), 385–399. https://doi.org/10.1086/281140

Room, P., Hanan, J., & Prusinkiewicz, P. (1996). Virtual plants: New perspectives for ecologists, pathologists, and agricultural scientists. *Trends in Plant Science, 1*(1), 33–38. https://doi.org/10.1016/s1360-1385(96)80021-5

Seleznyova, A. N., Saei, A., Han, L., & van Hooijdonk, B. M. (2018). From field data to modeling concepts: building a mechanistic FSPM for apple. In *2018 6th International Symposium on Plant Growth Modeling, Simulation, Visualization and Applications (PMA)*. https://doi.org/10.1109/pma.2018.8611582

Simon, L., & Steppe, K. (2019). Application of a functional-structural plant model on two different wheat varieties to enhance physiological interpretation. In *Master of Science in de bio-ingenieurswetenschappen:landbouwkunde*. https://lib.ugent.be/catalog/rug01:002791221

Smith, G. S., Curtis, J. P., & Edwards, C. M. (1992). A method for analyzing plant architecture as it relates to fruit quality using three-dimensional computer graphics. *Annals of Botany, 70*(3), 265–269. https://doi.org/10.1093/oxfordjournals.aob.a088468

Swinehart, D.F. (1962). The Beer-Lambert Law. *Journal of Chemical Education, 39*(7), p.333. https://doi.org/10.1021/ed039p333

Vos, J., Evers, J. B., Buck-Sorlin, G. H., Andrieu, B., Chelle, M., & de Visser, P. H. B. (2009). Functional–structural plant modeling: A new versatile tool in crop science. *Journal of Experimental Botany, 61*(8), 2101–2115. https://doi.org/10.1093/jxb/erp345

Yan, H.-P. (2004). A dynamic, architectural plant model simulating resource-dependent growth. *Annals of Botany, 93*(5), 591–602. https://doi.org/10.1093/aob/mch078

Zhang, Y., Henke, M., Buck-Sorlin, G. H., Li, Y., Xu, H., Liu, X., & Li, T. (2021). I am estimating canopy leaf physiology of tomato plants grown in a solar greenhouse: Evidence from simulations of light and thermal microclimate using a Functional-Structural Plant Model. *Agricultural and Forest Meteorology, 307*, 108494. https://doi.org/10.1016/j.agrformet.2021.108494

Remote Sensing for Mango and Rubber Mapping and Characterization for Carbon Stock Estimation—Case Study of Malihabad Tehsil (UP) and West Tripura District, India

S. V. Pasha, V. K. Dadhwal, and K. Saketh

Abstract The phytomass and soil carbon pools are the two largest pools that are directly influenced by anthropogenic activities and have large spatial heterogeneity. Orchards and plantation crops contribute significantly to terrestrial C-pools but have not received adequate attention. Remote sensing (RS) with vegetation discrimination and monitoring capacity is critical to describe spatial C-pool variability. The natural forest in India has been significantly disturbed by establishing large-scale commercial and horticulture crops which contributes significantly to the current forest and tree estimate of 81Mha by the Forest Survey of India (FSI). This study estimated the area under mango and rubber in two contrasting sites, i.e. Malihabad (Uttar Pradesh) and West Tripura (Tripura) districts. We used Sentinel-2 data and machine learning algorithms to discriminate target tree species. Multi-sensor-based geophysical product of biomass was analysed for aboveground biomass (AGB). Spaceborne Lidar data from the GEDI sensor was analysed to characterize the tree height. Additional characterization of tree density was carried out by counting tree canopies on high-resolution imageries. The results of phytomass and soil pools are comparable to published estimates under similar agroclimatic settings. The demonstrated approach of simultaneous high-resolution phytomass and soil mapping with geospatial techniques significantly enhances the capability to monitor and model terrestrial carbon pools in India.

Keywords Carbon pool · EO data · GEDI · Machine learning · Orchard · Plantation

S. V. Pasha · V. K. Dadhwal (✉) · K. Saketh
School of Natural Science and Engineering, National Institute of Advanced Studies (NIAS), IISC Campus, Bengaluru 560 012, India
e-mail: vkdadhwal@nias.res.in

S. V. Pasha
e-mail: drvazeedpasha@nias.res.in

K. Saketh
e-mail: saketh95@nias.res.in

© The Author(s), under exclusive license to Springer Nature Singapore Pte Ltd. 2023
S. Chaudhary et al. (eds.), *Digital Ecosystem for Innovation in Agriculture*,
Studies in Big Data 121, https://doi.org/10.1007/978-981-99-0577-5_9

1 Introduction

Current understanding of the global carbon cycle indicates that during 2011–2020, the atmospheric CO_2 increased by nearly 5.0 GtCyr-1, which represents nearly 47 per cent of fossil fuel and other anthropogenic emissions. The land CO_2 sink was 3.1 ± 0.6 GtCyr-1 during the 2011–2020 decade (29% of total CO_2 emissions). The role of the terrestrial carbon cycle is complicated due to human large uncertainty in agriculture, forestry and land use (AFOLU) net carbon emission on managed land is 1.1 ± 0.7 GtCyr-1 (Friedlingstein et al., 2022). Extensive studies are being conducted using remote sensing, flux monitoring, field data, carbon/biosphere models dynamic vegetation and earth system models and data-model fusion to enhance our understanding of the terrestrial carbon budget globally as well as at the national level (Ciais et al., 2022).

Indian carbon cycle emissions are reported in National Communication to UNFCCC (www.unfccc.int/NC7), and forest biomass and forest carbon pools are regularly reported in the Indian State of Forest Report by the Forest Survey of India (FSI) (www.fsi.nic.in/forest-report-2021). These estimates although based on remote sensing data for the forest, tree cover and field samples for forest biomass and soil carbon are aggregate numbers at district, state and national levels. Spatially, explicit and national scale estimates were made under the National Carbon Project (NCP) of ISRO. Recent spatial assessment at a high resolution of 250 m and using machine learning techniques have estimated the Indian forest aboveground pool as 3570.8 TgC (Fararoda et al., 2021) and soil organic carbon (SOC) pool to 100 cm depth as 22.72 ± 0.93 PgC (Sreenivas et al., 2016). Similarly, the forest cover change spatial analysis has been completed using multi-year remote sensing data and historical maps (Reddy et al., 2016).

The National Horticultural Board estimates that mango covered an area of 15.75 Mha in the country. The most extensive sampling of trees outside the forest is carried out by the Forest Survey of India, and in their 2021 assessment, it is reported that mango (*Mangifera indica*) has a total growing stock of 230.33 Mm^3 which is 12.94% of the national TOF growing stock, A national mango tree allometry and area statistics-based assessment by Ganeshamurthy et al. (2019) estimated total carbon pool of mango as 285 MtC.

The para rubber tree (*Hevea brasiliensis*) is one of India's top commercial plantation crops and has experienced significant area expansion in non-traditional regions. Currently, India is ranked the third-largest natural rubber (NR) producer in the world, after Malaysia and Thailand (TFDPC, 2018; www.ecostat.tripura.gov.in; www.rubberboard.org.in). The demand for natural rubber (NR) has increased in the recent past (Kou et al., 2018; Pradeep et al., 2020; Sethuraj & Jacob, 2012), and more areas have been brought under rubber in China, South and Southeast Asia leading to a significant change in the vegetation type. In the recent past, the rubber plantation is spreading to non-traditional areas such as Laos, Cambodia, northwest Vietnam, northeast Thailand and Yunnan province of China (Fox & Castella, 2013; Li & Fox, 2011). The rapid expansion of rubber monoculture is found in parts of Southwest China and

India (Chakraborty et al., 2018; Chen et al., 2016; Liu et al., 2006; Sethuraj & Jacob, 2012). In India, Tripura state is emerging as an important producer of NR, second only to Kerala.

Both mango and rubber plantations are listed as one of the major commercial crops in the Indian sub-continent, and these plantations are cultivated to enhance the tropical rural economy. The development of accurate and up-to-date maps of the spatial distribution of such plantations is necessary for agricultural monitoring and decision management (Luo et al. 2020). The role of remote sensing in horticulture has been reviewed by Usha & Singh (2013). Integrating geospatial databases such as land use land cover, Lidar canopy height and tree density would help in preparing the historical dynamics of plantations/orchards in the country. In this context, this study intends to distinguish two important plantation crops, i.e. mango and rubber. The current attempt would help in estimating plantation type and age-based dynamics and carbon stocks in the country.

The reduced forest and fuelwood shortage have led to the spread of agroforestry, especially popular, eucalyptus and fast-growing species. The role of orchards, plantations and trees in agricultural lands is increasing their area, and studying ecosystem services on their conservation, soil fertility, biomass supply and carbon sequestration is critical. In this context, Earth Observation (EO) data is critical for mapping generic categories also species-wise tree community discrimination using a wide range of advanced machine learning algorithms. The geophysical products of Vegetation Carbon Fraction (VCF) and machine learning-based forest biomass products at 1 km to 300 m are also being attempted. Case studies on mapping and change over two decades (2001–2020) use various pattern recognition indices. Validation of products with published literature in India. Use of newly available Global Ecosystem Dynamics Investigation (GEDI) tree height for discrimination as well as forest biomass estimation. The required analysis must include the type of plantation, area dynamics and age class along with EO-derived biomass/growing stock approaches. This study covers two plantations mango and rubber, their current distribution and land use land cover, carbon stocks and soil organic carbon.

Since the beginning of the EO satellite launch in 1972 (Landsat-1), many classification techniques have been used to classify pixels in satellite imagery. Classification methods range from parametric supervised such as maximum likelihood (MXL), and unsupervised algorithms, e.g. ISODATA and k-means clustering to machine learning algorithms such as artificial neural network (ANN), support vector machines (SVM) and ensembles classifiers (Kulkarni & Lowe, 2016). Random Forest algorithm has been frequently employed among machine learning classifiers in recent times. However, very few researchers used this algorithm for land cover classification. Chen et al. (1993) used dynamic learning neural networks for land use land cover (LULC) classification of multi-spectral data.

The Forest Survey of India (FSI) is mapping the national forest cover on a biennial basis at a 1:50,000 scale with four vegetation density classes and estimates the forest area. A spatial assessment of plantation type and its carbon stocks is crucial since there is no freely available spatial layer of the different plantation types. Regional level carbon cycles are significant to a region's unique ecosystems and helpful for

Table 1 Remote sensing studies in India on mango and rubber

Study area	RS data/(Spatial resolution)	Application/Objective	Analysis technique	References
Krishna District (AP)	LISS-II (36 m) LISS-III (23.5 m)	Acreage and production estimation	Maximum likelihood	Yadav et al. (2002)
Meerut District (UP)	Hyperion (20 m)	Discrimination and mapping	Spectral angle mapper	Paul et al. (2018)
Saharanpur District (UP)	Sentinel-1 (SAR)	Biophysical characterization	Time series	Sahu et al. (2020)
Malihabad Tehsil (UP)	LISS_IV (5.8 m)	Mango mapping	Object oriented	Hebbar et al. (2014), Ravishankar et al. (2022)
Malihabad Tehsil (UP)	LISS-IV (5)	Discrimination and mapping	Texture features + mapping	Nagori (2021)
Kerala	LISS-III (23.5 m)	Acreage and mapping	K-means clustering	Meti et al. (2016)

addressing regional environmental policy challenges. However, only a few studies used remote sensing to map the country's mango and rubber, and Table 1 provides information.

2 Study Area and Data Used

This study depicts two distinct landscapes: (1) Malihabad Tehsil in Uttar Pradesh and (2) West Tripura District in Tripura. The Malihabad Tehsil in Uttar Pradesh's (UP), Lucknow district, is bounded by 80.55 to 81.1 E longitude and 26.8 to 27.1 N latitude, and it is widely recognized for its mango orchards (Fig. 1). Agriculture and mango orchards are the most common land cover types. The West Tripura district in the north-eastern state of Tripura has a geographic area of 983.6 km^2 and is bounded by 91.23 to 91.55 E longitude and 23.74 to 24.1 N latitude. This study site is known for the expansion of rubber plantations (Fig. 2).

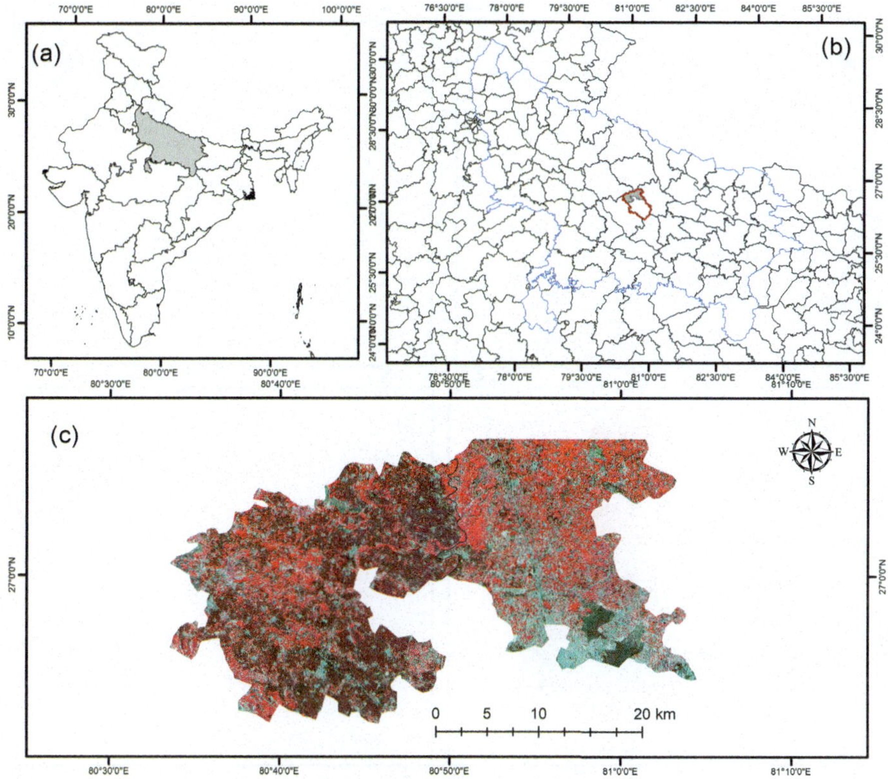

Fig. 1 Study area map of Malihabad **a** India states with Uttar Pradesh state, **b** Lucknow district and study site and **c** FCC of Sentinel-2 data of Malihabad

2.1 Remote Sensing Data Used in the Study

This study has utilized multi-source data sets which include.

(a) To compute and compare the canopy cover of mango and rubber trees, we used Sentinel-2A at 10 m resolution data (between March and April 2021–22) obtained from the Copernicus platform (www.scihub.copernicus.eu).
(b) The Normalized Difference Vegetation Index (NDVI) of the Moderate Resolution Imaging Spectroradiometer (MODIS) at 250 m resolution of the 16-day composite for six years was utilized as a predictor variable for modelling SOC in two study sites.
(c) Multi-sensor Forest/Tree biomass at 100 m resolution for the year 2018 was used for characterizing the mango and orchard from the identified pixels. This product used multi-temporal EO data as well as L and C band SAR data and machine learning algorithms applied to regional field data on tree biomass to

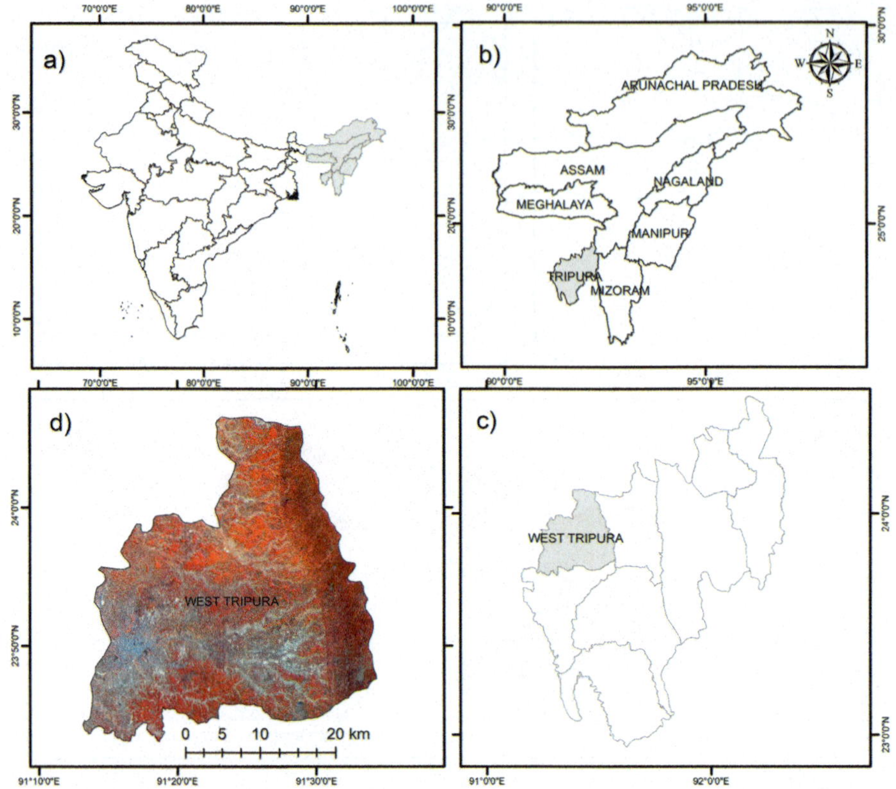

Fig. 2 Study area map showing West Tripura **a** India states with North East, **b** Tripura district and study site, **c** West Tripura district in Tripura state and **d** FCC of Sentinel-2 data of West Tripura

develop global data sets. These have also been extensively validated globally (Santoro et al., 2021).

(d) The vegetation canopy height from the Global Ecosystem Dynamics Investigation (GEDI) onboard the International Earth Station was applied to characterize the mango and rubber tree canopy heights.

(e) Google Earth Pro was used to visualize high-resolution (1/sub-metre class) information for online and on-screen mango tree counting (www.earth.google.com).

(f) Soil Health Card data was used to estimate SOC for the study sites.

3 Methodology

3.1 Estimating Area Under Mango and Rubber Plantations and LULC Using Machine Learning

In this study, the Random Forest machine learning algorithm in Q-GIS Software was used to identify plantations, orchards and other land cover. Later stage updated the LULC data of ESRI with 10 M resolution (Karra et al., 2021) using a hybrid classification approach.

The tree density for the mango and rubber plantations 1-ha sample plots were overlaid on the very high-resolution imageries (www.earth.google.com). A sum of 66 large-scale 1-ha sample plots was laid at individual plantations in the study sites, wherein 41 sites were used for calculating the tree densities in the Malihabad site and the remaining 25 plots in the West Tripura site. The Global Ecosystem Dynamics Investigation (GEDI) Lidar tree canopy height map integration for forest structure measurements and Landsat analysis-ready data time series used to discriminate between tree and crop (Potapov et al., 2021). The GEDI data is further classified into five height classes, i.e. < 2, 3–5, 6–10, 11–15 and > 16 for the two study sites. However, below 2 m height pixels were considered as non-trees/scrub and not included in the final analysis. The overall methodology adopted in the study depicted in Fig. 3.

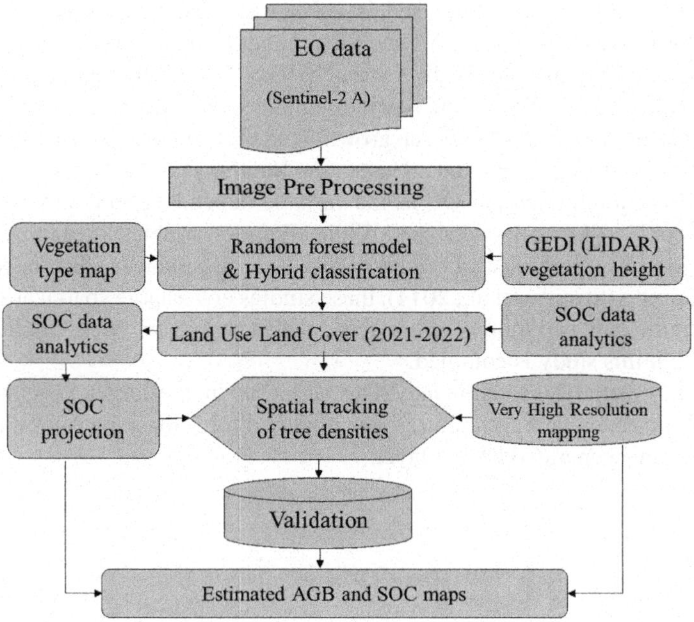

Fig. 3 Map depicted the overall methodology adopted in the study

3.2 Estimating Soil Organic Carbon (SOC)

SOC estimated using rubber, mango, tea and agricultural layer. The following steps were involved in the process: i. Soil data acquisition and quality analysis, ii. spatial data to be used as variables for predicting SOC, iii. overlay of vegetation type/land cover maps to extract SOC over targeted vegetation types and iv. conversion of the spatial layer to profile SOC density.

3.3 Data Acquisition and Analysis

Soil Health Card (SHC) data for the study area were acquired from the Department of Agriculture & Farmer Welfare SHC data portal (www.soilhealth.dac.gov.in/). The georeferenced point data set downloaded as comma-separated values (CSV) files had information on 12 different soil parameters including organic carbon (OC %) which was of interest to this study. Data quality check comprised of tests covering the OC per cent as well as the geographic location of samples. The latter test comprised checking if the coordinates were within the boundary of study districts and removal of observations with missing coordinates. Thus, while the original number of SHC observations was 20,673 and 16,618 for Malihabad and West Tripura, respectively, after the quality check, only 3069 and 9553 observations were used for further analysis.

The refined point data set along with MODIS NDVI at 250 m resolution of 16-day composite for six years was used as a predictor variable for modelling SOC in the two study sites. Six-year NDVI and the number of days with healthy vegetation cover over the six years (no. of days with NDVI > 2) were the parameters used for the prediction of SOC. The use of vegetation indices averaged over a period helps in better capturing the dynamics of crop growth with time as compared to single images with observations on a single date. Hence, the Random Forest regression algorithm was used to fit a model with SOC as the variable dependent on the two parameters derived from NDVI mentioned above. While climatic variables like temperature and precipitation have previously been recognized as important predictors of SOC in digital mapping (Srinivas et al., 2014), these studies cover large spatial area wherein there is significant variation of climatic variables within the area, unlike the sites within which this study is confined.

The final spatial layer showing SOC variation in the target vegetation types (rubber, mango, tea and agriculture) was obtained after masking out other regions using the above generated LULC layer.

3.4 Conversion of SOC Concentration to Profile SOC Density

The conversion of SOC concentration (%) to SOC density (Mg/ha) included two steps: (i) estimating the SOC density at a depth interval of 15–20 cm (from SHC data) and (ii) converting it to SOC density at a standard depth interval of 0–30 cm.

As SHC data does not include bulk density (BD), its estimate was used to convert SOC% to SOC density. Pedotransfer Functions (PTF) provide reliable estimates of unknown/difficult-to-measure soil properties from other known/easily measurable soil properties (McBratney et al., 2002). Many PTFs have been developed to estimate BD making use of different physical and chemical properties. Since the SHC observations do not include physical characterization, only OC-based PTFs were used. The estimated BD was the mean of four PTFs developed by Alexander et al. (1980) and Manrique & Jones (1991) each of which is given in Table 2.

For calculating the SOC density using OC concentration and BD, the following equation was used:

$$\mathrm{SOC} = \mathrm{OC} \times \mathrm{BD} \times d, \quad (1)$$

where

SOC Soil organic carbon in Mg/ha.
BD Bulk density in gm/cc.
d dept in cm.

In the second step, the layer SOC density was converted to SOC density at a depth interval of 0–30 cm using a computed conversion factor. This conversion factor, assuming a linear variation of SOC with depth, was derived using soil profile data from Tarun et al. (2018) which had soil data from areas within and surrounding our study sites.

Table 2 Description and source of the PTFs for estimating BD

PTF for estimating BD	References
$\mathrm{BD} = 1.66–0.308\ (\mathrm{OC}^{0.5})$	Alexander (1980)
$\mathrm{BD} = 1.72–0.294\ (\mathrm{OC}^{0.5})$	Alexander (1980)
$\mathrm{BD} = 1.510–0.113 \times \mathrm{OC}$	Manrique and Jones (1991)
$\mathrm{BD} = 1.660–0.318 \times \mathrm{OC}^{1/2}$	Manrique and Jones (1991)

4 Results and Discussions

4.1 Mapping Mango and Rubber Using Machine Learning

This research confirmed that multi-temporal, multi-source EO imagery had great potential for the identification of mango and rubber plantations in tropical areas. By integrating high-resolution LULC data (Karra et al., 2021) and open-source layers, the following conclusions can be drawn: ensemble RF and global data sets significantly improved the individual classification accuracy despite the limit of training samples.

For Malihabad site, vegetation class includes trees (mango and mixed vegetation), agriculture (crop and fallow), and category-II comprises non-vegetation classes such as (barren/fallow), built-up and water) (Fig. 4). Of the total land area, vegetation occupied 84.7% and the remaining area was under the non-vegetation category. The results reveal that agricultural land is occupied by about 49.4%. About 35.3% of the area is under tree vegetation 32.8% under mango orchards and 2.5% only under mixed vegetation. The western portion around Malihabad town represents the predominance of mango orchards. On the other hand, the eastern part of the study site is predominantly covered by agriculture. Optimal season data (peak greenness while other vegetation under leaf offstage) was used to discriminate mango orchards from other trees we used.

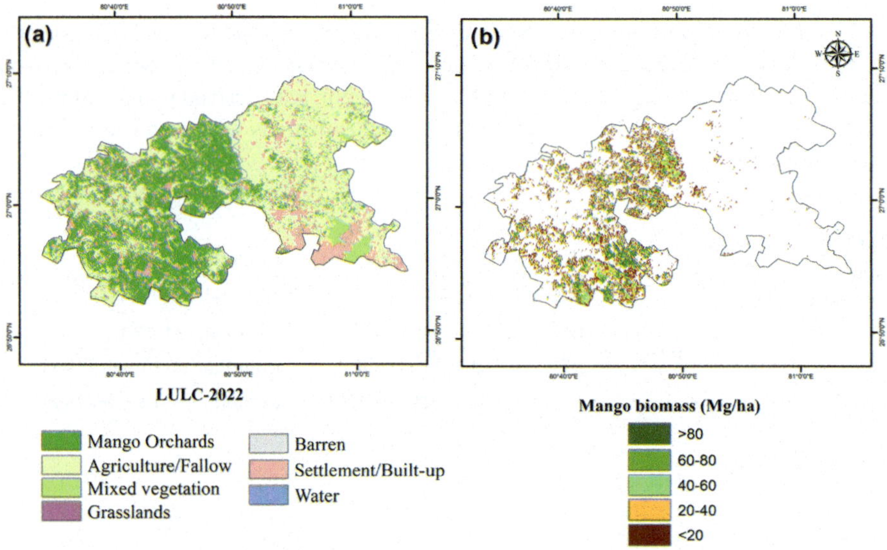

Fig. 4 Map showing the spatial distribution of **a** LULC map and **b** biomass of mango orchards in Malihabad, Lucknow and Uttar Pradesh

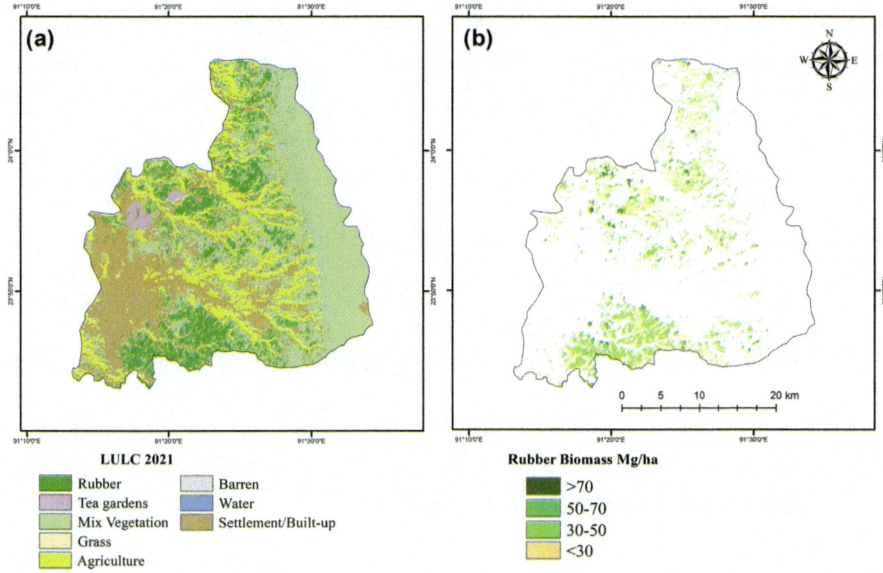

Fig. 5 Spatial distribution of **a** LULC map 2021 and **b** biomass map of rubber plantations in West Tripura district, Tripura

The West Tripura district is classified into nine LC classes, the vegetation class comprising rubber, mixed vegetation, agriculture, scrub, tea gardens and grasslands, while classes that do not have vegetation include barren, water and settlements (Fig. 5). According to calculations, mix-vegetation comprised 52% of the total area under the vegetation class, and rubber plantations comprised 15.8%.

4.2 Estimating Biomass and Carbon Densities for Mango and Rubber

In this study, a total of 16,104.6 tonnes of AGB (carbon stock at 7569.2) is determined for the Malihabad mango orchards. West Tripura biomass is estimated to be 13,409.6 tonnes. Between the two study sites, Malihabad had the highest AGB of 156 Mg/ha, with carbon stocks of 73.3 Mg/ha, whereas West Tripura had 104 Mg/ha and a carbon stock value of 48.9 Mg/ha (Table 3).

In the West Tripura site, the average biomass of rubber plantations ranges between 30 and > 70 Mg/ha. The majority of the biomass was found below 50 Mg/ha, with a few patches above 50 Mg/ha in the western portion of the study region. This variability could be attributed to planting year and site quality. According to Tripura's Rubber Board, more than 70% of the rubber was established in the recent past. The above

Table 3 Details of mango and rubber area, AGB, carbon stocks and tree density in the study sites

Vegetation class (Study site)	Plantation area in Ha	AGB Pool and (C-stock) (Mg)	Average AGB density and (C-density) (Mg/ha)	Average tree density (No of trees/ha)
Mango orchard (Malihabad)	28,184.7	16,104.6 (7569.2)	156.4 (73.5)	71.1
Rubber plantation (West Tripura)	13,098.9	13,409.6 (6302.5)	104.0 (48.9)	500.0

50 Mg/ha biomass patches imply increasing young and mature stages of rubber (Table 3).

4.3 Characterizing the Tree Density

The VHR data provided valuable information in two study sites to assess the tree densities for mango and rubber. In the case of rubber grown in West Tripura, visual counting of rubber trees was estimated by a sample of historical images at the planting/pre-planting stage e rubber is currently established. The results from counting pits provided an estimate of 500 (±20) trees. Currently, it is calculated tree density by counting the pits, whereas mango tree density in Malihabad was estimated by visual counting of individual trees in randomly placed plots mean density of 71.1 tree/ha with a standard deviation of 15.3 (Table 3). However, older plantations are difficult to calculate due to their complex canopy structure and coalescence which make it difficult to identify the tree densities. The tree density is also a critical parameter in determining tree biomass. However, in the case of mango where there are trends for establishing high tree-density plantations adopting early bearing new varieties (Singh & Nandi, 2021). As the study covered only two districts visual approach was adopted. However, a large number of remote sensing and automated analysis techniques have been established for automated tree counting (Ke & Quakenbush, 2011).

4.4 Tree Canopy Height Model

The Global Ecosystem Dynamics Investigation (GEDI) Lidar forest structure data was used to discriminate between trees, scrub and crops for the study sites (Potapov et. al., 2021). The height of the tree canopy ranged from 1 to 18 m in the Malihabad site, whereas it reached a maximum of 26 m in West Tripura. This analysis helps in discriminating various plantation and orchard types from the co-existing vegetation.

The Malihabad site represents a major mango orchard's height attributed to 6 to 10 m. About (63.8%) of the mango area is found in this category followed by 28.4%. in 3 to 5 m. On the other hand, the lowest area (7.8%) was attributed in > 11 categories. The second study site West Tripura shows the highest rubber area calculated under 11–15 m class at (74.4%) followed by 20.4% area under < 15 m. The remaining are found in the 3 m class. Height variations were estimated in both the mango and rubber plantations, and this disparity in height could be attributed to various age groups/plantation stages such as young, growing and mature.

4.5 Estimating Soil Organic Carbon (SOC)

Integrating the high-intensity field sampling data from SHC with remote sensing-derived vegetation indices helped capture the spatial variability of SOC within the major plantation types in our study area. The predicted values showed an acceptable correlation with the observed values ($R^2 = 0.54$ for Malihabad and $R^2 = 0.51$ for West Tripura).

The predicted values reveal the mean SOC value was higher for soils under tea (29.61 Mg/ha) and rubber plantations (27.01 Mg/ha) as compared to soils under mango plantations (19.22 Mg/ha)(Figs. 6 and 7). One of the explanations for the higher SOC in soils covered by rubber and tea plantations can be their respective geographic locations. Both study areas are characterized by different climatic conditions. Climate not only controls plant productivity but also affects decomposition rates, thereby affecting the quantity of organic carbon in soils (Martin et al., 2011). Hence, the lower temperatures and higher rainfall in West Tripura as compared to Malihabad can be attributed to the greater SOC accumulation in the soils under tea and rubber plantations in West Tripura. Similar trends of higher SOC stocks in Tripura and other north-eastern regions of India were found in previous national scale studies in the country (Bhattacharyya et al., 2000; Sreenivas et al., 2016). Also, SOC levels can vary with different management practices that weren't considered while estimating SOC in this study Table 4 shows the summary of various studies on AGB and SOC estimates for mango an rubber in India.

5 Conclusions

This study uses high-resolution Sentinel-2A data sets for vegetation, land cover mapping and estimated SOC. The important orchard and plantation crops were mapped, namely mango, rubber and tea gardens in two climatic zones, i.e. parts of Indo-Gangetic Plains and North East India. The RF algorithm provided acceptable results. The utility of remote sensing in further characterization of tree canopy with three biophysical parameters, tree density, tree height and aboveground biomass

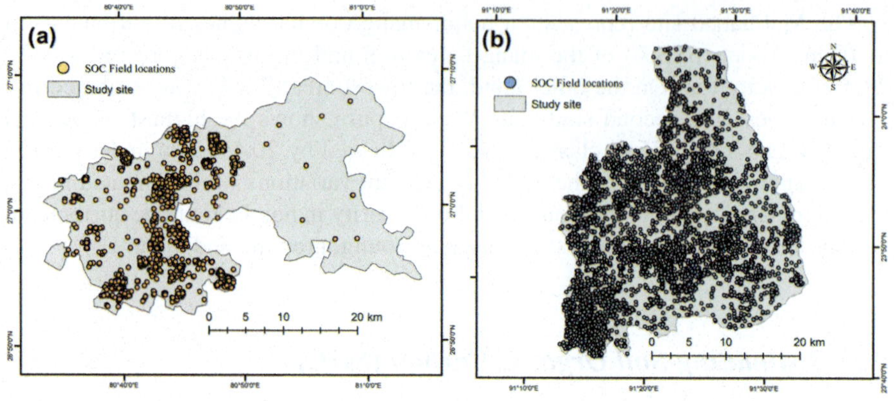

Fig. 6 Spatial distribution of the soil organic carbon field points on **a** Malihabad ($n = 3069$) and **b** West Tripura ($n = 9553$)

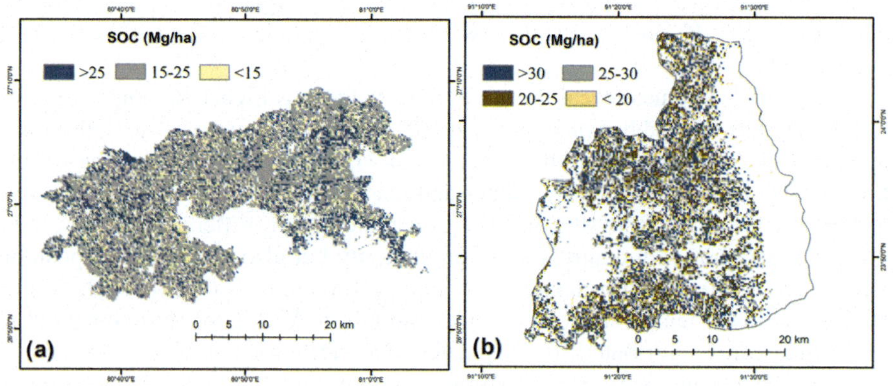

Fig. 7 Spatial distribution of predicted SOC (Mg/ha) for **a** Malihabad and **b** West Tripura

Table 4 Summary of AGB and SOC estimates for mango and rubber in India

Study area	AGB in tha^{-1}@	SOC in tha^{-1}$	Remarks	References
Mango studies				
Bengaluru	(21.1–24.8)	69.8–75.7(0–100)	Cultivation and weed management	Rupa et al. (2022)
Andhra Pradesh		61.3–66.5 (0–100)		Rupa et al. (2022)
Rubber studies				
Karimganj (Assam)	247–290 (D) 123–145(C)		35–40-year old	Brahma et al. (2018)

@ For Above Ground Biomass (C); Carbon tha-1; (D): Dry matter tha-1
$ For SOC (A): 0-100 cm; (B): 0–30 cm

was also carried out. The results are compatible with published field data over these study areas.

Soil organic carbon spatial variability could be captured by Soil Health Card data integrated with remote sensing derived vegetation canopy indices for developing a high-resolution SOC density map with help of a Random Forest machine learning algorithm. Space-based tree height data from Lidar onboard International Space Station (ISS), GEDI, could capture the broad differences among crops and trees as well as age-class differences. These results have an important bearing on the source of lands diverted to mango, rubber and tea gardens standing biomass assessment with clearly identified impacts on biodiversity and carbon pool and sequestration. New types of EO data, microwave and Lidar can provide additional cues for discrimination and characterizing rubber and mango for carbon pool estimation. This approach has great potential in supplementing the current field sample-based carbon assessment adopted by the Forest Survey of India.

Acknowledgements The authors gratefully acknowledge the encouragement from Prof. Shailesh Nayak, Director National Institute of Advanced Studies (NIAS). This research has been carried out under the Indian Terrestrial Carbon Cycle Assessment and Modelling (ITCAM) Project which has been financially supported by grants from Indira Gandhi Memorial Trust (IGMT), New Delhi to NIAS.

Declaration of Competing Interest The authors declare that we have no known competing financial interests or personal relationships that could appear to have influenced the work reported in this article.

Data Availability Statement The data sets generated during and analysed during the current study are available from the corresponding author upon reasonable request. The remote sensing data sets used in the study are publicly available from https://earthexplorer.usgs.gov/ and https://scihub.copernicus.eu/dhus/#/home.

References

Alexander, E. B. (1980). Bulk density of California soils in relation to other soil properties. *Soil Science Society of America Journal, 44*, 689–692. https://doi.org/10.2136/sssaj1980.03615995004400040005x

Bhattacharyya, T., Pal, D. K., Mandal, C., & Velayutham, M. (2000). Organic carbon stock in Indian soils and their geographical distribution. *Current Science*, 655–660.

Brahma, B., Nath, A. J., Sileshi, G. W., & Das, A. K. (2018). Estimating biomass stocks and potential loss of biomass carbon through clear-felling of rubber plantations. *Biomass and Bioenergy, 115*, 88–96. https://doi.org/10.1016/j.biombioe.2018.04.019

Chakraborty, K., Sudhakar, S., Sarma, K. K., Raju, P. L. N, & Das, A. K. (2018). Recognizing the rapid expansion of rubber plantation—A threat to native forest in parts of Northeast India. *Current Science, 114*, 207. https://doi.org/10.18520/cs/v114/i01/207-213.

Chen, K. S., Tzeng, Y. C., Chen, C. F., Kao, W. L., & Ni, C. L. (1993). Classification of multispectral imagery using dynamic learning neural network. In *Proceedings of IGARSS'93-IEEE international geoscience and remote sensing symposiuminternational* (pp. 896–898).

Chen, H., Yi, Z. F., Schmidt-Vogt, D., Ahrends, A., Beckschäfer, P., Kleinn, C., Ranjitkar, S., & Xu, J. (2016). Pushing the limits: The pattern and dynamics of rubber monoculture expansion in Xishuangbanna, SW China. *PloS One, 11*(2), e0150062.

Ciais, P., Bastos, A., Chevallier, F., Lauerwald, R., Poulter, B., Canadell, P., Hugelius, G., Jackson, R. B., Jain, A., Jones, M., & Zheng, B. (2022). Definitions and methods to estimate regional land carbon fluxes for the second phase of the Regional carbon cycle assessment and processes project (RECCAP-2). *Geoscientific Model Development, 15*(3), 1289–1316.

Dey, S. K. (2005). A preliminary estimation of carbon stock sequestrated through rubber (Hevea brasiliensis) Plantation in North Eastern Region of India. *Indian Forester, 131*, 2.

Fararoda, R., Reddy, R. S., Rajashekar, G., Chand, T. R. K., Jha, C. S., & Dadhwal, V. K. (2021). Improving forest above ground biomass estimates over Indian forests using multi source data sets with machine learning algorithm. *Ecology Information, 65*, 101392. https://doi.org/10.1016/j.ecoinf.2021.101392

Fox, J., Castella, J. C., & Ziegler, A. D. (2014). Swidden, rubber and carbon: Can REDD+ work for people and the environment in Montane Mainland Southeast Asia? *Global Environmental Change, 29*, 318–326.

Friedlingstein, P., Jones, M. W., O'Sullivan, M., Andrew, R. M., Bakker, D. C., Hauck, J., Le Quéré, C., Peters, G.P., Peters, W., Pongratz, J. & Zeng, J. (2022). Global carbon budget 2021. *Earth System Science Data, 14*(4), 1917–2005.

Ganeshamurthy, A. N., Ravindra, V., & Rupa, T. R. (2019). Carbon sequestration potential of mango orchards in India. *Current Science, 117*(12), 2006–2013.

Hebbar, R., Ravishankar, H. M., Trivedi, S., Subramoniam, S. K., Raj, U., & Dadhwal, V. K. (2014). Object oriented classification of high resolution data for inventory of horticultural crops. In *The international archives of the photogrammetry, remote sensing and spatial information sciences, presented at the ISPRS technical commission VIII Symposium*. ISPRS.

Karra, K. et al. (2021). Global land use/land cover with Sentinel-2 and deep learning. In *IGARSS 2021–2021 International Geoscience and Remote Sensing Symposium.*

Ke, Y., & Quackenbush, L. J. (2011). A review of methods for automatic individual tree-crown detection and delineation from passive remote sensing. *International Journal of Remote Sensing, 32*(17), 4725–4747.

Kou, W., Dong, J., Xiao, X., Hernandez, A. J., Qin, Y., Zhang, G., & Doughty, R. (2018). Expansion dynamics of deciduous rubber plantations in Xishuangbanna, China during 2000–2010. *Giscience Remote Sens, 55*(6), 905–925.

Kulkarni, A. D., & Lowe, B. (2016). *Random forest algorithm for land cover classification.* Computer Science Faculty Publications and Presentations. Paper 1 http://hdl.handle.net/10950/341.

Li, Z., & Fox, J. M. (2011). Rubber tree distribution mapping in Northeast Thailand. *Journal of Geochemical Exploration, 2*(04), 573.

Liu, W., Hu, H., Ma, Y., & Li, H. (2006). Environmental and socioeconomic impacts of increasing rubber plantations in menglun township, Southwest China. *Mountain Research and Development, 26*(3), 245–253.

Luo, H. X., Dai, S. P., Li, M. F., Liu, E. P., Zheng, Q., Hu, Y. Y., & Yi, X. P. (2020). Comparison of machine learning algorithms for mapping mango plantations based on Gaofen-1 imagery. *Journal of Integrative Agriculture, 19*(11), 2815–2828.

Manrique, L. A., & Jones, C. A. (1991). Bulk density of soils in relation to soil physical and chemical properties. *Soil Science Society of America Journal, 55*, 476–481. https://doi.org/10.2136/sssaj1991.03615995005500020030x

Martin, M. P., Wattenbach, M., Smith, P., Meersmans, J., Jolivet, C., & Boulonne, L. (2011). Arrouays D spatial distribution of soil organic carbon stocks in France. *Biogeoscience, 8*, 1053–1065. https://doi.org/10.5194/bg-8-1053-2011

McBratney, A. B., Minasny, B., Cattle, S. R., & Vervoort, R. W. (2002). From pedotransfer functions to soil inference systems, Geoderma, 109(1–2). *ISSN, 41–73*, 0016–7061. https://doi.org/10.1016/S0016-7061(02)00139-8

Meti, S., Pradeep, B., Jacob, J., Shebin, S. M., & Jessy, M. D. (2016). Application of remote sensing and GIS for estimating area under natural rubber cultivation in India. *Rubber Science, 29*, 7–19.

Nagori, R. (2021). Discrimination of mango orchards in Malihabad, India using textural features. *Geocarto International, 36*, 1060–1074. https://doi.org/10.1080/10106049.2019.1637467

Paul, N. C., Sahoo, P. M., Ahmad, T., Sahoo, R. N., Krishna, G., & Lal, S. B. (2018). Acreage estimation of mango orchards using hyperspectral satellite data. *Indian Journal of Horticulture, 75*, 27. https://doi.org/10.5958/0974-0112.2018.00005.1

Potapov, P., Li, X., Hernandez-Serna, A., Tyukavina, A., Hansen, M. C., Kommareddy, A., Pickens, A., Turubanova, S., Tang, H., Silva, C. E, & Hofton, M. (2021). Mapping global forest canopy height through integration of GEDI and Landsat data. *Remote Sensing of Environment, 253*, 112165.

Pradeep, B., Jacob, J., & Annamalainathan, K. (2020). Current status and future prospects of mapping rubber plantations in India. *Rubber Science, 33*, 127–139.

Ravishankar, H. M., Trivedi, S., Subramoniam, S. R., Ahamed, J. M., Nagashree, T. R., Manjula, V. B., Hebbar, R., Jha, C. S., & Dadhwal, V. K. (2022). Geospatial applications in inventory of horticulture plantations. In C. S. Jha, A. Pandey, V. M. Chowdary, & V. Singh (Eds.), *Geospatial technologies for resources planning and management, water science and technology library* (pp. 263–296). Springer International Publishing. https://doi.org/10.1007/978-3-030-98981-1_12.

Reddy, C. S., Jha, C. S., Dadhwal, V. K., Hari Krishna, P., Vazeed Pasha, S., Satish, K. V., Dutta, K., Saranya, K.R.L., Rakesh, F., Rajashekar, G, & Diwakar, P. G. (2016). Quantification and monitoring of deforestation in India over eight decades (1930–2013). *Biodiversity and Conservation, 25*(1), 93–116.

Rupa, T. R., Ganeshamurthy, A. N., Ravindra, V., Laxman, R. H., Rajeshwari, R., & Aruna, B. (2022). Carbon sequestration in mango orchards in seasonally dry tropical savanna climate under different management. *Communications in Soil Science and Plant Analysis, 53*(7), 862–871.

Sahu, H., Haldar, D., Danodia, A., & Kumar, S. (2020). Time series potential assessment for biophysical characterization of orchards and crops in a mixed scenario with Sentinel-1A SAR data. *Geocarto International, 35*(14), 1627–1639.

Santoro, M., Cartus, O., Carvalhais, N., Rozendaal, D., Avitabile, V., Araza, A., De Bruin, S., Herold, M., Quegan, S., Rodríguez-Veiga, P., & Willcock, S. (2021). The global forest above-ground biomass pool for 2010 estimated from high-resolution satellite observations. *Earth System Science Data, 13*(8), 3927–3950.

Sethuraj, M. R., & Jacob, J. (2012). Thrust areas of future research in natural rubber cultivation. *Natural Rubber Research, 25*(2), 123–138.

Singh, S. P., & Nandi, A. K. (2021). Investigate the socio-economic status of growers and determinants of mango yield in Lucknow district of Uttar Pradesh. *Journal of Crop and Weed, 17*, 86–92. https://doi.org/10.22271/09746315.2021.v17.i1.1409.

Sreenivas, K., Sujatha, G., Sudhir, K., et al. (2014). Spatial assessment of soil organic carbon density through random forests based imputation. *Journal Indian Social Remote Sensing, 42*, 577–587. https://doi.org/10.1007/s12524-013-0332-x

Sreenivas, K., Dadhwal, V. K., Kumar, S., Harsha, G. S., Mitran, T., Sujatha, G., Suresh, G. J. R., Fyzee, M. A., & Ravisankar, T. (2016). Digital mapping of soil organic and inorganic carbon status in India. *Geoderma, 269*, 160–173. https://doi.org/10.1016/j.geoderma.2016.02.002

Tarun, A., Vinod, K., Pandey, G. (2018). Soil physico-chemical and biological properties vis-à-vis Yield Gap Analysis in Mango cv. Langra Orchards in Lucknow. *Journal of Agricultural Science, 18*(2), 246–252, ISSN 0973-032X.

TFDPC. (2018). Tripura forest development and plantations corporation Ltd. Plan for Responsible Rubberwood and Bamboo Plantations Management 2013–14 to 2017–18.

Usha, K., & Singh, B. (2013). Potential applications of remote sensing in horticulture—A review. *Scientia horticulture, 153*, 71–83.

www.ecostat.tripura.gov.in/Tripura-At-a-Glance-2021.pdf. Accessed September 06, 2022.

www.earth.google.com. Accessed: September 04, 2022.

www.fsi.nic.in/forest-report-2021.
www.rubberboard.org.in/rbfilereader?fileid=526. Accessed September 24, 2022.
www.scihub.copernicus.eu/dhus/#/home. Accessed September 24, 2022.
www.soilhealth.dac.gov.in. Accessed January 10, 2022.
www.unfccc.int/NC7. Accessed September 10, 2022.
Yadav, I. S., Srinivasa Rao, N. K., Reddy, B. M. C., Rawal, R. D., Srinivasan, V. R., Sujatha, N. T., Bhattacharya, C., Nageswara Rao, P. P., Ramesh, K. S., & Elango, S. (2002). Acreage and production estimation of mango orchards using Indian Remote Sensing (IRS) satellite data. *Scientia Horticulturae, 93*, 105–123. https://doi.org/10.1016/S0304-4238(01)00321-1

Impact of Vegetation Indices on Wheat Yield Prediction Using Spatio-Temporal Modeling

Pragnesh Patel, Maitrik Shah, Mehul S. Raval, Sanjay Chaudhary, and Hasit Parmar

Abstract Precise yield prediction is necessary for any Government to design and implement agriculture-related policy. Usually, remotely sensed images are used for prediction, and it is a complex task with dependence on many parameters like weather, soil, and farm practices. The fusion of extra information can improve the prediction. Therefore, the chapter studies the impact of vegetation indices on wheat yield prediction using satellite images. The chapter uses convolutional neural network (CNN) to extract the spatial features, which are then fed into the long short-term memory (LSTM) to derive the temporal information. They are subsequently fed into a fully connected network (FCN) to predict the yield. The chapter demonstrates that adding information about vegetation indices improves yield prediction.

Keywords Convolutional neural network (CNN) · Long short-term memory (LSTM) · Moderate resolution imaging spectroradiometer (MODIS) · Prediction · Remote sensing · Satellites · Wheat · Yield

1 Introduction

Ensuring food security for the rising population is one of the UN's sustainable development goals. Due to the increase in population, rapid urbanization, and decreased agricultural land, climate changes are challenging factors for the food available in the coming years. By the year 2050, India will be the most populated country in the world, and providing food will be challenging. More than 50% of India's population is associated with agriculture to derive their livelihood (India, 2011). The majority of

P. Patel (✉) · M. Shah · M. S. Raval · S. Chaudhary
School of Engineering and Applied Science, Ahmedabad University, Ahmedabad, India
e-mail: pragnesh.p@ahduni.edu.in

M. Shah
e-mail: maitrik.s@ahduni.edu.in

H. Parmar
L. D. College of Engineering, Ahmedabad, India

© The Author(s), under exclusive license to Springer Nature Singapore Pte Ltd. 2023
S. Chaudhary et al. (eds.), *Digital Ecosystem for Innovation in Agriculture*,
Studies in Big Data 121, https://doi.org/10.1007/978-981-99-0577-5_10

the farmers do farming with their traditional knowledge and experience. Agriculture crop production depends on many factors, such as climate changes, drought, heavy rainfall, and farm practices. Due to rapid migration to cities and industrialization, the earth's temperature has increased, resulting in adverse climate changes that highly affect crop production. For assuring food security, precision agriculture is one of the solutions (Hakkim et al., 2016). According to (Sharma et al., 2020), we can classify agriculture problems into four clusters: preproduction, production, processing, and distribution. The issues where clustering is functional are as follows: crop classification, crop yield prediction, best practices to improve farm efficiency, supply chain management, and crop protection.

In this chapter, we have selected the crop yield prediction problem as its accurate and reliable prediction is vital for the farmers and the government. Crop yield prediction is a complex process affected by external factors such as rainfall, thunderstorm, soil, and weather. Traditional crop yield prediction uses manual surveys during the growing season, which requires substantial human hours and is difficult for a country like India. With the advancement in statistics, yield prediction is possible using statistical models (Bussay et al., 2015). Still, they require large amounts of soil data, weather data, and data related to farm practices at the finer level.

More than 30% of the population consumes wheat as the primary food and main supplement diet for protein and calories (Malik & Singh, 2010). Among the world's cereal production and consumption, wheat accounts for 26% and 44%, respectively (McGuire et al., 2015). India is the world's second-largest wheat producer (Food and Agricultural Organization (FAO), 2022). Therefore, reliable and accurate wheat yield prediction is essential. In this chapter, we use deep learning to study the impact of vegetation indices derived from satellite images on crop yield prediction. Unlike classical machine learning (ML) algorithms, CNN automatically extracts features, and it is one of the robust algorithms to solve tasks like detection, segmentation, and classification (Li et al., 2019). The extracted spatial features from CNN are fed into LSTM to learn the temporal variations across the crop growing season, which in turn trains FCN, which predict the wheat yield.

2 Related Work

In the past, Fisher has explored and identified the influence of weather on yield prediction (Thomas et al., 1997). Monitoring crop yield growth is critical in deciding agricultural practices and forecasting yield prediction. Recent advancements in spatial information technologies enable the assessment of spatial variability in crop yield. The vegetation property of the crop captures this spatial variability. Therefore, analysis and detection of changes in vegetation patterns give insight into crop growth captured by vegetation indices. They represent a combination of surface reflections of two or more wavelengths designed for the vegetation property. Intense red band energy absorption and strong near-infrared (NIR) band energy reflection indicate healthy plants (Thomas et al., 1997). The normalized difference vegetation index

(NDVI) is one of the most popular and widely used indices derived from multispectral data as a normalized ratio between the red and near-infrared bands (Arnon et al., 2010).

The following are a few approaches for yield prediction: field survey, crop growth model, statistical techniques-based approach, machine learning, and deep learning-based approach.

2.1 *Field Survey-Based Approach*

It is the easiest and oldest approach for crop yield prediction. In this approach, a manual field survey is performed, which is time-consuming and need the expertise to identify crop at the field level. In addition, this approach needs to scale better for a large country like India. Field surveys can validate the prediction results in collaboration with other systems.

2.2 *Crop Growth Model-Based Approach*

Crop models have a long history, during which their focus and application have altered in response to societal needs (Jones et al., 2016). Initial crop growth models include data from various economic and biological aspects of the crop. In the early days, (Fisher, 1925) the practical importance of weather on crop growth and (Baier, 1977) has formulated a model for crop yield prediction based on weather data. Keig introduced the soil water balancing (WATBAL) model for the crop yield model to link soil with plant growth (Keig et al., 1969). Keulen introduced the soil nitrogen (N) model for crop growth based on the water and nitrogen condition (Keulen et al., 1981). It was considered a base model for developing subsequent models. Wilkerson developed a CERES model based on the soil water, soil nitrogen and observed crop growth and predicted yield (Wilkerson et al., 1983). US Agency for International Development (USAID) funded and supported the International Benchmark Sites Network for Agrotechnology Transfer Project (IBSNAT) to develop a crop growth model. As a subsequent result, Decision Support System for Agrotechnology Transfer (DSSAT) was released for the crop simulation model in 1989.

DSSAT is widely used by agro-economists and provides the interface to work with the latest software, such as GIS. The majority of the crop growth models need crop data, soil data, weather data, environmental conditions, and management practices and predict crop growth with the help of one or more data types based on the structure of the model. These models need a significant amount of data at the finer level, which requires enormous work for data collection. Some examples of such models are Agricultural Production Systems Simulator (APSIM), Aquacrop, DSSAT, EPIC, WTGROWS, and WOFOST (Rauff & Bello, 2015).

2.3 Statistical Techniques-Based Approach

In this approach, models are designed and analyzed using various statistical techniques. For example, Hendricks and Scholl proposed a model for expressing the weather changes effect on yield as a function of correlation coefficients between yield and weather variables (Hendricks & Scholl, 1943). Crop growth models perform better crop yield prediction when information about climate, genotype, soil properties, and farm practices is available as it captures the physiological patterns of the experimented fields. However, its performance degrades when applied to the unseen area with the untested environmental condition as we need to validate with the farm data (Roberts et al., 2017). The statistical model can override this limitation of the crop growth models to a certain extent. One of the most experimented statistical techniques is regression, where one or more parameters are mapped through linear modeling (Birthal et al., 2014; Dkhar et al., 2017). The correlation between the various attributes and the dependent variable is identified in the regression technique, which plays an essential role in predicting output attributes. The linear model captures only linear relationships, and that too for a subset. However, some of these relationships are nonlinear and unable to grasp complex biological interactions among different variables affecting crop growth. Traditional linear methods such as autoregressive integrated moving average (ARIMA) are adequate for forecasting the same time series data. Still, it could not perform better during extreme changes in the time series data due to its assumption of normality (Petrică et al., 2016); hence, models capable of mapping nonlinearity patterns of the data are needed.

2.4 Machine Learning and Deep Learning-Based Approach

With the technology innovation, ML tools and techniques coupled with computing power give the ability to analyze big data (Basso and Liu 2019; Chlingaryan et al., 2018). Moreover, ML involves a helpful data-driven modeling approach to learning patterns and relationships from input data, and it offers promising results for improving yield forecasts (Willcock et al., 2018).

Researchers have applied ML algorithms along with soil data, climate data, farm practices, genome data, and predicted yield. Aghighi applied various advanced regression algorithms to predict the maize yield (Aghighi et al., 2018). The authors selected a 28,000 hectares area of maize field situated at Moghan Agro-Industrial and Animal Husbandry Company (MAIAHC) in Iran and collected data for 2013–15. The authors also collected vegetation indices and used support vector regression, Gaussian process regression, random forest regression model, and boosted regression tree. They found that boosted regression tree algorithm gives the best crop yield prediction compared to others, with an RMSE range from 8.5 to 11.10 and an average R-value higher than 0.87. Authors in (Gümüşçü et al., 2020) applied KNN, SVM, and decision tree to predict the planting date for the wheat crop in Turkey. Shahhosseini built metamodels based on random forest and multiple linear regression to predict maize yield and nitrogen loss from the US output data of the APSIM crop model (Shahhosseini et al., 2019). Jeong compared random forest and multiple linear regression and found random forest gives better results for yield predictions of wheat, maize, and potato (Jeong et al., 2016).

With the advancement in space technology, remote sensing data is helpful in different applications, such as weather prediction (Shin et al., 2020) and land-use-land-cover mapping (Nguyen et al., 2020). In addition, advances in modeling techniques and the availability of satellite images in the public domain led to the solution of many problems. In particular, multispectral images have information at different wavelengths, making them more feasible for agriculture. In the last few years, remote sensing-based approaches have been widely popular (Basso et al., 2013; Prasada et al., 2006) for applications such as weather prediction, large-scale crop maps (Belgiu & Csillik, 2018), and poverty estimation (Xie et al., 2016).

Many researchers worked on crop yield prediction using remote sensing images using machine learning and deep learning algorithms. Aghighi applied an advanced regression algorithm on the normalized difference vegetation index (NDVI) which derived from Landsat 8 and crop yield to predict silage maize crop yield (Aghighi et al., 2018). Kamir applied ML models on the MODIS data, rainfall and temperature data, observed yield map, and identified yield gap hotspot in wheat production (Kamir & Hochman, 2020). Cai mapped enhanced vegetation indices (EVI) derived from MODIS with climate data to predict wheat yield in Australia using machine learning approaches (Cai, 2019). X. Zhu has built to predict maize yield from the Agrometeorological Index and vegetation indices derived from the remote sensing images for the Jilin and Liaoning Provinces of China (Zhu et al., 2021). J. Han derived NDVI and EVI from MOD13Q1 products and integrated them with climate and soil data to predict winter wheat yield for North China (Han, 2020). The MODIS data helps monitor the growth of the crop and yield prediction (Domenikiotis et al., 2004; Dong, 2016). Therefore, we use the MODIS surface reflectance multispectral satellite images and crop yield as the input to train the model. We evaluate and validate our approach using state-level wheat data. The experimental results show that including vegetation indices increase crop yield prediction accuracy by 17% compared to prediction based on other parameters. From the literature study, we have observed that weather data—temperature, rainfall, humidity, soil data, and pH value—is a few critical parameters for crop growth. Most of the researches predict the yield based on the vegetation indices derived from the satellite images with a combination of climate and soil data. But, it is not easy to have reliable and accurate soil data at the district level. While using the integrated data, we need help finding the impact and importance of individual parameters. This chapter uses the actual surface reflectance values, temperature data, land cover data, and vegetation indices to predict the yield.

2.5 Vegetation Indices

Crop yield growth depends on many parameters and is also observed using vegetation indices. The vegetation index represents the spectral transformation of two or more bands that highlights vegetation properties and canopy structural variation of the plant (Huete et al., 2002). More than 50 vegetation indexes are available. Vegetation indices represent surface reflectance at two or more wavelengths and highlight the property of vegetation. In this chapter, we have taken NDVI and EVI to observe crop growth. The reflection value of the leaf varies across the band, and this principle is used for calculating vegetation indices. For example, in the blue and red bands, the reflected value is meager due to the high absorption of the photosynthetically active segment, while in the near-infrared (NIR) band, it is high due to scattering. Therefore, the difference between the red and NIR band is used as a measure of vegetation. As the plant grows, reflectance values will change, observed during the crop growing season with the help of temporal images.

2.6 Normalized Difference Vegetation Index (NDVI)

NDVI is one of the most widely used vegetation indices (Kriegler et al., 1969; Rouse et al., 1974). It is the normalized value of the difference between the reflectance of a given image's near-infrared (NIR) and red(R) band. It is expressed as follows:

$$\mathrm{NDVI} = \frac{(\mathrm{NIR} - \mathrm{RED})}{(\mathrm{NIR} + \mathrm{RED})} \tag{1}$$

Here, NIR and RED represent the reflectance value in the image's NIR and visible RED spectrum band, respectively.

NDVI values range from 1 to + 1, where -1 is mostly water bodies and + 1 is generally dense green–leafy vegetation. One of the advantages of NDVI is that it minimizes certain types of band-corrected noise (Didan et al., 2018). A drawback of NDVI is that it tends to amplify atmospheric noise in the red and NIR band and is very sensitive to background variation. In addition, NDVI is saturated at high biomass content, making it difficult to differentiate between moderately plant cover and very high plant cover.

2.7 Enhanced Vegetation Index

EVI (Huete et al., 1994; Jiang et al., 2008; Rocha & Shaver, 2009) is an improved and optimized vegetation index designed to overcome the problem of some atmospheric conditions and background noise, and it is more sensitive in highly dense vegetation areas. In addition to the red and near-infrared bands, EVI also incorporates the reflectance of the blue band. The equation for EVI is as follows:

$$\mathrm{EVI} = G.\frac{(\mathrm{NIR} - \mathrm{RED})}{(\mathrm{NIR} + C_1\mathrm{RED} - C_2\mathrm{BLUE} + L)} \tag{2}$$

where NIR, red, and blue are the total or partially atmospheric-corrected surface reflectance from NIR, red, and blue band, respectively, L is the canopy background adjustment, C_1 and C_2 are the coefficients, and G is the scaling factor. The drawback of the NDVI is covered with the use of EVI.

NDVI is an indicator of greenness that strongly correlates with green biomass; hence, crop state and its growth are quickly observed compared to other vegetation indices. However, NDVI suffers from the drawback of atmospheric effects and soil characteristics. On the other hand, EVI corrects atmospheric and soil conditions effects and is more responsive to canopy structural variation than another vegetation index. Hence, we selected it for our study.

3 Study Area and Dataset Description

The area of study is the state of Gujarat, one of India's central wheat-producing states. Gujarat is located at Latitude 23.00 N and Longitude 72.00 E. It has a significant coastal boundary and diverse weather patterns across different districts, making it more suitable for finding the impact compared to the Punjab, Haryana, or Rajasthan, which are top wheat producer states in India. In Gujarat, wheat is produced during the rabi season from September to March. We have collected wheat yield data for the years 2011–2020. The average yearly wheat production in Gujarat in the last ten years is shown in Table 1. The data is publicly available from the Crop Production Statistics Information System of the Department of Agriculture Cooperation, Government of India (Yield Statistics Data, 2022). In addition, the proposed work uses the following publicly available datasets from the MODIS sensor (Lp daac, 2022).

MOD09A1 (Vermote, 2015): It estimates the surface reflectance of Terra MODIS bands 1 through 7 at a spatial resolution of 500 m corrected for atmospheric conditions.

MYD11A2 (Wan et al., 2015): It provides land surface temperature and emissivity data at a spatial resolution of 1 km with eight days of temporal resolution.

MCD12Q1 (Friedl & Sulla-Menashe, 2015): It provides land cover types at a 500 m spatial resolution with a 12-month temporal resolution. This product identifies 17 different classes, including 11 natural vegetation classes, three human-altered classes, and three non-vegetated classes identified by IGBP (International Geosphere-Biosphere Program). We have used agriculture and urban built-up areas from these

Table 1 Average wheat area and yield of last ten years in Gujarat, India

Year	Average area (million hectares)	Production (million tones)	Average Yield (kgs/ha)
2010–11	1.58	5.01	75.95
2011–12	1.35	4.07	78.26
2012–13	1.02	2.94	73.05
2013–14	1.44	4.60	76.81
2014–15	1.17	3.29	90.17
2015–16	0.85	2.31	85.63
2016–17	0.99	2.75	87.74
2017–18	1.05	3.10	96.56
2018–19	0.79	2.40	96.35
2019–20	1.39	4.55	100.59

classes and incorporated them as synthetic binary bands, which we have used in training.

MOD13A1 (Didan et al., 2018): It provides values from the two primary layers such as NDVI and EVI at a 500 m spatial resolution with a 16-day temporal resolution.

4 Proposed Approach

We built a model based on deep CNN (Lecun et al., 1998) and long short-term memory (LSTM) in our study. Here, we have given a brief overview of these two models, followed by a description of the proposed architecture. The proposed CNN-LSTM-FCN architecture is shown in Fig. 1. It contains two parts—one of CNN and the second of LSTM operation. Detailed design architecture of both functions is shown in Figs. 2 and 3, respectively.

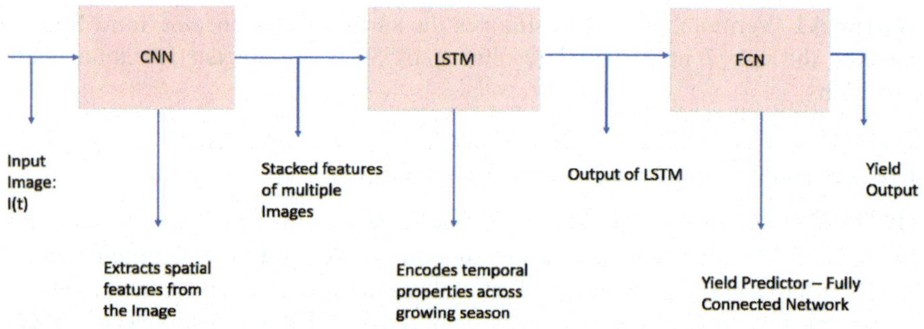

Fig. 1 Block diagram of the proposed approach

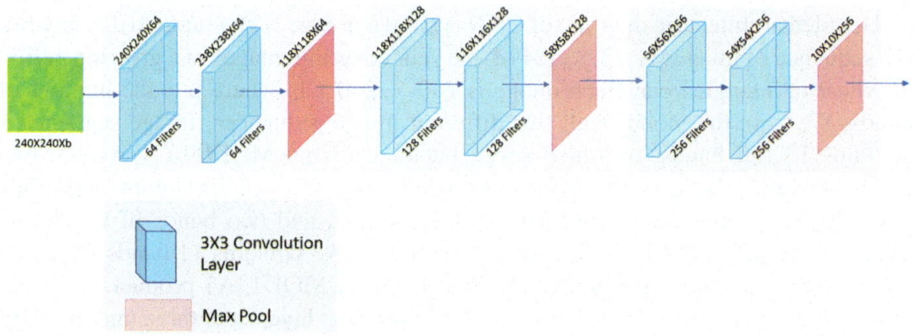

Fig. 2 CNN architecture used in the proposed study

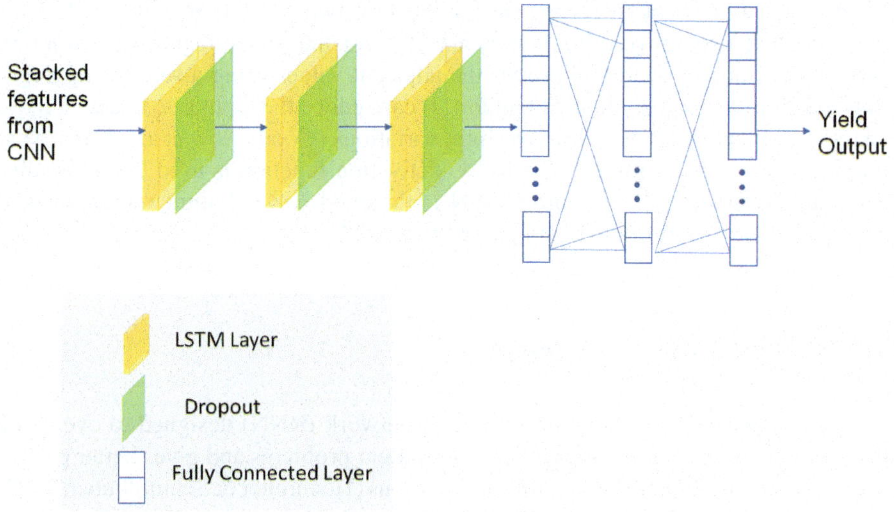

Fig. 3 LSTM block architecture

4.1 *Convolutional Neural Network (CNN)*

CNN is a widely used deep learning model that demonstrated excellent performance for image classification and segmentation. CNN can process the data available in different matrix formats, such as one-dimensional data like signals, two-dimensional data like images, and three-dimensional data like video. Moreover, it can map and learn hierarchical patterns of the data using nonlinear mapping (Lecun et al., 1998). The typical CNN model consists of several convolution and pooling operations followed by a small number of fully connected layers. During the convolution operation, a filter with the shared parameters is used and reduces the number of parameters.

Detailed architecture diagram of CNN is shown in Fig. 2. The input to the network is a sequence of images I1, I2, … I24 of one year covering images of a growing period of wheat in a single year. Each image is 240 × 240 x b, where b is the number of bands. We train the model with two different image sequences. Image sequence 1 contains 13 full bands, of which seven bands are from MOD09A1 product, two bands of MYD11A2 product, two binary bands comprising of agriculture and urban built-up classes generated from MCD12Q1 product, and two bands of vegetation indices from MOD13A1 product. Image sequence two contains 11 bands which do not include two bands of vegetation indices from the MOD13A1 product.

As shown in Fig. 2, the CNN part has 6 convolution layers and three max-pooling layers performed after every two convolution operations. CNN also has design parameters such as the number of filters, filter size, type of padding, and activation function. In our proposed model, we have used filters of size 3 × 3 in each convolution operation. The numbers of filters are 64 for the first two convolution layers, 128 for the following two convolution layers, and 256 for the remaining two convolution layers. The pooling operation groups the adjacent values using a selected aggregation function within a predefined window. It can reduce the input size to detect more coarse-level features to handle even slight variations (Goodfellow et al., 2016). Max pooling is used in our model. The ReLU activation function is used to add nonlinearity to the model. The output of CNN is flattened into a feature vector, stacked according to the timestamp T, and passed to LSTM.

4.2 Long Short-Term Memory

LSTM is a special kind of recurrent neural network (RNN) designed to overcome the problems of vanishing and exploding gradient problems and gives better performance in various sequence modeling applications (Hochreiter & Schmidhuber, 1997; Sherstinsky, 2020). Figure 3 shows the LSTM part of our proposed models. Input to the LSTM model is a flattened array. The LSTM model extracts temporal features across the wheat growing season. Our model consists of three LSTM layers, and a dropout layer follows each LSTM layer. The model consists of 13 trainable layers with 22 nodes with a ReLU activation function. Finally, the output of LSTM models is given to the fully connected neural network layer.

The FCN consists of 3 fully connected layers. The output of the yield predictor $\hat{y}$ is a wheat yield in kilograms per hectare for the input image sequence for the given year. We have used L2 loss functions, which find the loss between the predicted output and the actual output.

5 Results and Discussion

The images taken from the MODIS sensor are divided into patches of size 240 × 240. The model is trained on 4600 images and tested on 2500 images. The size of each image is (240, 240, 13) for image sequence one and (240, 240, 11) for image sequence two. The model is trained for 20 epochs with ReLU activation and a quadratic loss function. The model weights are optimized with Adam optimizer with batches of 100 images at a time. The model is implemented in TensorFlow and trained using a GPU equipped with × 86 based Intel processor, which uses NVIDIA Pascal architecture (P5000/6000).

The study uses root mean square error (RMSE) as a distance metric to measure the difference between the predicted outcome of the model yield prediction against ground truth values. As shown in Table 2, we trained models with two image sequences to identify the impact of vegetation indices. We found that the inclusion of vegetation indices reduced RMSE and gave better performance. Another input image sequence without VI shows higher RMSE when trained on the same model.

For learning temporal features and validating the usefulness of the proposed model, we also implemented two other variations of the models: the CNN–RNN model and the CNN–GRU model. In the CNN–RNN model, the architecture details of the model remain the same as the CNN–LSTM model, except simple RNN is used instead of LSTM. Recurrent neural networks (Mikolov et al., 2010) are a particular type of neural network designed for learning patterns from sequential data. RNN is suitable for modeling sequential data as it can recall an encoded representation of its past. In the CNN–GRU models, instead of LSTM, GRU is used and tested with the two different input image sequences. Another variation of RNN is gated recurrent unit (GRU), whose primary goal is to eliminate the information that is no longer meaningful and to govern how much information is carried over to the subsequent steps for remembering long-term information (Kyunghyun et al., 2014). The result for various deep learning architectures with two image sequences with and without vegetation indices is shown in Table 2. The RMSE of the proposed model is also higher when we train and test them without using vegetation indices. Table 2 shows that including vegetation indices, surface reflectance, and temperature bands increase the model's performance. Another observation is that the error rate is higher when it does not use VI and is approximately near the CNN–GRU-based model when used with VI indices. We observe RNN and GRU-based models capture patterns from the time series data, but their performance is lower than the LSTM model due to the simple structure of RNN and GRU compared to the LSTM. With vegetation indices,

Table 2 RMSE with and without vegetation indices

	RMSE (without VI bands)	RMSE (with VI bands)
CNN–LSTM	0.1015	0.0840
CNN–RNN	0.1136	0.0932
CNN–GRU	0.1124	0.0921

a 17% improvement has been seen using the CNN–LSTM-based model on yield prediction.

6 Conclusion

In the given chapter, we build a model to find vegetation indices' impact on crop yield prediction using satellite images. From the results, we conclude that the inclusion of vegetation indices increases the accuracy of wheat yield prediction. Furthermore, using the LSTM-based model, the vegetation indices collected from the different temporal images can capture crop growth more than the other features. We also observe that vegetation indices influence yield prediction during early crop growth. Crop yield prediction is a complex process that depends on many parameters like meteorological data, soil data, and farm practices. In future, we plan to study the impact of these parameters on crop yield prediction.

Acknowledgements We thank Dr. Srikrishnan Divakaran from Ahmedabad University for providing valuable input and guidance. In addition, the authors heartfully thank L D College of Engineering and Ahmedabad University for providing the computing facility.

References

Aghighi, H., Azadbakht, M., Ashourloo, D., Shahrabi, H., & Radiom, S. (2018). Machine learning regression techniques for the silage maize yield prediction using time-series images of landsat 8 OLI. *IEEE Journal of Selected Topics in Applied Earth Observations and Remote Sensing, 11*(12), 4563–4577. https://doi.org/10.1109/JSTARS.2018.2823361

Arnon, K., Nurit, A., Rachel, T., Martha, A., Marc, L., Garik, G., Natalya, P., & Alexander, G. (2010). Use of NDVI and land surface temperature for drought assessment: merits and limitations. *Journal of Climate, 23*(3), 618–633. https://doi.org/10.1175/2009JCLI2900.1

Baier, W. (1977). Cropweather models and their use in yield assessments. *WMO, 151*, 48.

Basso, B., & Liu, L. (2019). Seasonal crop yield forecast: Methods, applications, and accuracies. *154*, 201–255. https://doi.org/10.1016/bs.agron.2018.11.002

Basso, B., Cammarano, D., & Carfagna, E. (2013). Review of crop yield forecasting methods and early warning systems. In *Proceedings of the first meeting of the scientific advisory committee of the global strategy to improve agricultural and rural statistics*. FAO Headquarters (pp. 18–19). Rome, Italy.

Belgiu, M., & Csillik, O. (2018). Sentinel-2 cropland mapping using pixel-based and object-based time-weighted dynamic time warping analysis. *Remote Sensing of Environment, 204*, 509–523.

Birthal, P., Khan, T., Negi, D., & Agarwal, S. (2014). Impact of climate change on yields of major food crops in India: Implications for food security. *Agricultural Economics Research Review, 27*(2). https://ideas.repec.org/a/ags/aerrae/196659.html

Bussay, A., Velde, M., Fumagalli, D., & Seguini, L. (2015). Improving operational maize yield forecasting in Hungary. *Agricultural Systems, 141*, 94–106. https://doi.org/10.1016/j.agsy.2015.10.001

Cai. (2019). Integrating satellite and climate data to predict wheat yield in Australia using machine learning approaches. *Agricultural and Forest Meteorology, 274*, 144–159, https://doi.org/10.1016/j.agrformet.2019.03.010.

Chlingaryan, A., Sukkarieh, S., & Whelan, B. (2018). Machine learning approaches for crop yield prediction and nitrogen status estimation in precision agriculture: A review. *Computers and Electronics in Agriculture, 151*, 61–69. https://doi.org/10.1016/j.compag.2018.05.012

Didan, K., Munoz, A., Solano, R., Huete, A. (2018). MOD13A1 MODIS/Terra+Aqua vegetation indices 16-Day L3 Global 500m SIN Grid V006. *NASA EOSDIS Land Processes DAAC*. https://doi.org/10.5067/MODIS/MOD13A1.006

Dkhar, D., Feroze, S., Singh, R., & Ray, L. (2017). Effect of rainfall variability on rice yield in north eastern hills of India: A case study. *Agricultural Research, 6*(4), 341–346. https://doi.org/10.1007/s40003-017-0276-4

Dong, J. (2016). Mapping paddy rice planting area in northeastern Asia with Landsat 8 images, phenology-based algorithm and Google Earth Engine. *Remote Sensing Environment, 185*(SI), 142–154.

Domenikiotis, C., Spiliotopoulos, M., Tsiros, E., & Dalezios, N. (2004). Early cotton yield assessment by the use of the NOAA/AVHRR derived vegetation condition index (VCI) in Greece. *International Journal of Remote Sensing, 25*(14), 2807–2819.

Food and Agricultural Organization (FAO) (2022) India dataset. https://www.fao.org/india/fao-in-india/india-at-a-glance/en/. Accessed 16 Jan 2022.

Fisher. (1925). The influence of rainfall on the yield of wheat at Rothamsted. *Philosophical Transactions of the Royal Society, 213*, 89–142.

Friedl, M., & Sulla-Menashe, D. (2015). MCD12Q1 MODIS/Terra+Aqua land cover type yearly L3 global 500m SIN grid V006. *NASA EOSDIS Land Processes DAAC*. https://doi.org/10.5067/MODIS/MCD12Q1.006

Goodfellow, I., Bengio, Y., & Courville, A. (2016). *Deep learning*. MIT press.

Gümüşçü, A., Tenekeci, M. E., & Bilgili, A. V. (2020). Estimation of wheat planting date using machine learning algorithms based on available climate data. *Sustainable Computing: Informatics and Systems, 28*, 100308. https://doi.org/10.1016/j.suscom.2019.01.010

Hakkim, V., Joseph, E., Gokul, A., & Mufeedha, K. (2016). Precision farming: The future of Indian agriculture. *Journal of Applied Biology and Biotechnology, 4*(6).

Han. (2020). Prediction of winter wheat yield based on multi-source data and machine learning in China. *Remote Sensing, 12*(2, Art.2). https://doi.org/10.3390/rs12020236

Hendrick, W., & Scholl, J. (1943). Technique in measuring joint relationship. The joint effects of temperature and precipitation on crop yield. *North Carolina Agricultural Experimental Statistics Techniques Bulletin, 74*.

Hochreiter, S., & Schmidhuber, J. (1997). Long short-term memory. *Neural Computation, 9*(8), 1735–1780. https://doi.org/10.1162/neco.1997.9.8.1735

Huete, A., Justice, C., & Liu, H. (1994). Development of vegetation and soil indices for MODIS-EOS. *Remote Sensing of Environment, 49*, 224–234.

Huete, A., Didan, K., Miura, T., Rodriguez, E., Gao, X., & Ferreira, L. (2002). Overview of the radiometric and biophysical performance of the MODIS vegetation indices. *Remote Sensing of Environment, 83*(1–2), 195–213. https://doi.org/10.1016/S0034-4257(02)00096-2

India, G. (2011). F*aster, sustainable and more inclusive growth: an Approach to the 12th Five Year Plan Draft*. Planning Commission, Government of India.

Jeong, J. H., Resop, J. P., Mueller, N. D., Fleisher, D. H., Yun, K., Butler, E. E., & Kim, S. H. (2016). Random forests for global and regional crop yield predictions. *PLoS ONE, 11*(6), e0156571. https://doi.org/10.1371/journal.pone.0156571

Jiang, Z., Huete, A. R., Didan, K., & Miura, T. (2008). Development of a two-band enhanced vegetation index without a blue band. *Remote Sensing of Environment, 112*, 3833–3845.

Jones, J., Antle, J., Basso, B., Boote, K., Conant, R., Foster, I., & Wheeler, T. (2016). Brief history of agricultural systems modeling. *Agricultural Systems, 155*, 240–254.

Kamir, F., & Hochman, Z. (2020). Estimating wheat yields in Australia using climate records, satellite image time series and machine learning methods. *ISPRS Journal of Photogrammetry and Remote Sensing, 160*(124), 135.

Keig, M., Keig, J., & Mcalpine, W. (1969). WATBAL: A computer system for the estimation and analysis of soil moisture regimes from simple climatic data Tech CSIRO. *Division of Land Research, Canberra, 69.*

Van Keulen, N., Seligman, H., & Van Keulen, P. (1981). Simulation of Nitrogen behaviour of soil-plant systems. *Centre for Agricultural Publishing and Documentation Wageningen*, 192–220.

Kriegler, F., Malila, W., Nalepka, R., & Richardson, W. (1969). Preprocessing transformations and their effects on multispectral recognition. In *Proceedings of the sixth international symposium on remote sensing of environment*, (pp. 97–131). University of Michigan.

Kyunghyun, C., Bart van, M., Caglar, G., Dzmitry, B., Fethi, B., Holger, S., & Yoshua, B. (2014). Learning phrase representations using RNN encoder–decoder for statistical machine translation. In *Proceedings of the 2014 conference on empirical methods in natural language processing (EMNLP)* (pp. 1724–1734). Association for Computational Linguistics. https://doi.org/10.48550/arXiv.1406.1078.

Lecun, Y., Bottou, L., Bengio, Y., & Haffner, P. (1998). Gradient-based learning applied to document recognition. *Proceedings of the IEEE, 86*(11), 2278–2324. https://doi.org/10.1109/5.726791

Li, W., Liu, K., Yan, L., Cheng, F., Lv, Y., & Zhang, L. (2019). FRD-CNN: Object detection based on small-scale convolutional neural networks and feature reuse. *Science and Reports, 9*(1), 16294.

Lp daac. (2022). Available https://lpdaac.usgs.gov/. Accessed 25 Jan 2022

Malik, D., & Singh, D. (2010). Dynamics of production, processing and export of wheat in India. *Journal of Food Security, 1*, 1–12.

McGuire, I., Shelley, FAO, IFAD & WFP. (2015). The state of food insecurity in the world 2015: meeting the 2015 international hunger targets: taking stock of uneven progress. *Advances in Nutrition, 6*(5), 623–624.

Mikolov, T., Karafiát, M., Burget, L., Černocký, J.H., & Khudanpur, S. (2010). Recurrent neural network based language model. In *Proceedings of the 11th annual conference of the international speech communication association*, no. 9, ISSN 1990-9772.

Nguyen, T. T., Doan, T. M., Tomppo, E., & McRoberts, R. E. (2020). Land use/land cover mapping using multitemporal sentinel-2 imagery and four classification methods—A case study from Dak Nong, Vietnam. *Remote Sensing, 12*(9, Art.9). https://doi.org/10.3390/rs12091367.

Petrică, A., Stancu, S., & Tindeche, A. (2016). Limitation of ARIMA models in financial and monetary economics. *Theoretical and Applied Economics, XXIII*(4(609), Winter), 19–42, https://ideas.repec.org/a/agr/journl/vxxiiiy2016i4(609)p19-42.html.

Prasada, A., Chai, L., Singha, R., & Kafatos, M. (2006). Crop yield estimation model for IOWA using remote sensing and surface parameters. *International Journal of Applied Earth Observation and Geoinformation, 8*, 26–33.

Rauff, O., & Bello, R. (2015). A review of crop growth simulation models as tools for agricultural meteorology. *Agricultural Sciences, 6*(9), 1098–1105. https://doi.org/10.4236/as.2015.69105

Roberts, M., Braun, N., Sinclair, T., Lobell, D., & Schlenker, W. (2017). Comparing and combining process-based crop models and statistical models with some implications for climate change. *Environmental Research Letters, 12*(9), 095010. https://doi.org/10.1088/1748-9326/aa7f33

Rocha, A. V., & Shaver, G. R. (2009). Advantages of a two band EVI calculated from solar and photosynthetically active radiation fluxes. *Agricultural and Forest Meteorology, 149*, 1560–1563.

Rouse, J., Haas, R. H., Schell, J. A., & Deering, D. W. (1974). Monitoring vegetation systems in the great plains with ERTS. In *Proceedings of Third ERTS symposium* (pp. 309–317). NASA.

Shahhosseini, M., Martinez-Feria, R. A., Hu, G., & Archontoulis, S. V. (2019). Maize yield and nitrate loss prediction with machine learning algorithms. *Environmental Research Letters*. https://doi.org/10.1088/1748-9326/ab5268

Sharma, R., Kamble, S., Gunasekaran, A., Kumar, V., & Kumar, A. (2020). A systematic literature review on machine learning applications for sustainable agriculture supply chain performance. *Computers and Operations Research, 119*, 104926. https://doi.org/10.1016/j.cor.2020.104926

Sherstinsky, A. (2020). Fundamentals of recurrent neural network (RNN) and long short-term memory (LSTM) network. *Physica D: Nonlinear Phenomena, 404*, 132306. https://doi.org/10.1016/j.physd.2019.132306

Shin, J. Y., Kim, K. R., & Ha, J. C. (2020). Seasonal forecasting of daily mean air temperatures using a coupled global climate model and machine learning algorithm for field-scale agricultural management. *Agricultural and Forest Meteorology*, 281(107858).

Thomas, G., Taylor, J., & Wood, G. (1997). Mapping yield potential with remote sensing. *Precision Agriculture*, 713–720.

Vermote, E. (2015). MOD09A1 MODIS/Terra surface reflectance 8-Day L3 global 500m SIN Grid V006. *NASA EOSDIS Land Processes DAAC*. https://doi.org/10.5067/MODIS/MOD09A1.006

Wan, Z., Hook, S., & Hulley, G. (2015). MYD11A2 MODIS/Aqua land surface temperature/emissivity 8-Day L3 global 1km SIN Grid V006. *NASA EOSDIS Land Processes DAAC*. https://doi.org/10.5067/MODIS/MYD11A2.006

Wilkerson, G., Jones, W., Boote, J., Ingram, T., & Mishoe, W. (1983). Modeling soybean growth for crop management. *Transactions of the American Society of Agricultural Engineers, 26*, 63–73.

Willcock, S., Javier, M., Danny, P., Kenneth, J., Stefano, B., Alessia, M., Carlo, P., Saverio, S., Giovanni, S., Brian, V., Ferdinando, V., James, M., & Ioannis, N. (2018). Machine learning for ecosystem services. *Ecosystem Services, 33*, 165–174. https://doi.org/10.1016/j.ecoser.2018.04.004

Xie, M., Jean, N., Burke, M., Lobell, D., & Ermon, S. (2016). Transfer learning from deep features for remote sensing and poverty mapping. In *Thirtieth AAAI conference on artificial intelligence*.

Yield Statistics Data (2022). https://aps.dac.gov.in/APY/Index.htm Accessed 20 Feb 2022

Zhu, X., Guo, R., Liu, T., & Xu, K. (2021). Crop yield prediction based on agrometeorological indexes and remote sensing data. *Remote Sensing, 13*(10, Art.10). https://doi.org/10.3390/rs13102016.

Farm-Wise Estimation of Crop Water Requirement of Major Crops Using Deep Learning Architecture

Mihir Dakwala, Pratyush Kumar, Jay Prakash Kumar, and Sneha S. Kulkarni

Abstract Each crop has different cultivation practices with various phases including seed treatment, soil management, land preparation, sowing of seeds, irrigation, application of fertilizers, etc. Irrigation is a very important phase of any crop cultivation practice. Irrigation scheduling, water management, crop forecasting, and demands precise crop-specific water requirements (CWR) which nowadays become extremely important for various crops grown under irrigation, especially in arid and semi-arid regions. This study will enable efficient use of water and better irrigation practices like scheduling as the supply of water through rainfall is limited in some areas. In growing crops, irrigation scheduling is a critical management input to ensure optimum soil moisture status for proper plant growth and development as well as for optimum yield, water use efficiency, and economic benefits. Operational CWR methods in India are mainly based on sparsely located in situ measurements and high-resolution remote sensing data, which limit the overall precision. To overcome the mentioned challenge, the deep learning architecture and soil moisture techniques have been used in this study to generate high-resolution farm boundaries, followed by the generation of crop maps and then generated soil moisture at the parcel level using our company's own algorithms to estimate the farm-specific CWR. Over most of the farms, a direct positive relationship has been observed between the crop growing period and its particular CWR. The irrigation scheduling module of the Agrogate platform developed is currently used in many states by different stakeholders for the proper management of water resources.

Keywords Crop water requirement · Deep learning · Reference evapotranspiration (ETo) · Crop coefficients (Kc) · Cultivation practices · Irrigation

M. Dakwala (✉) · P. Kumar · J. P. Kumar · S. S. Kulkarni
Amnex Infotechnologies Pvt. Ltd. Ahmedabad, Gujarat 380054, India
e-mail: mihir.d@amnex.com

© The Author(s), under exclusive license to Springer Nature Singapore Pte Ltd. 2023
S. Chaudhary et al. (eds.), *Digital Ecosystem for Innovation in Agriculture*,
Studies in Big Data 121, https://doi.org/10.1007/978-981-99-0577-5_11

1 Introduction

Irrigation is the key to the production of crops. India being the second largest irrigated country needs to apply an efficient irrigation system. Growing demand for food for the world population expects more land under cultivation and irrigation. Efficient application of water on the farm using satellite images is one of the other challenges for researchers (Gupta et al., 2022). The term such as precision irrigation water management has been coined to meet the application of optimum water at the right time and in the right manner. India is in famous for its flood irrigation system which is not only reduces the crop yield but also increases the chances of pest and disease attacks (Brahmanand and Singh, 2022). In India, challenges such as irrigation water pricing (IWP) done based on the area of the farm are irrelevant to water application. Monitoring and estimating the irrigation water pricing based on the user-based, volumetric, and crop can reduce the burden on the farmers. Also, remote sensing-based CWR methodology can estimate the water requirement in a particular area well in advance and can help the government bodies to plan the same (Upadhyaya et al., 2022). Database management systems developed based on spatial data such as rainfall, evaporation, weather, and the crop can lead to create the best decision support system for the individual or an organization (Pandey & Mogarekar, 2022).

Scientists made different attempts to develop the net and gross water requirement using FAO-CROPWAT 8.0 showing better water management system adoption for lower water needs crops (Gabr, 2022). Assessment of crop water requirements during droughts is yet another challenge for the scientific community with limited water resources for optimum yield production. Different models such as CROPWAT[1] and AquaCrop-GIS[2] have been studied for limited crops such as corn and soybean. Overall robust irrigation management for better yield is required (Deb et al., 2022). The crop growth monitoring system plug-in of AquaCrop includes spatiotemporal variability of crop yield, evapotranspiration, and water productivity to check with the water stress. But the study is limited to the maize grain (Ahmadpour et al., 2022).

New development in the era of big data analytics and machine learning has been a tool to explore the domain of irrigation water management, crop water requirement, and scheduling of irrigation (Saggi & Jain, 2022). Traditional canal water distribution without information on the demand or soil water status can create an irregular supply of water leading to reduced total production and loss of water (Zhang et al., 2022). Crop maps generated using the time series satellite data can tell the ground status of the crops grown in the area. Open-source remote sensing data such as Sentinel-1 and -2 can help us to monitor the crop at regular intervals. Different machine learning (ML) algorithms such as random forest (RF) and support vector machine (SVM) give the crop type mapping using Sentinel-1 and -2 data in time series (Chakhar et al., 2021). Crop water estimation directly depends on the available soil moisture. Recent development in soil moisture estimation using Sentinel-1 and Sentinel-2

[1] https://www.fao.org/land-water/databases-and-software/cropwat/en/.

[2] https://www.fao.org/aquacrop/software/en/.

data has given us scope for near real-time satellite-based soil moisture for analysis. Synthetic aperture radar (SAR) technology and backscattering models have been used to retrieve soil moisture for crop planning and hydrological modeling (Parida et al., 2022). The flow of the water depends upon catchment area development, and canal flow directly depends upon the slope of the contour. Our previous work on object-based crop classification, where parcel-level farm has been generated using image processing techniques and crop classification has been done using random forest, has motivated us for the next level of work of CWR at the farm level (Kumar et al., 2022).

2 Objective

This chapter study aims to address the following important and critical things which have not been addressed so far at a larger scale in India for the development of comprehensive irrigation water management practices which are essential to improve water management and eliminate the associated problems.

1. To generate the farm boundary at the farm level using proprietary algorithms based on deep learning and remote sensing.
2. To determine the crop water requirement and irrigation scheduling of major crops in India at the farm for optimal resource allocation to increase yield and water productivity.
3. Providing near real-time parameters that are affecting the CWR on our analytical Amnex's Agrogate platform. Parameters such as temperature, accumulated rainfall, and soil moisture.

The present study demonstrated the capabilities of Amnex's Agrogate platform with respect to creating an end-to-end pipeline. This platform starts the process with generation of farm boundaries, then generation of crop map using multi-temporal high-resolution drone and satellite images. Once the crop map is generated, the platform will identify soil moisture at individual farm level, and based on all these data, farm level crop water requirements will be estimated. This platform successfully demonstrates the crop water requirement and irrigation scheduling of major crops in India using proprietary algorithms based on deep learning and remote sensing for optimal resource allocation and to increase yield and water productivity.

3 Study Area

The study area is located at 22.7502 latitude and 72.6764 longitude in the Kheda District of Gujarat state (Fig. 1). Kheda district is one of the thirty-three districts of Gujarat state in western India. Kheda district is primarily an agricultural district with tobacco and paddy as the predominant crops. The other major crops cultivated

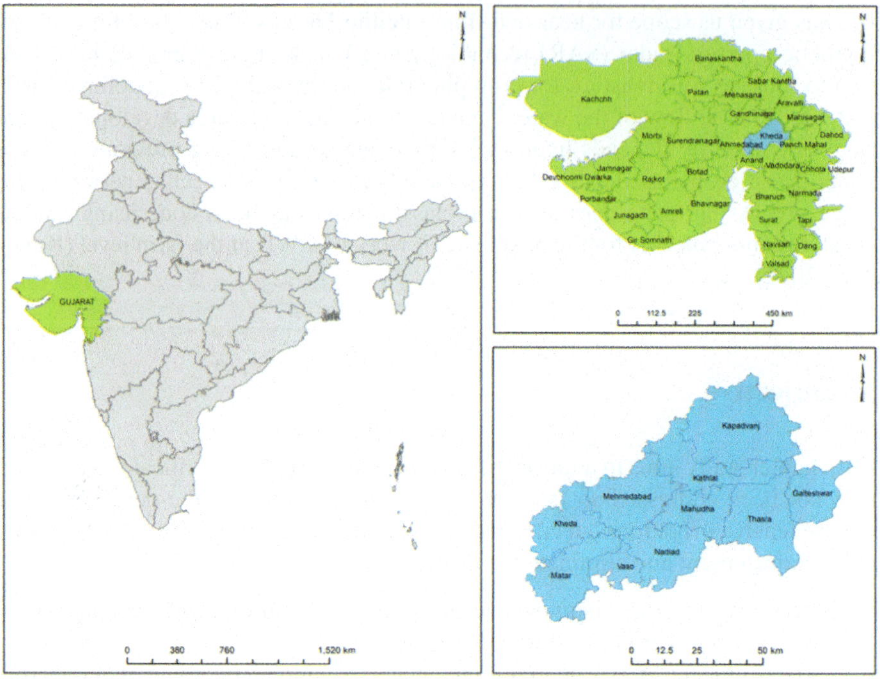

Fig. 1 Study area map

are wheat, bajra, maize, cotton, and groundnuts. About 21.39% of land holdings are with small and marginal farmers, and the average size of the holdings is 1.20 Ha. The mean annual rainfall of Kheda is 690 mm. The farming patterns are irregular, and the majority are small in size and frequently change cropping patterns.

4 Amnex's Agrogate Platform

Amnex has a list of solutions which can help entire agriculture sectors to overcome the challenges persisting. Amnex proposes its agriculture product "Agrogate and CropTrack" which can provide valuable near real-time insights on crop condition. AGROGATE[3] is a Web GIS-based application developed to provide integrated Agri intelligence services for various stakeholders involved in the decision-making process. It offers a holistic view of critical information on agro ecosystems such as crop-wise area, detection of crop stress due to water, nutrients, and diseases, and crop loss assessment due to natural calamities. This vital information helps in guiding the decision-makers in formulating optimal strategies for planning, distribution, price

[3] https://www.agriculture.amnex.com/agrogate/

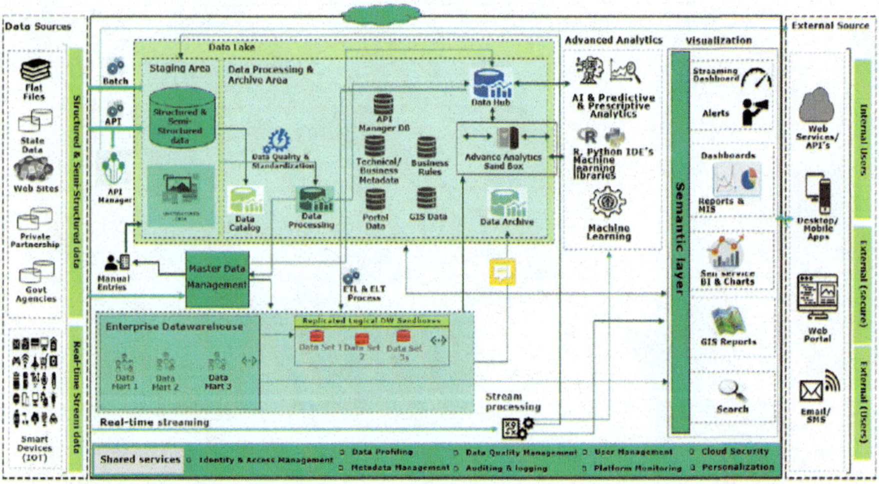

Fig. 2 Agrogate platform architecture

fixation, procurement, transportation, and storage of essential agricultural products. The architecture of the product is represented in Fig. 2

5 Methodology

The detailed methodology used in the study is given below.

5.1 Dataset

Farm boundary or parcel-level data has been generated with the help of Google satellite maps at 0.5-m resolution with the help of QGIS. Our internal team at Amnex has created the labels for the same which were shown in Fig. 3.

Classified crop map of study area has been used from our previous study area (Kumar et al., 2022). Temperature data of the GLDAS Noah Land Surface Model has been used for land surface temperature data. Atmospherically corrected Level-2A cloud-free images of Sentinel-2 satellite data were used for deriving the NDVI layer. Monthly climate hazards group infrared precipitation with station (CHIRPS) data has been used for rainfall monitoring in our study area. Soil moisture has been generated using Sentinel-1 data using our own algorithm which has obtained a copyright on this particular task ***(diary number 4608/2022-CO/SW)***. Crop coefficients (K_c) used for each crop in the study area were referred from a research paper for Gujarat. Reference evapotranspiration (ET_o) was calculated using Penman Monteith method described

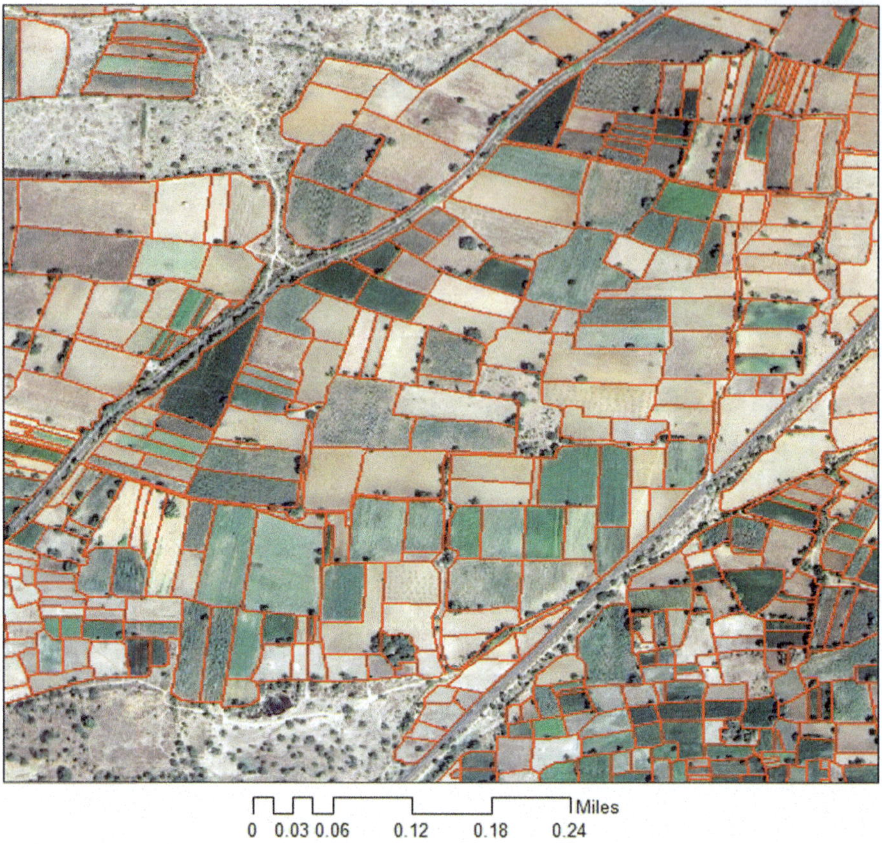

Fig. 3 Training datasets for farm boundary generation

in FAO irrigation and drainage paper (Allen et al., 1998). Also, to understand the direction of the flow, slope was calculated based on hydrological modeling using SRTM data. The below are the details of the input data used in study (Table 1).

5.2 Process Flow

The data described above is processed as per the flow diagram to get the analytical dashboard for CWR as shown in Fig. 4.

Table 1 Details of the datasets used for the result generation

Data	Period	Temporal scale	Data source
Rainfall	Rabi season (2019–20)	Monthly	https://www.data.chc.ucsb.edu/products/CHIRPS-2.0/global_monthly/tifs/
Land surface temperature (LST)	Rabi season (2019–20)	Monthly	https://www.disc.gsfc.nasa.gov/datasets/GLDAS_NOAH025_M_2.1/summary?keywords=GLDAS
Soil moisture (SM)	Rabi season (2019–20)	Fortnightly	https://www.sentinel.esa.int/web/sentinel/missions/sentinel-1
NDVI	Rabi season (2019–20)	Fortnightly	https://www.sentinel.esa.int/web/sentinel/missions/sentinel-2
Slope			https://www.usgs.gov/centers/eros/science/usgs-eros-archive-digital-elevation-shuttle-radar-topography-mission-srtm-1
Farm boundary			Google Satellite using QGIS
Crop map	Rabi season (2019–20)	Growing Season	https://www.sentinel.esa.int/web/sentinel/missions/sentinel-2

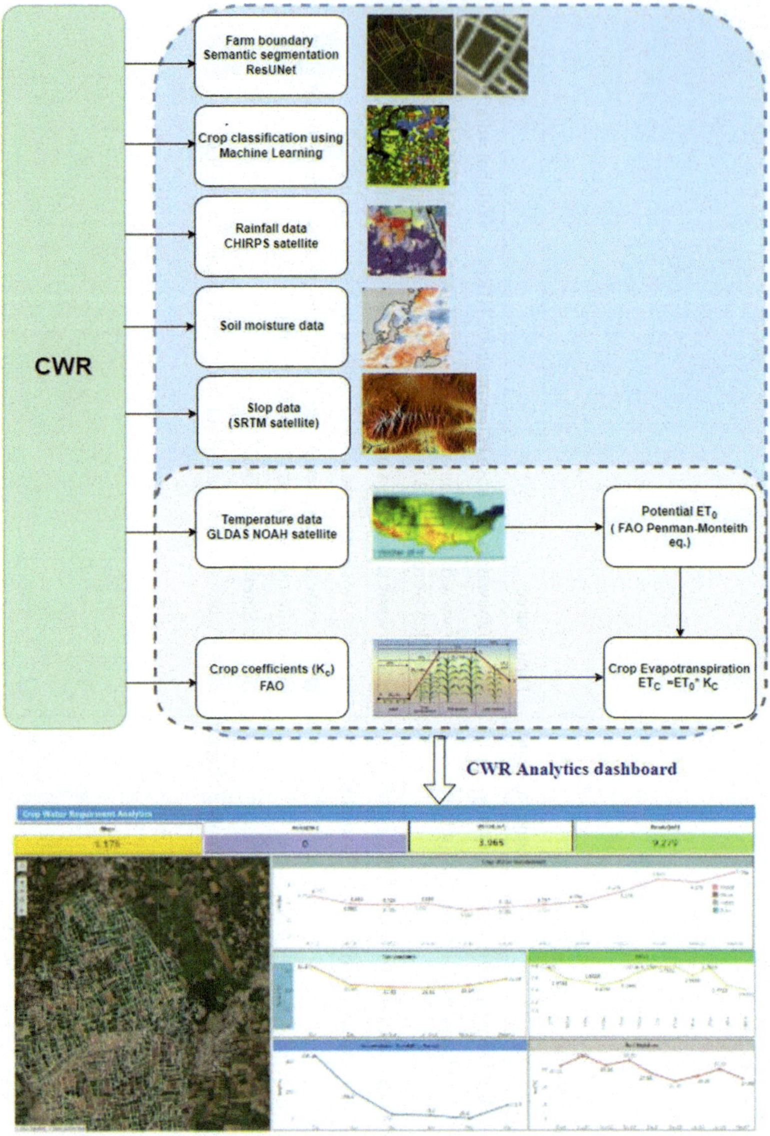

Fig. 4 Methodology followed to calculate crop water requirements

5.3 Farm Boundary Delineation

Google satellite maps are exported using QGIS at 0.5 m resolution and are used for farm boundary generation. The team at Amnex has created the training datasets for the model. Data cleaning and pre-processing were performed on training datasets.

Fig. 5 Generated farm boundary map of study area using ResUNet

We have filtered the cloud from images. Also, non-agriculture areas, e.g., roads, residential areas, and forest, were removed. Pre-processing such as data standardization and data augmentation was performed. The accuracy of deep learning models largely depends on the quality, quantity, and contextual meaning of training data. However, data scarcity is one of the most common challenges in building deep learning models. In production use cases, collecting such data can be costly and time-consuming. Data augmentation is a process of artificially increasing the amount of data by generating new data points from existing data. This includes adding minor alterations to data or using machine learning models to generate new data points in the latent space of original data to amplify the dataset. Data augmentation such as rotations at 90, 180, and 270 degrees was performed. Horizontal flip, vertical flip, increase illumination, and decrease illumination wer introduced in the data to create more different training datasets. Semantic segmentation models such as modified ResUNet with loss binary cross entropy and F1 loss were selected as the best result. Post-processing was done on the final raster using QGIS software to generate a clean vector file. The final output can be seen in Fig. 5

5.4 Classification Output

Classified maps generated using machine learning algorithm-based random forests were used for parcel-level crop type mapping (Kumar et al., 2022). The final output of classified crops can be seen in Fig. 6.

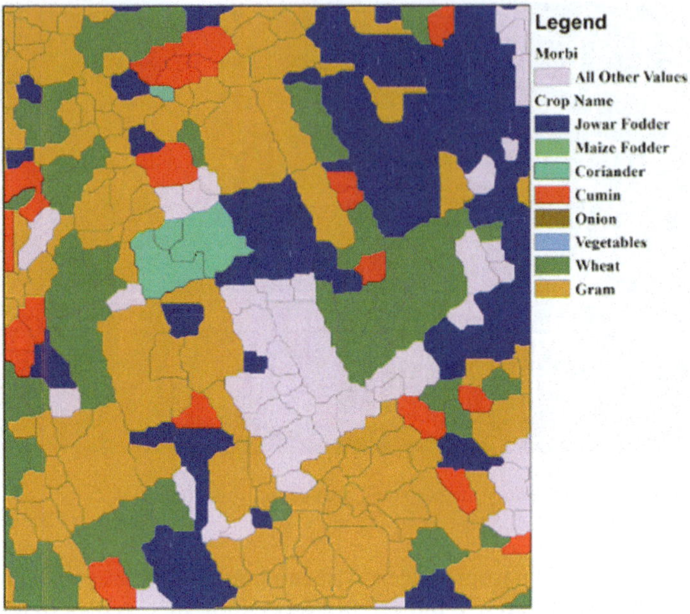

Fig. 6 Object-based crop classification output. Different crop classes were symbolized, and crop labels are seen in the legend section of the map

5.5 *Soil Moisture*

Retrieving soil moisture (SM) at the agriculture field level from remote sensing is a very challenging job. The SAR data has tremendous potential in soil moisture-related studies. Our company has developed a soil moisture workflow with Sentinel-1 SAR data, which will enable the user to generate SM maps every 12 days with 10 m spatial resolution. Amnex Infotechnologies Pvt. Ltd. has obtained a copyright on this particular task ***(diary number 4608/2022-CO/SW)***. The SM variation at the sub-parcel level was shown in the following Fig. 7.

5.6 *Temperature, Rainfall, Kc, ETo*

Land surface temperature is the key for CWR. Land surface temperature works as an inter-linkage between soil moisture and evapotranspiration for CWR estimation. The monthly LST data was used for Rabi (2019–2020). As rainfall directly affects the CWR, monthly CHIRPS rainfall data helps us to understand farm level water requirement. NDVI derived from Sentinel-2 data helped us to understand the growth pattern of a particular farm. This gives the idea about the growing stage of the crop as well as this helps us in crop identification of areas of interest. Crop coefficient (Kc) derived for the reference is used to calculate crop evapotranspiration (ET). Reference evapotranspiration (ETo) is calculated using the Penman Monteith method (Fig. 8).

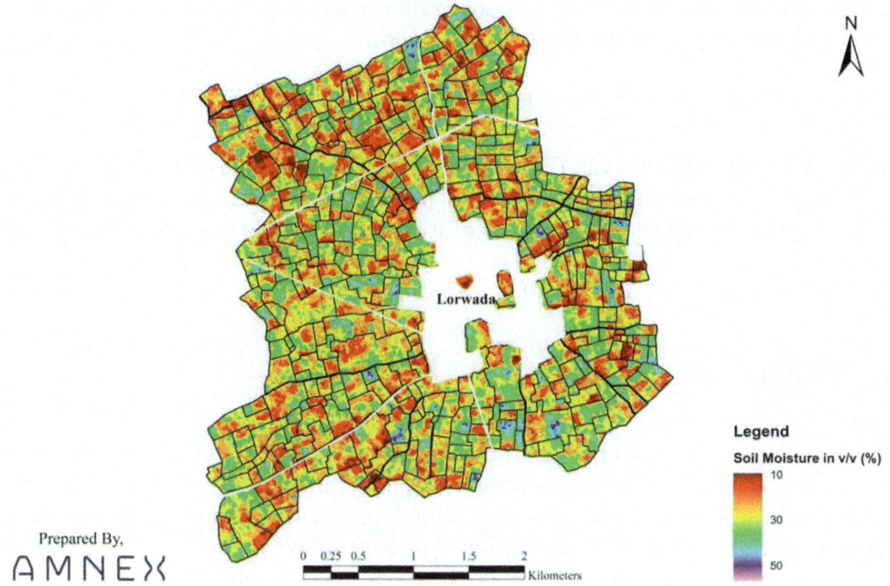

Fig. 7 Map of field-wise soil moisture variation of Lorwada Village

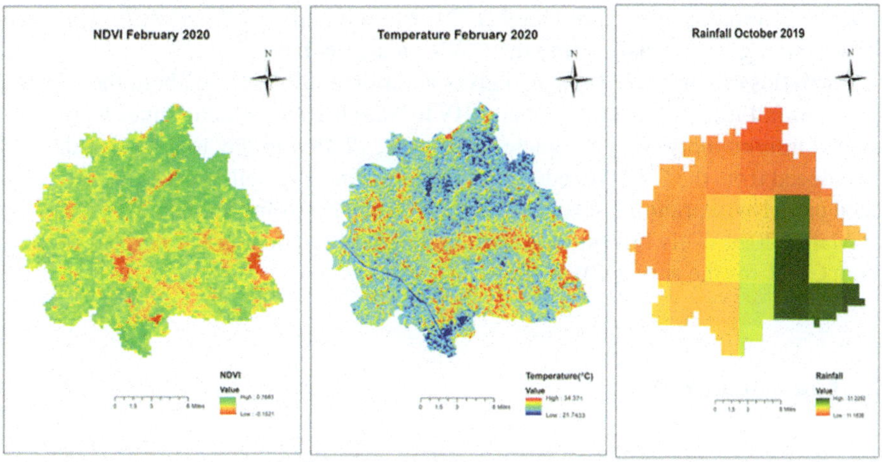

Fig. 8 Map of field-wise **1** NDVI, **2** temperature, and **3** rainfall variation of study area

5.7 CWR

CWR is the amount of water that is lost from a cropped field by the evapotranspiration and is expressed by the rate of evapotranspiration in mm/day. Crop evapotranspiration (ET) is derived by using the following equation.

$$\mathrm{ET} = K_c \times \mathrm{ET}_0 \tag{1}$$

where K_c = Crop coefficient, ET_o = Reference evapotranspiration

In addition to theoretical ET, additional parameters such as soil moisture, temperature, vegetation indices, and rainfall of a parcel allow the analysis of the ground situation.

6 Result and Discussion

Amnex's Agrogate platform incorporates all parameters affecting the CWR. As shown in Fig. 9, all input and estimated output parameters slope, crop name, crop area, CWR, temperature, accumulated rainfall, and soil moisture of the individual farm level can be seen on the platform. Figure 9 shows the analysis of CWR for selected Maize farms. Area of the maize farm as seen is 0.3196 ha having land slope of 1.859%. A zoom view of the graphs can be seen in Fig. 10, where we can see each parameter at the parcel level for the Rabi season from 2019 to 2020. The average temperature ranges from 29.618to 28.074 °C. Total CWR varies from 4.9 to 7.7 mm/day from mid-October 2019 to the end of March 2020. This basic information gives us insight into the CWR on a daily basis.

Analytics on our dashboard also gives additional information about the accumulated rainfall monthly from October 2019 to March 2020, which ranges from 374.4 to 80.4 mm/monthly for a complete parcel, i.e., 12.48 mm/day to 2.68 mm/day will be reduced from the CWR. In addition to this information, soil moisture is also calculated parcel-wise, ranging from 35.25 to 24.73 mv(%) from October 2019 to the end of March 2020. This information will be utilized by reducing the percent of water from CWR for the actual CWR (Figs. 11 and 12).

7 Conclusion

Amnex Agrogate platform for CWR study is of its own case and gives the detailed analysis of CWR in near real-time. Farm boundaries were generated using deep learning architecture. Post-processing using QGIS was done for raster vector operation. Once parcels were generated, classified crop maps at the farm level were incorporated. Soil moisture layers generated using the Amnex algorithm were tabulated

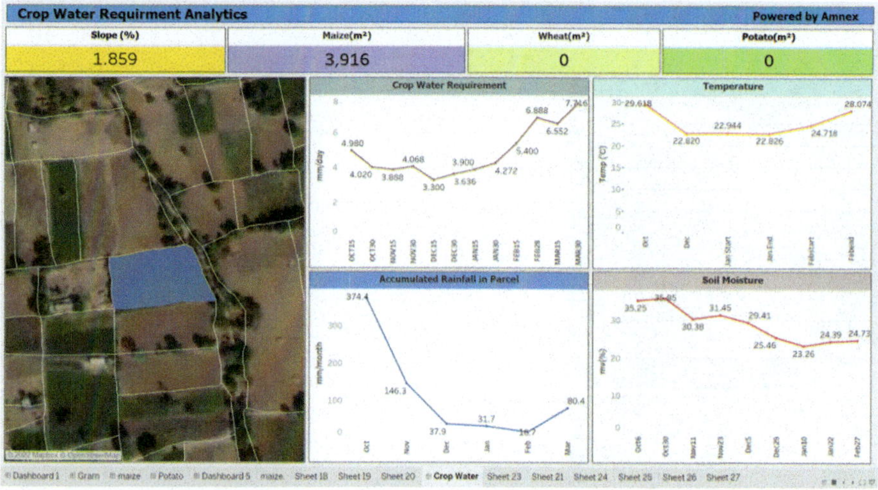

Fig. 9 Crop water requirement dashboard for randomly selected maize crop field on Amnex Agrogate platform

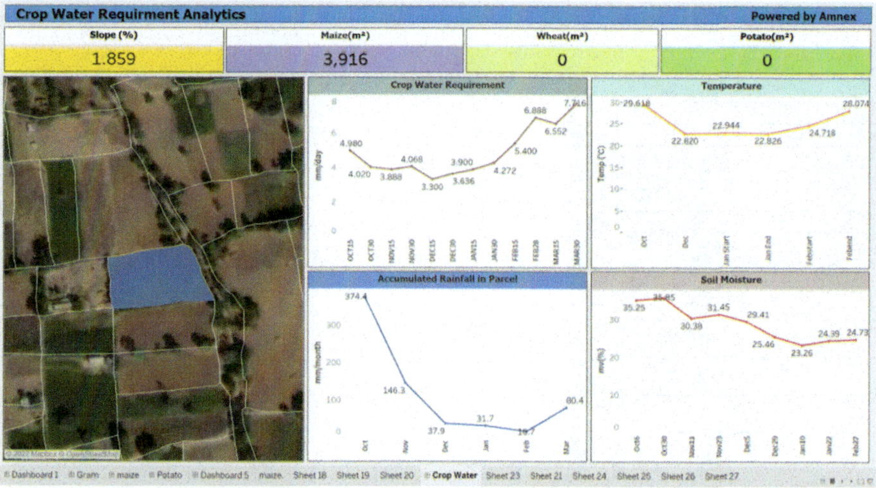

Fig. 10 Graphical analytics of parameters which affects maize crop water requirements

using QGIS. Parameters such as rainfall and temperature were also tabulated. Once all input per parcel was ready, CWR based on K_c and ET_0. It has been seen that this methodology to estimate near real-time CWR can be helpful for irrigation scheduling, hence increasing the water use efficiency and crop yield. Future work includes the implementation of high-resolution and temporal satellite data for enhanced CWR estimation. Open source or Indian Meteorological Department (IMD)-based weather

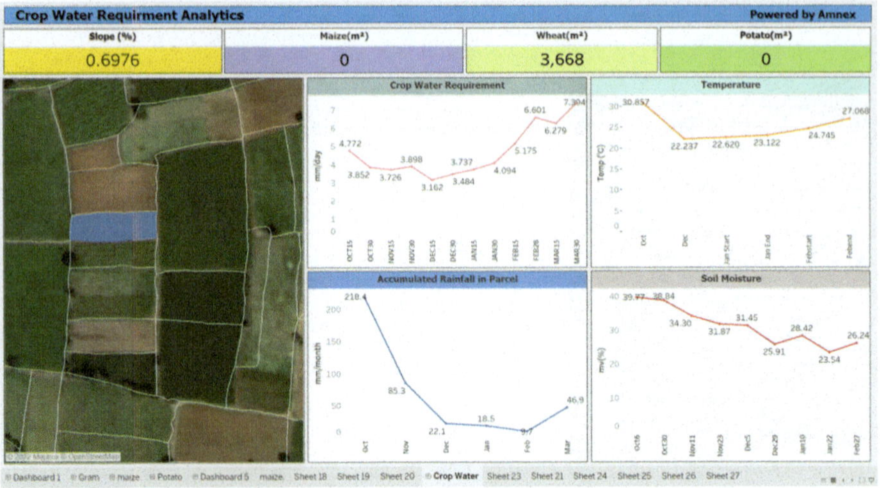

Fig. 11 Crop water requirement dashboard for randomly selected wheat crop field on Amnex Agrogate platform

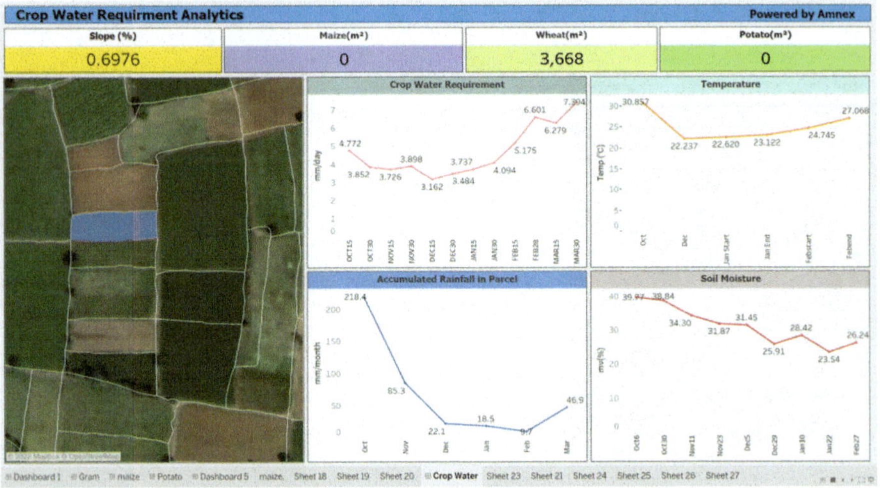

Fig. 12 Graphical analytics of parameters which affects wheat crop water requirements

data can be utilized on a daily basis for a better understanding of the effect of these parameters on CWR. Crop stage-wise crop coefficient and NDVI can be incorporated to make CWR more precise.

Acknowledgements The authors would like to thank the Google Earth Engine (GEE) team for providing cloud-computing resources for data visualization and European Space Agency (ESA) for providing open-access data. We would also like to thank the USGA, NASA. The authors thank

Amnex Infotechnologies Pvt. Ltd. for providing all the needed support and environment to work on this project. Additionally, the authors would like to thank anonymous reviewers for their helpful comments and suggestions.

References

Ahmadpour, A., FarhadiBansouleh, B., & Azari, A. (2022). Proposing a combined method for the estimation of spatial and temporal variation of crop water productivity under deficit irrigation scenarios based on the AquaCrop model. *Applied Water Science, 12*(7), 1–19.

Allen, R. G., Pereira, L. S., Raes, D., & Smith, M. (1998). Crop evapotranspiration- Guidelines for computing crop water requirements-FAO Irrigation and drainage paper 56.Fao. *Rome*, 300(9), D05109.

Brahmanand, P. S., & Singh, A. K. (2022). Precision irrigation water management-current status, scope and challenges. *Indian Journal of Fertilisers, 18*(4), 372–380.

Chakhar, A., Hernández-López, D., Ballesteros, R., & Moreno, M.A. (2021). Improving the accuracy of multiple algorithms for crop classification by integrating sentinel-1 observations with sentinel-2 data. *Remote Sensing, 13*(2), 243.

Deb, P., Moradkhani, H., Han, X., Abbaszadeh, P., & Xu, L. (2022). Assessing irrigation mitigating drought impacts on crop yields with an integrated modeling framework. *Journal of Hydrology, 609*, 127760.

Gabr, M. E. (2022). Modelling net irrigation water requirements using FAO-CROPWAT 8.0 and CLIMWAT 2.0: a case study of Tina Plain and East South ElKantara regions, North Sinai, Egypt. *Archives of Agronomy and Soil Science, 68*(10), 1322–1337.

Gupta, A., Singh, R. K., Kumar, M., Sawant, C. P., & Gaikwad, B. B. (2022). On-farm irrigation water management in India: Challenges and research gaps. *Irrigation and Drainage, 71*(1), 3–22.

Kumar, J. P., Singhania, D., Patel, S. N., & Dakwala, M. (2022). Crop classification for precision farming using machine learning algorithms and sentinel-2 Data. In: *Data science in agriculture and natural resource management* (pp. 143–159).Springer, Singapore.

Pandey, A., & Mogarekar, N. (2022). Development of a spatial decision system for irrigation management. *Journal of the Indian Society of Remote Sensing*, 50(2), 385–395.

Parida, B. R., Pandey, A. C., Kumar, R., & Kumar, S. (2022). Surface soil moisture retrieval using sentinel-1 SAR data for crop planning in Kosi River Basin of North Bihar. *Agronomy, 12*(5), 1045.

Saggi, M. K., & Jain, S. (2022). A survey towards decision support system on smart irrigation scheduling using machine learning approaches. *Archives of Computational Methods in Engineering*, 1–24.

Upadhyaya, A., Jeet, P., Singh, A. K., Kumari, A., & Sundaram, P. K. (2022). Efficacy of influencing factors in the decision-making of irrigation water pricing: a review. *Water Policy*.

Zhang, F., He, C., Yaqiong, F., Hao, X., & Kang, S. (2022). Canal delivery and irrigation scheduling optimization based on crop water demand. *Agricultural Water Management, 260*, 107245.

Hyperspectral Remote Sensing for Agriculture Land Use and Land Cover Classification

MuraliKrishna Iyyanki and Satya Sahithi Veeramallu

Abstract Food production is accountable for about 20–30% of anthropogenic greenhouse gas emissions, with the agricultural sector becoming the dominant source of these emissions. Land use information is important for agriculture management, the information about which can be obtained by hyperspectral (HyS) remote sensing. The high spectral information from hyperspectral sensors can help in differentiating various LU/LC classes. In LULC, focus is to be laid in classification of closely resembling classes which is possible only from HyS RS. This requires development of specific algorithms. A review of current algorithms for processing of HyS datasets is carried out in this article. This includes validating various atmospheric correction (AC) models, dimensionality reduction techniques (DR), and classification methods. Results show that FLAASH absolute AC model gave a higher resemblance with the ground spectra with higher correlation for agriculture and built-up classes. Classification is performed using seven per pixel classifiers and one ensemble classifier. Support vector (SVM) and ensemble classifiers for both Hyperion and AVIRIS-NG HyS images have shown higher accuracy with accuracy percentage ranging between 90 and 95%. Accordingly, the case studies for delineation of LU/LC under different scenarios facilitate a feasible and viable overall carbon sequestration.

Keywords Classification · Hyperion · Hyperspectral data · LU/LC classification · HyS data processing methods

1 Introduction

Land use is a pivotal sector in taking around augmented carbon sequestration globally. Agricultural land provides the largest share of food supplies and guarantees different

M. Iyyanki (✉)
DRDO, Delhi, India
e-mail: iyyanki@berkeley.edu

S. S. Veeramallu
Jawaharlal Nehru Technological University, Hyderabad, India

© The Author(s), under exclusive license to Springer Nature Singapore Pte Ltd. 2023
S. Chaudhary et al. (eds.), *Digital Ecosystem for Innovation in Agriculture*,
Studies in Big Data 121, https://doi.org/10.1007/978-981-99-0577-5_12

ecosystem services. Land use management assists the control and the allocation of land for specific uses. Food production is found to be equally a significant activity accountable for about 20–30% of anthropogenic greenhouse gas emissions, with agricultural sector becoming a pre-dominant source of these emissions. Thus, there is increased emphasis to significantly reduce the emissions from agriculture sector and also augment the carbon sequestration under different sectors including the land use sector. Reay (2020) mentions that this observation is most clear in the drive for "net zero" where unavoidable emissions, such as those from food production, are balanced by more sequestration via land use change. It has been highlighted that successful land use policy for net zero will require extremely demanding levels of integration and spatial resolution, and that the research community has a vital role to play in providing a robust evidence base for this. Accordingly, this study lays emphasis on the application of advanced tools of remote sensing, viz. hyperspectral remote sensing, as well as case studies for delineation of land use under different scenarios essential for overall carbon sequestration program design and implementation.

The upgradations in sensor technology have led to the invent of spaceborne sensors that can monitor the earth at a very narrow and contiguous spectral bands, thus providing land cover information at a very high spectral resolution. Some of such spaceborne HyS sensors include EO-1 Hyperion of 242 bands and 30 m spatial resolution, compact high resolution imaging spectrometer (CHRIS) of 18 bands and 17 m spatial resolution, advanced very high resolution imaging spectrometer (AVIRIS) of 224 bands and 20 m spatial resolution, and HYDICE. The application of spaceborne and airborne HyS remote sensing sensors started from identifying various vegetation species and rock types and has extended in various branches of earth system studies like oil slick contamination, target detection, vegetation stress monitoring, mineral mapping, and urban roof type mapping. In the field of agriculture, this narrow bandwidth high spectral resolution images can help in accurate crop acreage estimation, crop stress monitoring, crop type and health identification, and many other advanced studies (Ravikanth et al., 2017). Most of the confusing class pairs in LU/LC classification like urban roofs and sand, swamps and shrubs, plantations and forests, crop and grass, wetlands and shallow water bodies, barren lands and fallow lands, etc., can be differentiated using the HyS datasets. Although the utility of HyS images is diverse, the processing procedure of these huge datasets is complex and needs certain special algorithms or methods to map the crop species, forest types, minerals, or other natural resources.

The utilization of HyS data cannot be confined only to producing an accurate or detailed classification map but it is also used for many important analyzes like optimum band or optimum wavelength identification for various crop types, development of spectral indices for vegetation or crop stress studies, developing a spectral library for various dominant varieties of crop, vegetation, soil, and minerals. The studies on spaceborne HyS datasets are also extended to improving the qualities or spatial resolutions of the medium or coarse resolution HyS images using an external multispectral image or microwave image. Keeping all the above utilities of HyS data, the general processing flow of spaceborne and airborne HyS data for any mapping application is broadly presented in Fig. 1. The data received from sensor is first radiometrically corrected and then atmospherically corrected. Once the data is converted

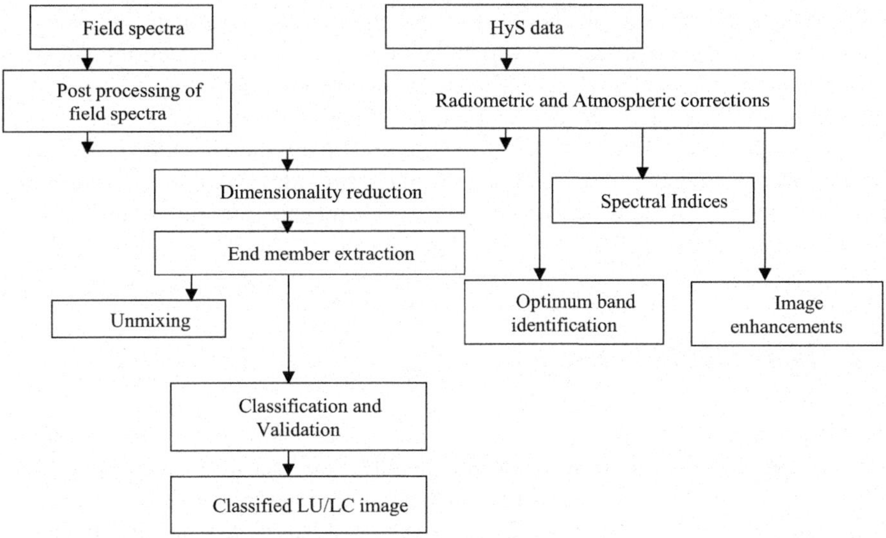

Fig. 1 General working flow of HyS data

to reflectance using ACs, it is then used for DR and endmember extraction, and finally, classification is performed over the image.

Due to the spectral richness of the HyS data, intense care must be taken while collecting the training samples for classification. There are three different ways of collecting the endmembers/pure pixels for HyS data classification. These include

1. Collecting the ground spectra using a spectroradiometer.
2. Carefully picking the pure pixels directly from the image through keen observation.
3. Pixel purity index and n-d visualization method of collecting the endmember spectra from the image.

This chapter further provides an overview on pre-processing of HyS data, atmospheric correction (AC) methods followed by the dimensionality reduction (DR) techniques. Later, some classification methods commonly used in HyS data classification are reviewed after which the spectral unmixing methods, optimum band identification, spectral indices, and image fusion were reviewed. The chapter also presents few results from processing of Hyperion spaceborne HyS data.

2 Pre-processing—Radiometric and Atmospheric Corrections

The HyS images are significantly effected by the radiometric and atmospheric parameters. Improper functioning of the sensors, scattering, and absorption of the reflected radiation reaching the sensor, greatly affect the quality of the HyS image. Hence, HyS data requires a special processing which includes radiometric corrections, ACs, and unwanted noise and smile effects. The calibration of HyS data to reflectance is the most important step of image processing. Once the data is converted to reflectance, it is then used for DR and endmember extraction, and finally, classification is performed over the image. Endmember extraction is the process of identifying some of the purest pixels from the entire image for each LU/LC class and which are further used as input for classification.

2.1 Radiometric Corrections

The data recorded in the sensor has certain abnormalities which are caused due to mal functioning of the sensor or earth curvature or other radiometric errors (Jensen, 2005). These cause a bad column or black lines in the image which are corrected by replacing them with an average of the neighboring values. Also, certain no information or noisy bands in the blue, short-wave infrared (SWIR) region, and few bands in the near infrared (NIR) region caused because of the absorption due to water vapor and other atmospheric gases need to be removed. Also, certain negative pixels observed in the image need to be replaced with an average of the surrounding pixel values.

2.2 Atmospheric Corrections

The amount of radiation reflected from the earth's surface and reaching the sensor is prone to various atmospheric effects like absorption, transmission, reflection, and scattering which are caused due to factors like—temporal changes in solar illumination, varying Earth–Sun geometry with the time of the year, and geographically changing atmospheric conditions. Such atmospheric effects must be accounted for in order to achieve a reliable reflectance spectrum of features on the ground. The AC procedure helps in obtaining an image with reflectance close to the ground reflectance by removing the atmospheric noises like adjacency effects, scattering, and water vapor. AC methods are used for converting a HyS or multispectral image from a raw digital number (DN) to reflectance can be of two types:

- Relative or image-based AC methods.
- Absolute or model-based AC methods.

The relative AC methods also commonly known as statistical or empirical methods are mere calibration techniques and solely depend on the diversity of features available in an image. They do not require any prior information about the ground and can be applied to any image that is free from cloud and other radiometric errors. These statistical techniques come into play when there is very little information about the ground and are used to adjust or normalize the raw radiance data. Some of the relative

AC methods include—internal average relative reflectance, flat field correction, log residuals, empirical line method, etc.

Absolute AC methods require a detail on the atmospheric conditions on the ground that existed at the time of acquisition of the image from the sensor. These physics-based atmospheric correction modules derive the spectral radiance at satellite sensor (Lλ) based on satellite ephemeris parameters and atmospheric constituents. The atmospheric constituents can be derived from the metadata of the image or in-situ measurements. The output of these absolute AC methods is a reflectance image that matches the reflectance of the ground spectra with a maximum estimated error of 10%, if atmospheric profiling is adequate enough. The description, advantages and limitations of various AC methods is given in Table 1.

Table 1 Method, description, advantages, and limitations of various AC methods

	Method	Description	Advantages	Limitations
Relative AC methods	Flat field correction (Gao et al., 2009)	– Derives an average radiance spectrum from a large flat area within the scene as reference spectrum and then divides pixel in the image with the reference spectrum	– No ground truth is required – Suitable for dry regions with less vegetation	– Requires a large flat area within the scene – Not suitable for heterogenous areas – Strongly depends on the composition of features in the image and susceptible to artifacts (Ben-Dor et al., 2004)
	Internal average relative reflectance (IARR) (Gao et al., 2009)	– Takes an average spectrum for the entire scene and divides each pixel using this as the reference spectrum to calculate the apparent reflectance	– No ground information required	– Depends on the range of DN values in the image (landscape) (Ben-Dor et al., 2004)
	Empirical line (Ben-Dor et al., 2004)	– Calculates a regression line between the ground and image reflectance spectra of two objects in the image (one bright and one dark). The gain and offset of this regression line applied to the radiance spectra for each pixel to produce apparent reflectance on a pixel-by-pixel basis	– Considered to be equivalent to removing the solar radiance and atmospheric path radiance from the measured signal	– Requires certain field information – Difficult to apply for large and inaccessible areas (San & Suzen, 2010)

(continued)

Table 1 (continued)

	Method	Description	Advantages	Limitations
	Log residuals	– Input spectrum is divided by the spectral geometric mean, then divided by the spatial geometric mean (Green & Craig, 1985)	– Does not require any ground truth – Spectral mean removes the solar irradiance, while the spatial mean compensates the atmospheric transmittance, instrument gain, topographic effects, and albedo effects from radiance data – Useful for analyzing mineral-related absorption features	– Depends on the range of DN values in the image (landscape) (Ben-Dor et al., 2004)
	QUAC (Quick AC)	– Determines AC parameters directly from the observed pixel spectra in a scene (Guo & Zeng, 2012)	– Does not require any ancillary information except sensor type – Produces accurate results close to absolute AC methods	– Requires the intervention of trained HyS data analyst (Kale et al., 2017)
Absolute AC methods	Fast line of sight atmospheric attenuation for spectral hypercubes (FLAASH)	– Uses MODTRAN-based AC model (Kawishwar, 2007)	– Adjacency correction and spectral polishing are added features – Automated wavelength calibration – Significant results can occur under moist and hazy conditions	– Need for ancilliary information sometimes makes it difficult – Produces anomalies for mineralogical applications. (Cetin et al., 2017)
	ATCOR (AC)	– Uses MODTRAN-based AC model – ATCOR2 is used for flat terrains with two geometric degrees of freedom (DOF), while ATCOR-3 is useful for rugged terrain and takes the third dimension -terrain height into consideration. The ATCOR 4 model is useful for airborne multispectral and HyS datasets	– Useful for VNIR as well as thermal region – Useful for flat, rugged terrains – Takes care of adjacency correction	– Defining a new sensor is difficult – Aerosol model and atmospheric models are fixed – Fixed visibility values
	6S method (second simulation of a satellite signal in the solar spectrum)	– A basic RT code used for the calculation of look-up tables in the MODIS	– Most widely used and easily adaptable	– More useful for ocean and crop studies using multispectral data – More prone to spectral artifacts

(continued)

Table 1 (continued)

	Method	Description	Advantages	Limitations
	LaSRC	– A level-2 dataset produced and released as a product by USGS, essentially to support terrestrial remote sensing applications (Ilori et al., 2019) – Generated using a dedicated Landsat Surface Reflectance code (USGS 2017) – Uses a radiative transfer model for the inversion of atmospheric parameters such as aerosol and water vapor	– Validated and assessed for land applications, and has a dedicated aerosol retrieval algorithm for pixels over water – Dominantly used for MODIS and Landsat data	– Surface reflectance is available for first seven spectral bands and only for scenes with a solar zenith angle less than 76° – OLI bands at 443 and 482 nm are not suitable for analysis as they are utilized for aerosol inversion tests within the model

2.3 *Dimensionality Reduction Techniques*

The AC process leaves the HyS data with reflectance values but retains the large number of bands in HyS data which leads to the problem of redundancy of information between the bands and increased processing time. Reducing the redundancy and extracting the important features from this huge data is an important pre-processing step for improving the accuracy of classification, clustering, and other comprehensive models. DR techniques transform the data into a new domain or a feature space where the data is projected in such a way that the bands are uncorrelated with each other based on a certain criterion (Sahithi et al., 2016). This phenomenon is otherwise called as curse of dimensionality where the data requires a higher dimensional feature space in order to extract the exact information and to further classify and analyze the data (Kakarla et al., 2019).

DR techniques can be mainly classified in two categories—feature extraction methods and feature selection techniques. The feature extraction process alters the raw data to a higher dimensional feature space to define a new subset of the features containing a varied range of information from the original data. The feature selection techniques on the other hand identifies or selects a subset of the original features which cover the most useful information from the highly correlated and redundant features (Pal & Foody, 2010). The feature extraction techniques are further classified as linear and non-linear methods. Fejjari et al. (2018) made a comparative study on linear and non-linear feature extraction methods using Indiana Pine data and concluded that the non-linear techniques have higher classification accuracy than linear methods, while the linear methods are computationally faster. The most popularly known DR feature extraction techniques are principal component analysis (PCA), minimum noise fraction (MNF), and independent component analysis (ICA) methods. Each of these techniques work on a unique principle and have their own advantages and disadvantages. The PCA techniques works on the principle of data variance while MNF sorts the information based on the SNR. The ICA assumes each band to be a linear mixture of some independent hidden components and thus applies a linear unmixing procedure to extract the independent features. Table 2 gives an overview of three well-known DR methods.

Kale et al. (2017) made a review on various HyS DR techniques which is stated that the selection of one single technique is very hard to handle the dimensionality issue, while the PCA method extracts some useful information and ignores the noise in the image while MNF considers noise whitening as an important step and works with second-order statistics to reduce the noise. ICA has the capability to withhold some minute or inherent information which cannot be retained by PCA and MNF (second-order statistics-based methods). The selection of the number of components in DR process is dependant on the application of the user and is indirectly related to the final classification accuracy.

Table 2 Method, characteristics, advantages, and limitations of various DR methods

DR method	Characteristics	Advantages	Limitations
PCA	– Based on a mathematical principle known as eigenvalue decomposition of the covariance matrix – Transforms the data on to a new coordinate axis by maximizing the variance – As the number or order of components increases, the variance decreases, and the components are ordered in the decreasing order of their eigen value – First PCA contains more information than the second component, and the second PCA has more information than the third and so on (Panwar et al., 2014)	– Sorts the components based on the information content – Computationally faster	– Noise filtering is not performed, and hence, the feature of interest will be buried in the noisy bands – In certain cases, like aircraft data, few lower order components may contain some significant information – Determining the number of components to retain is an issue (Wang & Chang, 2006)

(continued)

Table 2 (continued)

DR method	Characteristics	Advantages	Limitations
MNF	– An unsupervised DR technique with A two-step process where the first step includes noise whitening – The second step involves a PCA – Transformed MNF data is highly decorrelated and have zero mean and a unit noise variance	– The noise factor that is not considered in PCA is resolved in MNF transformation – MNF-based methods achieve higher SNR than PCA-based methods for signal-dependent noise (Luo & Chen, et al., 2016)	– Determining the number of components to retain is an issue (Wang & Chang, 2006)
ICA	– Based on the non-Gaussian assumption of the independent sources, and uses higher-order statistics to reveal interesting features (Yusuf & He, 2011) – It assumes that each band is a linear mixture of independent hidden component and extracts the independent feature using a linear unmixing operation (Panwar et al., 2014)	– Can distinguish features of interest even when they occupy only a small portion of the pixels in the image	– Fails in non-linear mixing models – Cannot sort the components based on the information (Hyvarinen & Oja, 2000)
Linear discriminant analysis (LDA)	– A supervised DR method that computes the linear discriminants (directions) with maximum separation between multiple signal sources (Chang, 2013)	– Less errors are possible due to maximum separation	– Needs ground information

3 Field Spectra Collection and Post-processing of Field Spectra

Field spectroscopy collects ground-based information about a feature and gives detail about the spectral properties of the feature and the relation with its material properties. The endmembers collected from the image may sometimes be falsely labeled due to the unavailability of ground truth information. The availability of field information and ground spectra can compensate this error. While collecting the field spectra using a spectroradiometer, parameters like solar illumination angle, time of the day, illumination over the target/incidence angle, field of view (FOV) of the fiber optics, shadow, calibration of the instrument, etc., effect the purity of the collected spectra. The spectral gun used for collecting the spectra needs to be held at a height of 60 cm from the target, and a FOV is to be set at 25°. All the spectra need to be collected between 10:00 and 14:30 h to reach the necessary illumination conditions (Arun Prasad et al., 2015). After collecting the spectra from ground using a spectroradiometer, certain post-processing steps need to be performed to remove any anomalies from the ground spectra. These post-processing steps include—splice correction, removal of noisy and water vapor regions, spectral smoothening, and library building.

3.1 Splice Correction

Spectral profiles collected from spectroradiometer exhibit three steps at the joining point or splice of the three transition regions such as one at 1000 nm—end of VNIR, second at 1800 nm—transition between SWIR1 to SWIR2 region, and the other at the end of SWIR2. This step or loop like effect in the spectra needs to be corrected in the initial step.

3.2 Removal of Noisy and Water Vapor Regions

Due to the atmospheric absorption and scattering effects, the incoming solar radiation in the wavelength regions below 450 nm, 1350–1425 nm, 1800–1955 nm, and beyond 2350 nm range gets afflicted and contains some noise. This noise cannot be corrected or smoothened as the reflected radiation is highly attenuated and hence needs to be removed.

3.3 Spectral Smoothening

The narrow bandwidth of the sensor and reflections from multiple objects in the field causes certain inherent noise in the spectra which appears like an unwanted speckle like noise in the curve. This needs to be smoothened, and the inherent noise needs to be removed. Many smoothening filters were used for spectral curves like moving average filter which takes the average of neighboring values and smoothens the curve, weighted average filter, which uses a weighted value along with the moving window. However, these averaging filters were not very successful as they could not handle the outliers and the edges in the data (starting and ending points of the curve). Savitzky-Golay (S-G) is a well-known smoothening filter in image and signal processing that can produce smoother curves from the data without losing the important information. The outliers and edges are also well handled using a S-G smoothening filter. The selection of filter size and order of the polynomial are crucial while applying the S-G filter as this may lead to over-smoothening and loss of spectral shape (Savitzky & Golay, 1964; Vaiphasa, 2006).

3.4 Building the Spectral Library

The smoothened spectra are then added to a spectral library along with the associated metadata like name of the feature, latitude/longitude, image, and other secondary details.

4 Classification Methods of Hyperspectral Data

Classification of HyS image requires certain special algorithms that can handle its huge volume, redundancy, mixed pixel effect, and heavy processing time. Various factors like AC methods, DR approaches, spectral response of the target material, parameters used for classification, training sample size, purity of the training sample, etc., play an important role in obtaining an accurate classified map. Hence, many advanced classification methods have come up that can aid in overcoming these limitations and accurately classifying the voluminous HyS datasets. The classifiers can be parametric, non-parametric, supervised, unsupervised, per pixel methods, sub-pixel method, conceptual methods, single classifiers or ensemble classifiers, and characteristics of them which are presented in Table 3.

Some of the widely classification methods in the recent studies are presented in Table 4.

Moughal (2010) made a comparison of SVM classifier with two well-known ML and SAM classifiers to classify seven different LU/LC classes using HYDICE data. A prior MNF transform was applied on the data before classification. It was concluded

Table 3 Various hyperspectral data classification approaches

Criteria	Methods of classification	
Based on availability of training samples	Supervised	Training samples are available, and prior knowledge about the study area is available **Eg**: maximum likelihood (ML), spectral angle mapper (SAM), artificial neural networks (ANN), support vector machines (SVMs), etc
	Unsupervised	No prior information about the study area and image is available. No training samples are given as input for classification. **Eg**: Iso data, K-means
Based on assumptions on distribution of data	Parametric	Gaussian assumption of data. **Eg**: ML classifier (MLC)
	Non-parametric	Non-Gaussian assumption of data. **Eg**: SVM, ANN
If spatial data is included during classification	Object-based classification method	Uses the spatial or contextual information apart from the spectral information. **Eg**: Texture-based or contextual classifier or watershed segmentation
Based on pixel information and within pixel information extraction	Hard classifiers or per pixel classifiers	Land cover continuity is not shown. Classification is done based on the information available in one pixel. **Eg**: SVM, ANN, SAM, MLC, etc.
	Soft classifiers or sub pixel classifiers	Continuity of land cover classes is seen in the classifier. It considers the per pixel information for classifying the data. **Eg**: Unmixing—linear and non-linear, mixture tuned matched filtering (MTMF), etc.
Whether a single classifier is used or multi-classifiers are used	Single classifier	Only a single classifier is used to produce the classification results. **Eg**: All per pixel classifiers
	Ensemble classifiers	More than one classifiers are used for classification. **Eg**: Random forests, ensemble methods, etc.

Table 4 Some of the well-known hyperspectral data classifiers

Method of classification	Classifier	Principle of working and key parameters
Per pixel classifiers	SVM (Zhang et al., 2001)	– Constructs a hyper plane that maximizes the margin between two classes – It is best suitable for binary classifiers and uses kernels for multiclass classification problems – Its major advantage is that it yields better accuracy even with minimum number of training samples – Key parameters: penalty parameter, kernel type, gamma value, training sample
	ANN (Zagajewski and Olesiuk 2009)	– Typically follows the model of the brain neuron system – Input, hidden layers, and output are the major components of a neural network – The back propagation ANN method uses any weights assigned to the hidden layer, and the difference between output and the estimated output is back propagated to the input neuron until the output and the estimated output are the same – Time consuming technique but gives very accurate results when properly trained – Key parameters: hidden layers, number of iterations, and threshold values
	SAM (Sahithi and Krishna 2019)	– SAM calculates the cosine angle between the unknown spectra (pixel under consideration) and target spectra and assigns the pixel to the target class if the angle is less than the threshold angle (user specified) – Prominently used for vegetation species mapping because of its insensitivity to illumination angle – Key parameters: threshold angle (α)

(continued)

Table 4 (continued)

Method of classification	Classifier	Principle of working and key parameters
	Spectral feature fitting (SFF)	– Absorption-feature-based method for matching image spectra to reference end members SFF uses a continuum removed profile to observe the dominant absorption regions in a profiles – Based on the absorption regions, reference endmember spectrum will be scaled to match the unknown spectrum
Decision tree classifier	Decision tree classifiers	– A subclass of non-parametric approaches, which can be used for both classification and regression. During the construction of a decision tree, the training set is progressively split into an increasing number of smaller, more homogeneous groups
Ensemble methods	Ensemble methods	– Integrates more than one classifiers based on the weights and yields an improved classification output – Key parameter: weights for each classifier
	Random forest classifier	– For each new training set that is generated, one-third of the training samples are randomly left out, called the out-of-bag (OOB) samples. The remaining (in the-bag) samples are used for building a tree – Votes for each case are counted every time the case belongs to the OOB samples – A majority vote will determine the final label
Sub pixel classifiers	Linear spectral unmixing	– Assumes that the radiation incident on the earth interacts with only a single material before it reaches the sensor – Two important constraints – Non-negativity—which requires all the fractions from a pixel to be positive – Sum to one—where sum of fractions of all the features should be equal to one

(continued)

Table 4 (continued)

Method of classification	Classifier	Principle of working and key parameters
	Non-linear unmixing	– Used for problem of unmixing HyS images, when the light suffers multiple interactions among distinct endmembers – The non-linear model will consider the second-order scattering interactions, which are assumed to be the most significant
	Mixture tuned matched filtering	– Filters the input image for good matches to the chosen target spectrum by maximizing the response of the target spectrum within the data and suppressing the response of everything else – Yields two outputs—MF score and infeasibility image
Deep learning methods	Deep learning methods and deep belief systems	– A kind of neural network with multi-layers, typically deeper than three layers, tries to hierarchically learn the features of input data
Mathematical morphology-based	Morphological profiles	– Various morphological operators like erosion, dilation, opening, and closing are used with a structuring element to extract the information from the image

that a combination of SVM with MNF transform can help in reducing the complexity of classification and improves the overall classification accuracy. The impact of including multiple DR methods in multiple classifier system (MCS) architecture for the supervised LU/LC classification of a HyS image has been assessed by Damodaran and Nidamanuri (2014). It was concluded that utilization of multiple DR techniques along with multiple classifiers in MCS has increased the overall accuracy by 5% than the SVM classifier when considered alone.

5 Some Results from Hyperion Spaceborne Datasets

Studies were carried out using EO-1 Hyperion spaceborne HyS image, with a spatial resolution of 30 m and swath width of 7.5 m and a spectral resolution of 242 bands with a 10 nm interval. The following were the results of processing on Hyperion datasets.

5.1 Radiometric Correction of Hyperion Data

In the considered Hyperion image, the bands under 416 nm that had no information and the overlapping bands between 851 and 1053 nm were removed. An example of bad line and column correction using an average of the neighboring pixels in Hyperion data is shown in Figs. 2 and 3.

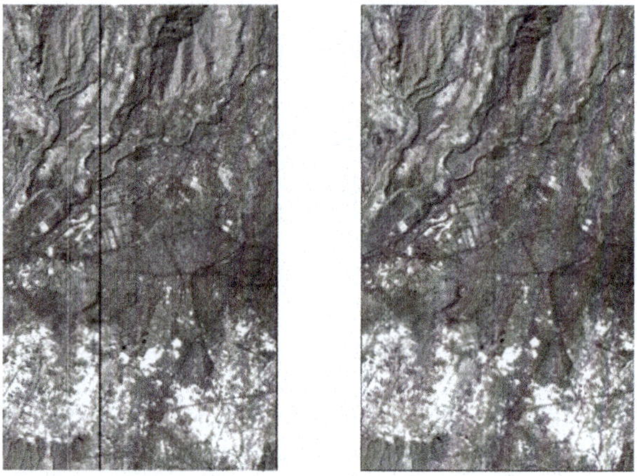

Fig. 2 Radiometric correction of Hyperion data

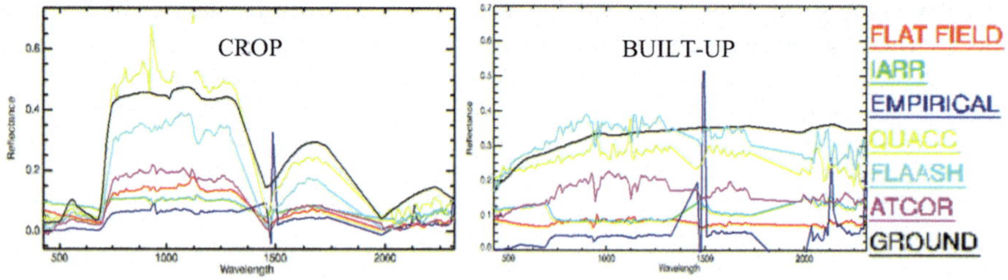

Fig. 3 Reflectance profiles of crop and built-up classes with six different AC methods along with the ground spectra

5.2 *Atmospheric Correction of Hyperion Data*

For atmospheric correction, four relative and two absolute AC methods, namely flat field, IARR, empirical line and QUACC, FLAASH, and ATCOR methods were used to analyze the best method for Hyperion data. The comparative analysis was carried out using Hyperion data as it was an inbuilt sensor in both FLAASH and ATCOR modules. In order to quantitatively asses the performance of the considered AC methods, regression analysis was performed, and regression coefficients were calculated between ground spectra and corresponding image spectra. Figure 4 shows the correlation between ground and atmospherically corrected spectra for crop and built-up class.

Upon conducting a regression analysis, it was observed that spectra of image from FLAASH module had a closer resemblance with the ground spectra with R^2 values of 0.966 for crop class and 0.74 for built-up class. For certain classes like mixed vegetation and barren land classes, the ATCOR and QUAC models gave an edge over FLAASH model. However, FLAASH was observed to give a constant higher R^2 values for the considered LULC classes. Figure 3 gives a glimpse of reflectance profiles of crop and built-up classes with six different AC methods along with the ground spectra.

In order to quantitatively asses the performance of considered AC methods, regression analysis was performed, and regression coefficients were calculated between ground spectra and corresponding image spectra. Figure 4 shows the correlation between ground and atmospherically corrected spectra for crop and built-up class. The ATCOR module has certain fixed parameters that cannot be fine-tuned and has certain complications in defining new sensors which may not be suitable for all atmospheric conditions and all HyS images. Quantitative results also proved that image-based QUAC method can be a reliable method when there is no prior information about the image or the ground.

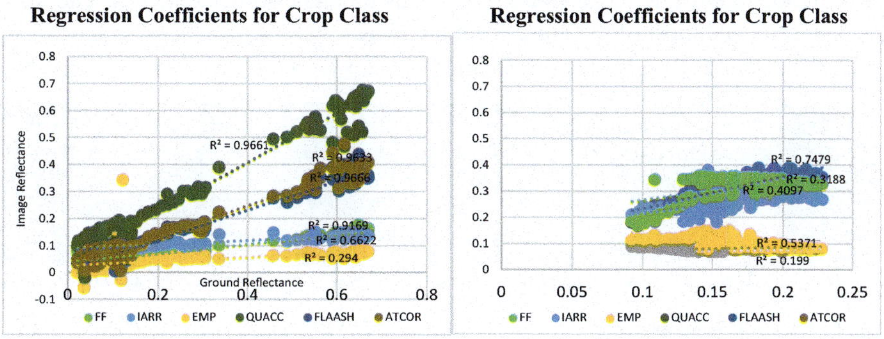

Fig. 4 Correlation between ground and atmospherically corrected spectra for crop and built-up class

5.3 Dimensionality Reduction (DR)

In this step, the three unsupervised DR techniques, namely PCA, MNF, and ICA, were tested on Hyperion data. Figure 5 shows the case study of DR of Hyperion image of Rishikesh and Dehradun area of Uttarakhand. The image was radiometrically and atmospherically corrected and then used for analyzing the three well-known DR methods. The resúltant PCA, MNF, and ICA components was visually analyzed for extraction of any hidden features.

The principal component analysis transform was applied on the 137 band atmospherically corrected data. Then, an inverse PCA was applied using the first 9 bands. The first 9 bands were chosen based on the eigen value curve which represent the noise percentage in the corresponding PCA components. For the considered Hyperion data, the eigen value plot falls sharply after the first 9 eigen values and flattens out which indicates that the information is noisy after those components.

Unlike the PCA transform, ICA components are not sorted in the order of their information content. ICA1 appeared to have higher information, while ICA2 was observed to have some random white noise. The ICA 3–7 had relevant sorted information, and bands after ICA 9 had very low signal to noise ratio. It was noticed that ICA 1, 3, 4, and 5 had good amount of information and had considerate SNR, while 6 and 7 had moderate information, and the rest had more noise and did not contain any useful information. The first 9 ICA components are presented in Fig. 5.

MNF transform can help in removing the smile effect in the image which was observed in band 3. Most of the important information was stored in MNF 3–7, and hence, these bands were used for analyzing the land use features. Figure 6 shows the MNF components from 3 to 7.

These highly informative components from each DR method were observed through the RGB guns to observe any visual improvements in interpretations which is presented in Fig. 7. It was observed that the PCA method brought out all the major information in the first five components, ICA method identified the minute/hidden information that is ignored as noise by PCA using higher-order statistics. The MNF method removed most of the noise from the image.

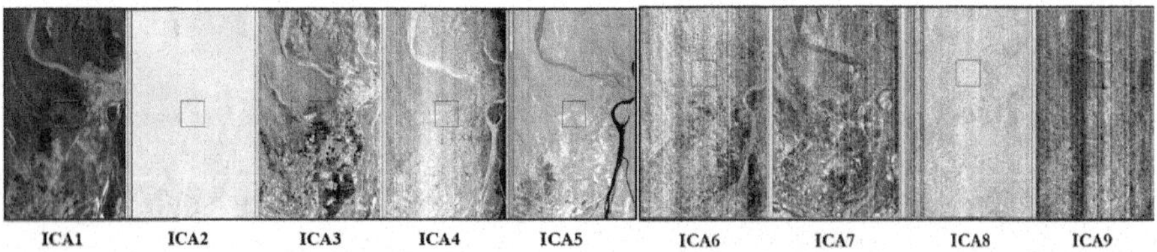

Fig. 5 First 9 ICA components

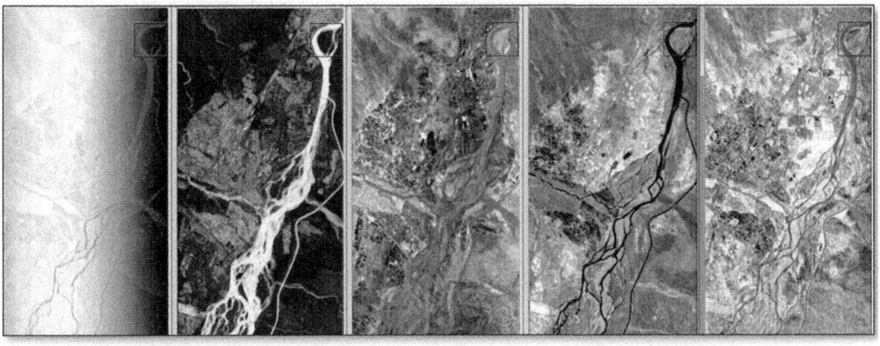

MNF3 MNF4 MNF5 MNF6 MNF7

Fig. 6 MNF components from 3 to 7 containing highest information

5.4 Post-processing of Field Spectra

The ground spectra collected from the field contain certain unwanted noise like splices/step like appearances in the VNIR-SWIR transition zone and absorption errors in water vapor absorption regions. Hence, certain post-processing steps are applied before building the spectral library. In this section, the field spectra collected using ASD field spec 3 spectroradiometer within the 400–2500 nm range were post-processed using the following steps, and the results of each step are as follows.

5.4.1 Splice Correction

The inherent variations in the detector sensitivity to thermal cooling at times lead to a drift in the spectra in the VNIR and SWIR regions and especially at their transition points. Hence, correcting the drift at these splice regions is important to remove the unwanted variations in the sensor signal. In the present spectra, these drifts were mostly observed in VNIR-SWIR joint at 1000 nm region only. The results of splice corrected spectra are presented in Fig. 8.

5.4.2 Removal of Noisy/Non-illuminated Regions

The reflectance values in the absorption regions between 1350 and 1425 nm, 1800–1955 nm, and beyond 2350 nm range are ranging between − 2.5 and 3.10 which is not in the range of normal surface reflectance (0–1). These values were removed in order to properly analyze the curves in further steps for smoothening and classifications. Along with this, the regions before 400 nm which were attenuated due to atmospheric haze were also removed, and the resultant curve can be seen in Fig. 9.

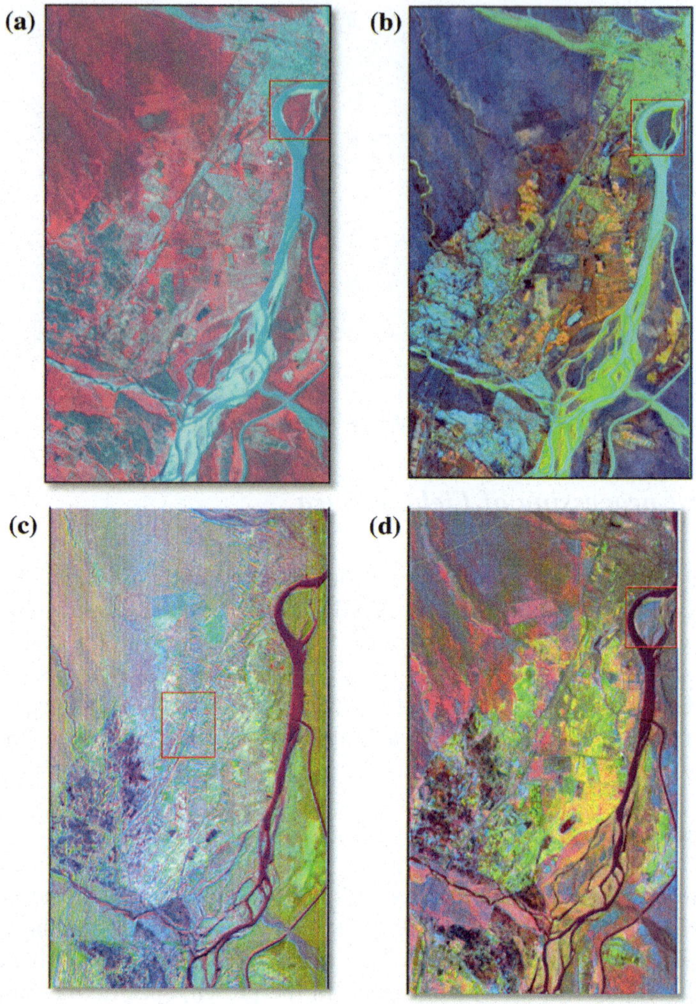

Fig. 7 **a** RGB of atmospherically corrected image. **b** RGB of PCA 1, 2, 3. **c** RGB of ICA 3, 5, 6. **d** RGB of MNF 4, 5, 6

5.4.3 Spectral Smoothening

Savitzky-Golay filter was used with a filter size of 15 and a polynomial order of 2. It not only removed the noise from the curve but also retained the shape and important information from the original curve. Figure 10 shows the smoothened spectra using S-G filter.

The final smoothened spectra after removing atmospheric noise are then built to a spectral library with the associated meta data.

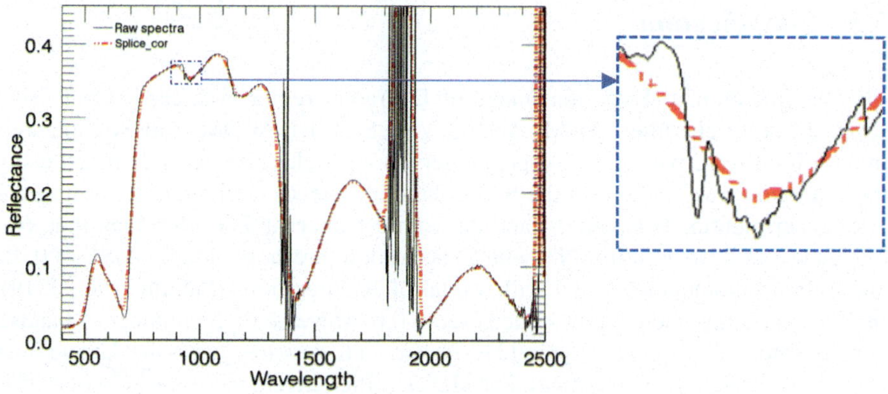

Fig. 8 Splice correction of spectral profiles of vegetation class

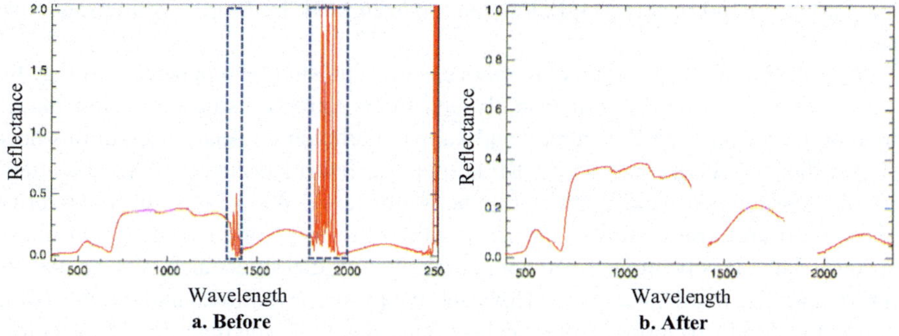

Fig. 9 Removal of noisy and water vapor absorption regions

Fig. 10 Smoothened spectra using savitzky-Golay filter

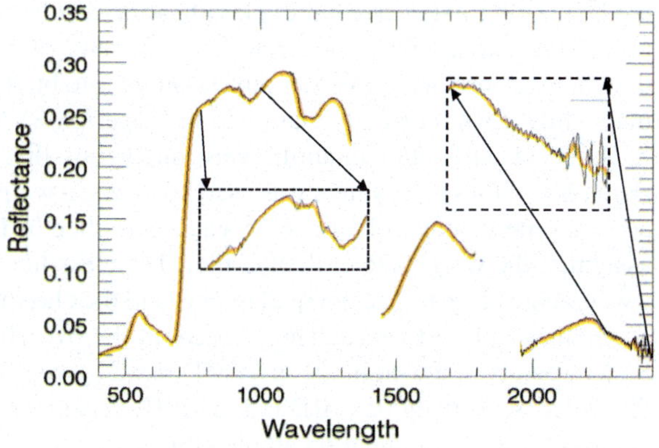

5.5 *Classification*

The classification of Hyperion image of Dehradun region was carried out using seven different classifiers SAM, ANN, SVM, SID, BE, MDM, and ML methods in ENVI software package. The parameters in each classifier are fine-tuned using the trial-and-error method. In the SAM classifier, the spectral angle was set to 0.2 after experimenting between 0.1 and 0.8 with 0.1 interval. For ANN classifier, two hidden layers were used in ANN with an activation threshold of 0.5, RMSE of 0.01, and training momentum of 0.5 with a total of 100 iterations (Sahithi et al., 2019). In SVM classifier, the gamma value is set to 0.07 (inverse of the number of bands), penalty value of 100 and the pyramid levels are set to zero to process the data without compression (Sahithi et al., 2022). For SID classifier, divergence threshold of 0.05 is used and for MLC, and a probability threshold of 0.01 is used (Sahithi et al., 2021). Ten different classes were identified in the study area—Sal Forest, grass, croplands, tea garden, mixed vegetation, fallow land, barren land, built-up, riverine sand, and river.

The results of all the seven classifiers used for the study are presented in Fig. 11. The analysis was carried out visually and quantitatively using confusion matrix method based on overall accuracy and kappa coefficient. Visual observations have shown that SVM classifier with RBF kernel outperformed the other six classifiers. SVM classifier gave a clear delineation between class pairs like built-up and riverine class, crop and grass, shrubs, and crop which are most often confused in multispectral data classification. Artificial neural network classifiers also gave an edge to SVM classifier in certain classes. However, the most conventional and reliable MLC method yielded very poor results in classifying the Hyperion data. However, before settling to any one classifier for a particular image with a specific application, it is necessary to fine tune the parameters to obtain accurate results. In this case study, the Hyperion image gave better classification results for LULC classification for the given study area with a moderate heterogeneity of features. It can be clearly seen that the riverine sand class was misclassified into built-up class in LISS IV classified map while it was correctly classified in Hyperion SVM classified map.

After visually and quantitatively analyzing the results of classifications, an ensemble of these classifiers was used by giving a weightage to each one. The overall accuracy percentage obtained for each classifier in Fig. 11 is taken as a base for assigning the weights to each classifier. The classifier with highest OA will be given a weightage close to one or equal to one, and the classifier with least OA will be given a weightage close to zero. Thus, based on the overall accuracy percentage criteria, SVM was given a weightage of 1, MDM was given 0.7, ANN was given 0.8, SAM 0.6, MLC was given 0.3, SID 0.5, and BE was given a least weightage of 0.1. The results of SVM classification along with the improved classified map from ensemble classification are presented in Fig. 12. Visual observations from Fig. 12 indicate that classes like tea plantations, fallow and barren lands, built-up, and riverine sand were well classified when using an ensemble of various methods. Certain overfitting problems observed in SVM were resolved when using an ensemble of seven classifiers.

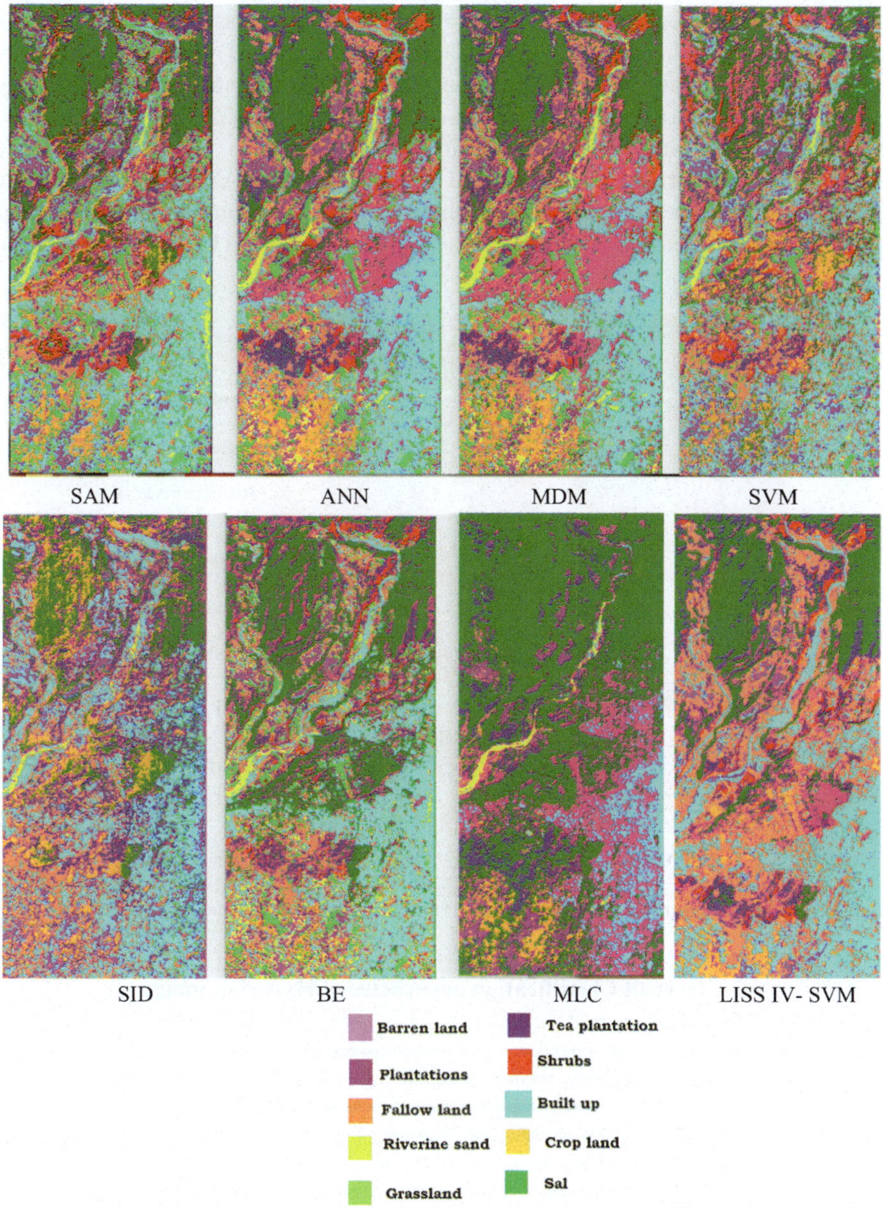

Fig. 11 Classification results of seven different classifiers and LISS IV(SVM)

The overall classification accuracies of all the seven base classifiers and ensemble classifier are given in Fig. 13.

Another study was carried out using the AVIRIS-NG airborne hyperspectral image of Udipi region, Karnataka. The AVIRIS-NG data contains 425 continuous spectral bands between 350 and 2500 nm range with a spatial resolution of 7.6 m. Radiometric and atmospheric correction were carried out using the FLAASH AC model, and a MNF transform was applied. Endmembers were collected from the image and used as input for classification. The study area is mostly dominated by various plantation

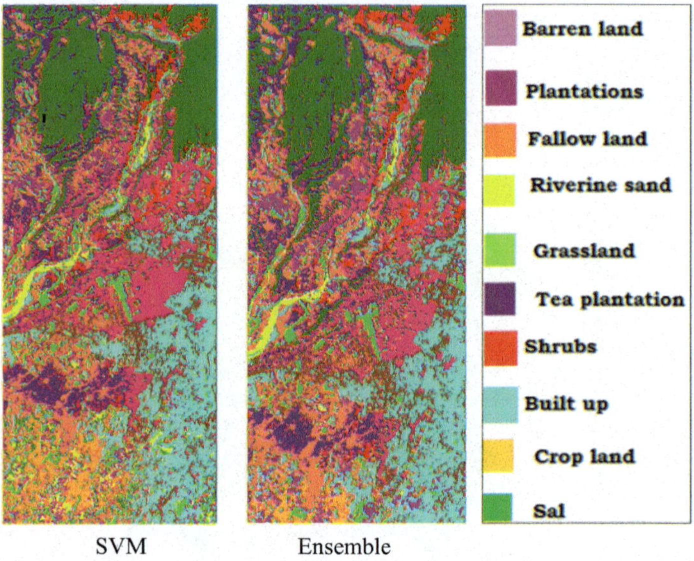

Fig. 12 SVM and ensemble classified maps of Hyperion image

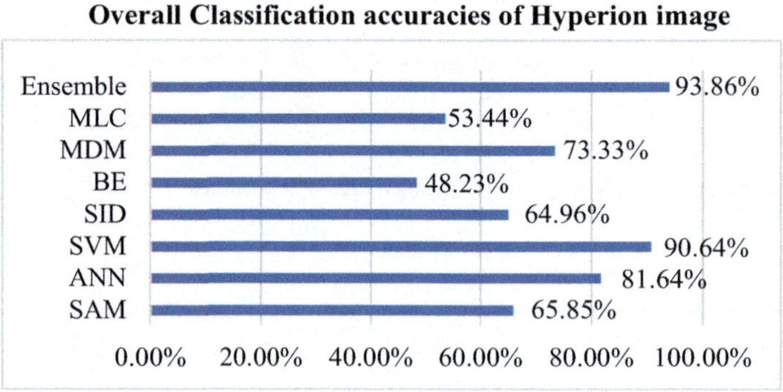

Fig. 13 Graph showing the overall accuracies of LULC classified maps from seven classifiers and ensemble method for Hyperion image

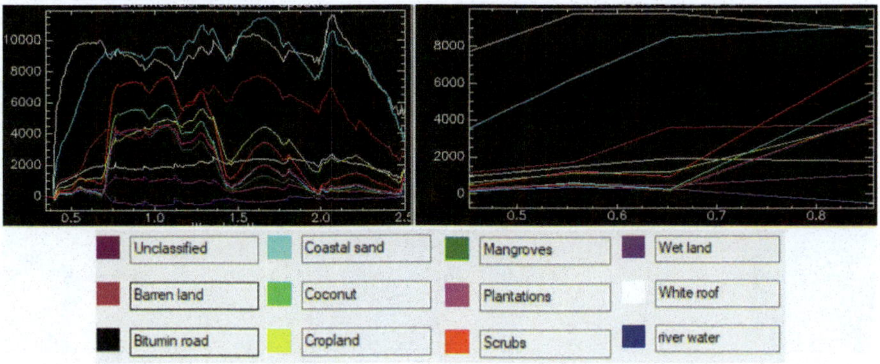

Fig. 14 Spectral profiles of endmembers from AVIRIS-NG (left) and LISS IV (right) datasets, respectively

types like coconut, mangroves, arecanut, paddy crop, fallow lands, sea, and river waters. In this study, classification was carried out using SVM and ensemble classifiers on AVIRIS-NG HyS image as well as four band LISS IV multispectral (MX) images. The differences in the classified outputs and the overall accuracies were observed. Figure 14 shows the spectra profiles of 11 endmember classes collected from HyS and MX datasets simultaneously. The x-axis is the wavelength, and y axis represents reflectance (with a scale factor of 10,000).

Classification results from ANN, SVM, and ensemble classification method for both AVIRIS-NG and LISS IV are presented in Fig. 15.

It was observed that the low spectral resolution of LISS IV image could not clearly differentiate between coconut plantations and mangroves, crop, and scrubs class pair. These classes were differentiated well in HyS image. The ANN classifier failed to classify all the 11 classes in LISS IV image due to confusion in spectral properties of classes. The same parameters as mentioned for Hyperion image were used for AVIRIS-NG and LISS IV images. However, a higher spatial resolution of the hyperspectral image can further improve the classification results. This was evident due to the improved classification accuracy with AVIRIS-NG image as compared to Hyperion image (Fig. 16).

5.5.1 Utility of HyS Data for Agricultural LU/LC Applications

An accurate LU/LC map is the basis for modeling any climatic or environmental parameters including the agricultural productivity. While producing the LU/LC maps, it is very important to obtain a delineation between the closely resembling vegetation classes like—crop, grass, scrubs, plantations, mangroves, fallow lands, and wetlands. Automatic classification of multispectral datasets can lead to confusion between grass areas and crops, plantations, and mangroves which can in-turn effect the final

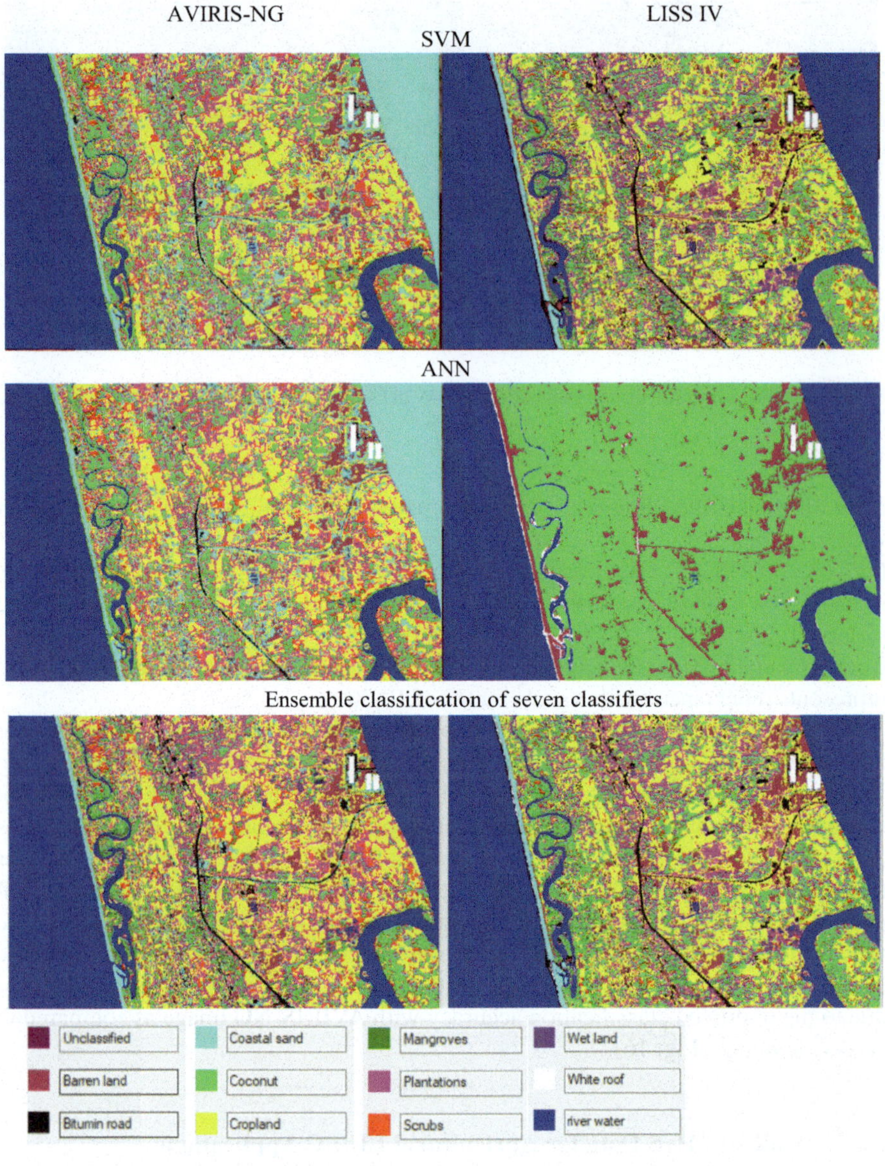

Fig. 15 Classification of 11 LU/LC classes in AVIRIS-NG and LISS IV images

coverage statistics. The utility of this high spectral information data thus not only helps in obtaining improved and accurate LU/LC classified maps but also helps in-

- In differentiating various crop types using automatic methods.
- Differentiating the fallow lands, grass lands, and scrubs from croplands.

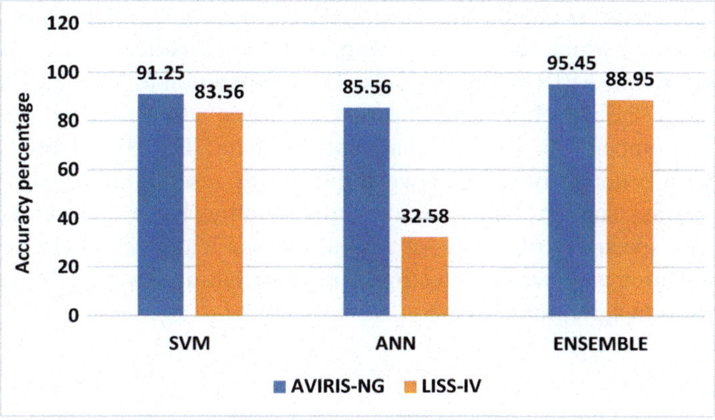

Fig. 16 Overall accuracies of AVIRIS-NG and LISS IV datasets

- Providing an important input, i.e., accurate LULC map for crop and drought modeling.
- Improving the statistics of agricultural lands in inaccessible areas.

6 Conclusions

Hyperspectral datasets are being widely used in various fields like crop classification and acreage estimation, vegetation species identification, mineral studies, water quality assessment studies, ocean studies, land use/land cover identification, etc. In spite of the field of application, the narrow spectral interval or bandwidth of 3–5 nm can help in extracting certain useful information that is specific to its high spectral properties. However, handling these voluminous HyS datasets require certain special algorithms or techniques which may be again marginally same for any field of application. The present chapter gives an aerial view on various steps involved in processing of HyS datasets, some commonly used algorithms and techniques in HyS data processing and classification methods. The chapter also includes few results from processing of spaceborne Hyperion and airborne AVIRIS-NG HyS data. Results show that FLAASH AC model gave a higher resemblance with the ground spectra with a correlation coefficient R^2 value ranging between 0.966 for crop class and 0.74 for built-up class. In the DR methods, it was observed that the PCA method sorts all the major information of the data into its first few components, while ICA method identifies the minute/hidden information that is ignored as noise by PCA using higher-order statistics. Classification performed using seven base classifiers on Hyperion image proved that SVM classifier can outperform the other methods with an overall accuracy of 90.64%. The classification of LISS IV multispectral image with SVM has given an accuracy of 84.53%, thus showing an improvement in classification results upon using the HyS data. An ensemble classification method was implemented which further improvement in the classification results by 3–4% with Hyperion and AVIRIS-NG datasets. Thus, utilization of HyS data has proved to differentiate the closely resembling LULC classes with improved classification accuracy as compared to LISS IV multispectral data. The HyS datasets can help in obtaining a better delineation between the closely resembling vegetation classes like—crop, grass, scrubs, plantations, mangroves, fallow lands, and wetlands. Application of this spectrally rich data for agriculture application can help in automatic crop classification and disease detection, thus contributing toward digital ecosystem for innovation in agriculture sector.

References

Arun Prasad, K., Gnanappazham, L., Selvam, V., Ramasubramanian, R., & Kar, C. S. (2015). Developing a spectral library of mangrove species of Indian east coast using field spectroscopy. *Geocarto International, 30*(5), 580–599.

Ben-Dor, E., Kindel, B., & Goetz, A. F. H. (2004). Quality assessment of several methods to recover surface reflectance using synthetic imaging spectroscopy data. *Remote Sensing of Environment, 90*, 389–404.

Cetin, M., Musaoglu, N., & Kocal, O. H. (2017). *A comparison of AC methods on Hyperion imagery in forest areas*. http://hdl.handle.net/11452/12207. Last accessed September, 2022.

Chang, C. I. (2013). *HyS data processing: Algorithm design and analysis*. Wiley.

Damodaran, B. B., & Nidamanuri, R. R. (2014). Assessment of the impact of DR methods on information classes and classifiers for HyS image classification by multiple classifier system. *Advances in Space Research, 53*(12), 1720–1734.

Fejjari, A., Saheb Ettabaa, K., & Korbaa, O. (2018) Feature extraction techniques for HyS images classification. In *International Workshop Soft Computing Applications* (pp. 174–188). Springer.

Gao, B. C., Montes, M. J., Davis, C. O., & Goetz, A. F. (2009). Atmospheric correction algorithms for hyperspectral remote sensing data of land and ocean. *Remote sensing of environment, 113*, S17–S24.

Green, A. A., & Craig, M. D. (1985). Analysis of aircraft spectrometer data with logarithmic residuals. In *JPL Proceedings of the Airborne Imaging Spectrometer Data Analysis Workshop*.

Guo, Y., & Zeng, F. (2012). AC comparison of SPOT-5 image based on model FLAASH and model QUAC. *International Archives of the Photogrammetry, Remote Sensing and Spatial Information Sciences, 39*(7), 21–23.

Hyvarinen, A., & Oja, E. (2000). Independent component analysis: Algorithms and applications. *Neural Networks, 13*(4), 411–430.

Ilori, C. O., Pahlevan, N., & Knudby, A. (2019). Analyzing performances of different atmospheric correction techniques for Landsat 8: Application for coastal remote sensing. *Remote Sensing, 11*(4), 469.

Jensen, J. R. (2005). *Introductory digital image processing: A remote sensing perspective* (3rd ed.). Pearson Prentice Hall.

Kakarla, S., Gangula, P., Singh, C., & Sarma, T. H. (2019). DR in HyS images using auto-encoders. In *International Conference on Advances in Computational Intelligence and Informatics* (pp. 101–107). Springer.

Kale, K. V., Solankar, M. M., Nalawade, D. B., Dhumal, R. K., & Gite, H. R. (2017). A research review on HyS data processing and analysis algorithms. *Proceedings of the National Academy of Sciences, India Section A: Physical Sciences, 87*(4), 541–555.

Kawishwar, P. (2007). *AC models for retrievals of calibrated spectral profiles from Hyperion EO-1 data*. Master's Thesis submitted to ITC, Netherlands.

Luo, G., Chen, G., Tian, L., Qin, K., & Qian, S. E. (2016). Minimum noise fraction versus principal component analysis as a preprocessing step for hyperspectral imagery denoising. *Canadian Journal of Remote Sensing, 42*(2), 106–116.

Moughal, T. A. (2010). HyS image classification using support vector machine. In *Journal of Physics: Conference Series* (Vol. 439, No. 1, p. 012042). IOP Publishing.

Pal, M., & Foody, G. M. (2010). Feature selection for classification of HyS data by SVM. *IEEE Transactions on Geoscience and Remote Sensing, 48*(5), 2297–2307.

Panwar, A., Singh, A., & Bhaduria, H. S. (2014). *International Journal of Emerging Technology and Advanced Engineering, 4*(5), 701–705. ISSN 2250-2459, ISO 9001:2008.

Panwar, S. S., & Raiwani, Y. P. (2014). Data reduction techniques to anayze NSL-KDD dataset. *International Journal of Computer Engineering & Technology, 5*(10), 21–31.

Ravikanth, L., Jayas, D. S., White, N. D., Fields, P. G., & Sun, D. W. (2017). Extraction of spectral information from hyperspectral data and application of hyperspectral imaging for food and agricultural products. *Food and Bioprocess Technology, 10*(1), 1–33.

Reay, D. S. (2020). Land use and agriculture: Pitfalls and precautions on the road to net zero. *Frontiers in Climate, 2*. https://doi.org/10.3389/fclim.2020.00004

Sahithi, V. S., Iyyanki, M., & Giridhar, M. V. S. S. S. (2016). Performance evaluation of DR techniques on chris hyperspectral data for surface discrimination. *Journal of Geomatics, 10*(1), 7–11.

Sahithi, V. S., Iyyanki, M., & Giridhar, M. V. S. S. S. (2021). Hyperspectral data classification algorithms for delineation of LULC classes. In *Proceedings of the International Conference on Industry* (Vol. 4).

Sahithi, V. S., Iyyanki, M., & Giridhar, M. V. S. S. S. (2022). Analysing the sensitivity of SVM kernels on hyperspectral imagery for land use land cover classification. *Journal of Image Processing and Artificial Intelligence 8*(2), 15–23.

Sahithi, V. S., Subbiah, S., & Agrawal, S. (2019). Comparison of support vector machine, artificial neural networks and spectral angle mapper classifiers on fused hyperspectral data for improved LULC classification. In *2019 8th International Conference on Modeling Simulation and Applied Optimization (ICMSAO)* (pp. 1–6). IEEE.

San, A. B. T., & Suzen, B. M. L. (2010). Evaluation of different AC algorithms for EO-1 Hyperion imagery. *International Archives of the Photogrammetry, Remote Sensing and Spatial Information, 38*, 392–397.

Savitzky, A., & Golay, M. J. (1964). Smoothing and differentiation of data by simplified least squares procedures. *Analytical Chemistry, 36*(8), 1627–1639.

USGS. (2017). *Product guide: Landsat 8 surface reflectance code (LASRC) product*. Available online: https://doi.org/10.1080/1073161X.1994.10467258. Accessed on February 18, 2019.

Vaiphasa, C. (2006). Consideration of smoothing techniques for HyS remote sensing. *ISPRS Journal of Photogrammetry and Remote Sensing, 60*(2), 91–99.

Wang, J., & Chang, C. I. (2006). Independent component analysis-based DR with applications in HyS image analysis. *IEEE Transactions on Geoscience and Remote Sensing, 44*(6), 1586–1600.

Yusuf, B. L., & He, Y. (2011). Application of HyS imaging sensor to differentiate between the moisture and reflectance of healthy and infected tobacco leaves. *African Journal of Agricultural Research, 6*(29), 6267–6280.

Zagajewski, B., & Olesiuk, D. (2009). SAM and ANN classification of HyS data of seminatural agriculture used areas. In *Remote Sensing for a Changing Europe* (pp. 505–510). IOS Press.

Zhang, J., Zhang, Y., & Zhou, T. (2001). Classification of HyS data using support vector machine. In *Proceedings 2001 International Conference on Image Processing (Cat. No. 01CH37205)* (Vol. 1, pp. 882–885). IEEE.

Computer Vision Approaches for Plant Phenotypic Parameter Determination

Alka Arora, Tanuj Misra, Mohit Kumar, Sudeep Marwaha, Sudhir Kumar, and Viswanathan Chinnusamy

Abstract Climate change and the growing population are major challenges in the global agriculture scenario. High-quality crop genotypes are essential to counter the challenges. In plant breeding, phenotypic trait measurement is necessary to develop improved crop varieties. Plant phenotyping refers to studying the plant's morphological and physiological characteristics. Plant phenotypic traits like the number of spikes/panicle in cereal crops and senescence quantification play an important role in assessing functional plant biology, growth analysis, and net primary production. However, conventional plant phenotyping is time-consuming, labor-intensive, and error-prone. Computer vision-based techniques have emerged as an efficient method for non-invasive and non-destructive plant phenotyping over the last two decades. Therefore to measure these traits in high-throughput and non-destructive way, computer vision-based methodologies are proposed. For recognition and counting of number of spikes from visual images of wheat plant, a deep learning-based encoder-decoder network is developed. The precision, accuracy, and robustness (F_1-score) of the approach for spike recognition are found as 98.97%, 98.07%, and 98.97%, respectively. For spike counting, the average precision, accuracy, and robustness are 98%, 93%, and 97%, respectively. The performance of the approach demonstrates that the encoder-decoder network-based approach is effective and robust for

A. Arora (✉) · S. Marwaha
ICAR-Indian Agricultural Statistics Research Institute (IASRI), New Delhi, India
e-mail: Alka.Arora@icar.gov.in

S. Marwaha
e-mail: Sudeep.Marwaha@icar.gov.in

T. Misra
Teaching Cum Research Associate, Rani Lakshmi Bai Central Agricultural University, Jhansi, UP, India

M. Kumar
Computer Application, ICAR-Indian Agricultural Statistics Research Institute (IASRI), New Delhi, India

S. Kumar · V. Chinnusamy
ICAR-Indian Agricultural Research Institute (IARI), New Delhi, India
e-mail: viswa.chinnusamy@gmail.com

© The Author(s), under exclusive license to Springer Nature Singapore Pte Ltd. 2023
S. Chaudhary et al. (eds.), *Digital Ecosystem for Innovation in Agriculture*,
Studies in Big Data 121, https://doi.org/10.1007/978-981-99-0577-5_13

spike detection and counting. For senescence quantification, machine learning-based approach has been proposed which segments the wheat plant into different senescence and greenness classes. Six machine learning-based classifiers: decision tree, random forest, KNN, gradient boosting, naïve Bayes, and artificial neural network (ANN) are trained to segment the senescence portion from wheat plants. All the classifiers performed well, but ANN outperformed with 97.28% accuracy. After senescence segmentation, percentage of senescence area is also calculated. A GUI-based desktop application, m—Senescencica has been developed, which processes the input images and generates output for senescence percentage, plant height, and plant area.

Keywords Computer vision · Deep learning · Image analysis · Machine learning · m—Senescencica · Plant phenomics

1 Introduction

Food grain production must be doubled by 2050 to meet the demand of growing population (Ray et al., 2013). This is a major challenge due to climate change induced stresses and slow rate of genetic gain in conventional crop improvement programs. High-quality crop varieties are crucial to counter the challenges. Plant phenotypic trait measurement is necessary to develop improved crop varieties. In this connection, plant phenomics refers to studying the plant's morphological and physiological characteristics. Plant phenotypic traits quantification of germplasm lines and mapping population in a given environment is necessary for gene mapping and trait pyramiding. Manual recording of physiological traits is used in traditional methods, which is time-consuming and labor-intensive, and may be error-prone when recording a large number of genotypes. Computer vision-based techniques have recently emerged as an efficient framework for non-destructive plant phenotyping. Plant phenotypic traits like number of spikes/panicle in cereal crops and senescence quantification play an important role in assessing functional plant biology, growth analysis, and net primary production. Conventional techniques of measuring these traits are tedious, time-consuming, and error-prone for phenotyping large dataset. In this chapter, computer vision-based approaches for measuring the phenotypic traits (number of spikes/panicle in cereal crops and senescence quantification) are presented.

2 Recognizing and Counting of Spikes in Visual Images of a Wheat Plant

Wheat spike is the grain-bearing organ, and spike number is the key measure in yield determination of the plant. Manual or conventional technique of counting number of spikes based on naked-eye observation is tedious and time-consuming to record large number of genotypes. Recently, computer vision (integration of image analysis and machine learning techniques)-based technologies acquire strong attention in recognition and counting of spikes through image processing. A computer vision-based approach is presented in this chapter for identifying spikes in visual images (VIS) of wheat plants.

2.1 *Image Acquisition*

In this study, visual images of the plant were taken using 6576 $\times$ 4384 pixel RGB camera from three different side view directions (angles: 0°, 120°, and 240°) with respect to the initial position of the plant. To reduce the issue of spikes overlapping, three side view directions were taken into consideration. This LemnaTec imaging facilities (LemnaTec GmbH, Aachen, Germany) are installed at Nanaji Deshmukh Plant Phenomics Center at Indian Agriculture Research Institute (IARI), New Delhi (Misra et al., 2020). Hundred wheat plants were grown in pot under controlled environmental condition with recommended cultural practices. Images were captured during reproductive stage of the plant by maintaining a uniform background for better image processing and stored in PNG format. After image acquisition, spikes number per plant pot was recorded manually to validate the developed approach.

2.2 *Architecture of the Deep Learning Approach*

The developed approach consists of two deep learning networks: Patchify Network (PN) and Refinement Network (RN). PN extracts spatial and local contextual features at patch level (a small area of the image that overlaps), while RN refines the segmented output of PN (as it sometimes contain segments spikes inaccurately) (Fig. 1). The convolutional encoder-decoder deep learning network with hourglasses serves as the approach's backbone and bottleneck network for pixel-by-pixel segmentation of objects (here, spikes). For retrieving the feature map depiction clutching the spatial and factual information from the instruction digital representation, an encoder network with three encoder blocks (each encoder block consisting of two convolution layers followed by a ReLU and Max-pooling layer of window size 2 $\times$ 2) was constructed. Three encoder blocks each had 16 filters, 64 filters, and 128 filters to encode the features. Three decoder blocks make up the decoder network, and each block's two-transpose convolution layers are, used to upsample the incoming feature maps. In contrast to the encoder network used to renew the features, there were 128, 64, and 16 filters in each of the three-decoder blocks, correspondingly. For better localization of the segmented features, the upsampled feature maps are combined with the appropriate encoded feature maps. The hourglass network is made of three hourglasses, each of which is made of a series of residual blocks that have three

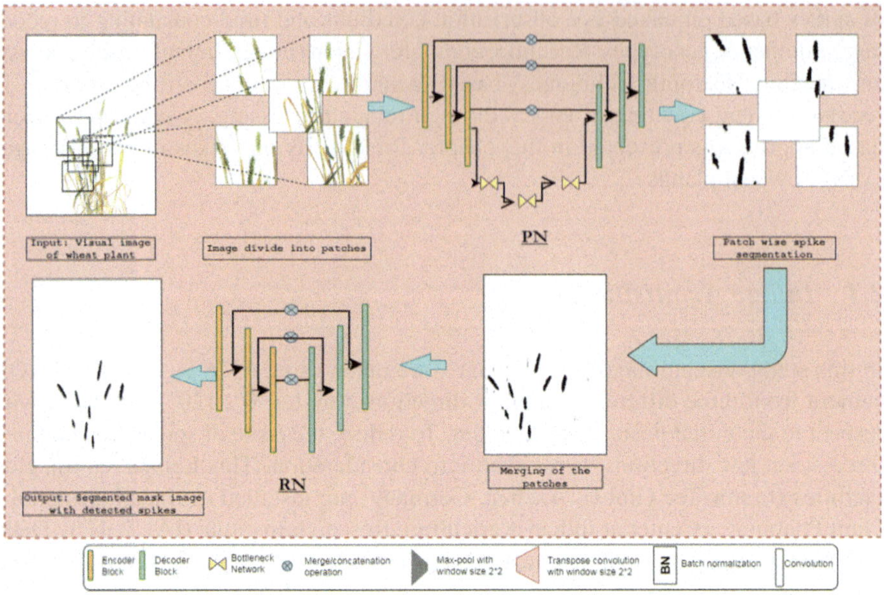

Fig. 1 Input image is split into patches before entering into PN. Patch-wise segmented mask images are the output of PN and are concatenated which contains some erroneous segmentation of spikes. The output are then refined using RN network. Refined mask image contains spike regions only

convolution layers with the filter sizes 1×1, 3×3, and 1×1. This allows for more confident spike segmentation by focusing on the key features that are affected by scale, viewpoints, and occlusions. Based on the optical performances, the number of encoders, decoders, and hourglasses was estimated on empirical evidence.

2.3 *Training of the Deep Learning Model*

For developing the deep learning model, image dataset comprising 300 images (3 direction images of 100 plants) was divided randomly into training dataset and test dataset at 85% and 15% ratio, respectively. The deep learning model was developed on Linux operating system with 32 GB RAM and NVIDIA GeForce GTX 1080 Ti graphics card (with memory 11 GB). The training dataset consists of 255 images (*i.e.,* 85% of the total dataset). The images were divided into patches before entering into the PN. The popular optimizer "Adam" with learning rate 0.0005 was used to update the weight of the network, and "binary cross-entropy" (Misra et al., 2021, 2022) loss function was utilized to for predicting the spikes and non-spikes pixels. As it is a binary class classification problem (spike pixels or not), "binary cross-entropy" is used to calculate the loss function at pixel level. Both the networks (PN and RN) were trained separately for 100 epochs with batch size 32 (due to the system

Table 1 List of hyperparameters

Optimizer	Learning rate	Epoch	Batch size	Loss function
Adam	0.0005	100	32	Binary cross-entropy

Table 2 Performance of SpikeSegNet in spike segmentation

E1	E2	JI	Accuracy	Precision	Recall	F-measure
0.0016	0.0487	0.9982	0.9807	0.9897	0.9889	0.9897

constraints) and then merged to confine a single network. The hyperparameters used in developing the model are given in Table 1.

2.4 *Result*

SpikeSegNet, which consists of both networks (PN and RN), was trained sequentially. The pixel-wise segmentation performance of the developed model was measured by the performance metrics (Type I classification error (E_1), Type II classification error (E_2), Jaccard index (JI), accuracy, precision, recall, and F-measure) presented in Table 2.

Precision = TP/(TP + FP); Recall = TP/(TP + FN); Accuracy = (TP + TN)/(TP + TN + FP + FN); TP: True positive; TN: True negative; FP: False positive; FN: False negative.

E_1 indicates that a very small number of pixels were incorrectly identified. The created model's accuracy is almost 99%, and spikes can be identified with an average precision of 98.97%. For spike counting, the "analyze particles"—function of imageJ (Abràmoff et al., 2004)—was applied on the output of SpikeSegNet network, i.e., binary image containing spike region only. For spike counting, the average precision, accuracy, and robustness are 98%, 93%, and 97%, respectively. The performance of SpikeSegNet indicates that it is an important advance forward toward high-throughput phenotyping of wheat plant.

In the next section, another approach has been presented for the senescence quantification.

3 Machine Learning-Based Plant Senescence Quantification

Senescence is the last stage of the wheat crop cycle, and it is at this time that nutrients start to flow back into the developing grain from the plant. The first and most significant change in wheat senescence is the damaging of chloroplasts, which result in breakdown of photosynthetic pigments such as chlorophyll in leaf (Nikolaeva et al., 2010). Due to damage in chlorophyll, color of the leaf changes from the usual deep green to yellow and finally brown (Fig. 3).

Measuring of plant senescence is an important aspect as this helps to select the best genotypes tolerable to senescence in the stressed conditions. Conventionally, senescence is measured by manual scoring, in which an expert assign senescence score by observing the plant. But this method has many drawbacks, first of all it is subjective in nature and highly biased. This method is time-consuming as in any breeding program, there is a large population of grown plants. Manually measuring senescence for such a large population of plants is time-consuming and prone to errors. With the availability of image data, image-based measuring of plant phenotypic parameters is gaining the interest of researchers. It is high throughput and non-destructive in nature.

Here, a computer vision-based approach has been proposed for plant senescence quantification (Kumar, 2020). This is the plant pixels classification problem to classify plant pixels into each of the defined classes. Six classes were decided for the senescence dry, yellow, pale yellow, dark green, and light green and one background class by observing the senescence pattern in Fig. 2. Around 1000 pixels values were sampled from the image data. Sampled dataset was divided in two training and test sets with 75–25% ratio, respectively. Six machine learning-based classifiers (ANN, naïve Bayes, random forest, gradient boosting classifier, decision tree, K-nearest neighbors) were trained on training data by using scikit-learn library. In order to get the best models parameters, tuning with tenfold cross-validation was used to select the best-performing models.

Fig. 2 Changes in leaf color due to senescence. Initially, the leaf is green, but due to senescence, it results in change in leaf color. It changes to pale yellow to yellow and at last to brown

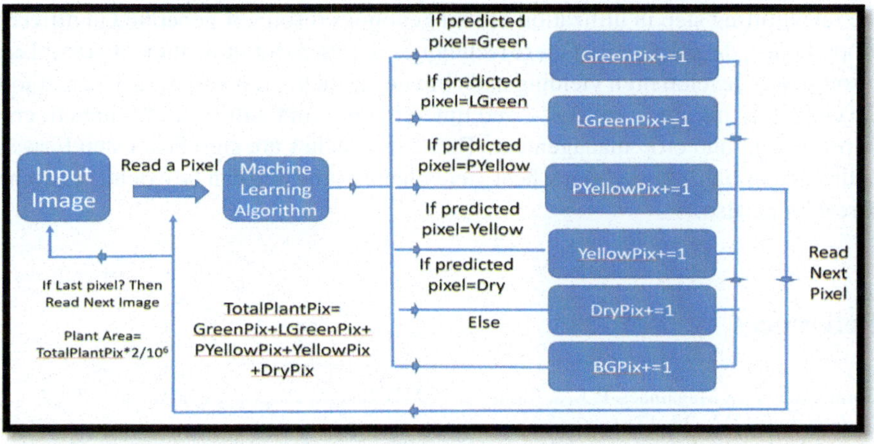

Fig. 3 Flowchart for senescence quantification

Precision, recall, and F1-scores were measured on test data. Among all the trained classifiers, ANN outperformed with 97.28% test accuracy. After pixel classification, the total number of pixels in each class was counted. Sum of all the pixels provides the total number of plant pixels (Fig. 3). Upon dividing total pixels in each class by total plant pixels gives the percentage of pixels in each class. Division by zero in Python causes zero division error, and this exception is handled by using try and catch block.

Among the six defined classes, yellow, pale yellow, and brown account for senescence classes. Hence, the percentage sum for those three classes gave the senescence percentage.

In this study, four approaches have been presented based on artificial intelligence in the area of plant phenomics. All these approaches gave promising results in the area of phenomics. Artificial intelligence techniques have tremendous potential in determination of other plant phenomics parameters with the ultimate goal for yield estimation.

4 Conclusion

In the era of modern plant phenotyping, computer vision-based technologies are much needed to counter the challenges that exist in traditional plant phenotyping of recognizing and counting spikes and senescence quantification in wheat plant. In this study, a deep learning approach has been developed for recognizing and counting spikes form visual images of wheat plant with satisfactory precision, accuracy, and robustness performances. Besides, machine learning-based approaches also performs a promising result in plant senescence quantification. As conventional phenotyping is the rate limiting step in utilization of vast genomics resources generated in different crops, hence, development of these techniques is critical for utilization of germplasm resources to develop high yielding and climate-resilient crop varieties. The methods developed in this study are cost—and time-effective and will be useful in both crop improvement and crop management. These approaches are significant step forward in the area of high-throughput wheat yield phenotyping and can be extended to other cereal crops also.

References

Abràmoff, M. D., Magalhães, P. J., & Ram, S. J. (2004). Image processing with ImageJ. *Biophotons Intention, 11*(7), 36–42.

Kumar, M. (2020). Wheat Plant Senescence Quantification using Machine Learning Algorithms (Master's thesis). Indian Agricultural Statistics Research Institute, IARI, New Delhi.

Misra, T., Arora, A., Marwaha, S., Chinnusamy, V., Rao, A. R., Jain, R., ... & Goel, S. (2020). SpikeSegNet-a deep learning approach utilizing encoder-decoder network with hourglass for spike segmentation and counting in wheat plant from visual imaging. *Plant Methods, 16*(1): 1–20.

Misra, T., Arora, A., Marwaha, S., Jha, R. R., Ray, M., Varghese, E., Kumar, S., Nigam, A., Sahoo, R. N., & Chinnusamy, V. (2021). Web-SpikeSegNet: Deep learning framework for recognition and counting of spikes from visual images of wheat plants. *IEEE Access, 9*, 76235–76247.

Misra, T., Arora, A., Marwaha, S., Ranjan Jha, R., Ray, M., Kumar, S., & Chinnusamy, V. (2022). Yield-SpikeSegNet: An extension of SpikeSegNet deep-learning approach for the yield estimation in the wheat using visual images. *Applied Artificial Intelligence, 36*(1), 2137642.

Nikolaeva, M. K., Maevskaya, S. N., Shugaev, A. G., & Bukhov, N. G. (2010). Effect of drought on chlorophyll content and antioxidant enzyme activities in leaves of three wheat cultivars varying in productivity. *Russian Journal of Plant Physiology, 57*(1), 87–95.

Ray, D. K., Mueller, N. D., West, P. C., & Foley, J. A. (2013). Yield trends are insufficient to double global crop production by 2050. *PloS one, 8*(6), e66428.

The manufacturer's authorised representative in the EU is Springer Nature Customer Service Centre GmbH, Europaplatz 3, 69115 Heidelberg, Germany. If you have any concerns regarding our products, please contact ProductSafety@springernature.com

Printed and bound by CPI Group (UK) Ltd, Croydon, CR0 4YY
01/10/2026
02225870-0003